Nonlinear Spatio-Temporal Dynamics and Chaos in Semiconductors

Nonlinear transport phenomena are an increasingly important aspect of modern semiconductor research. *Nonlinear Spatio-Temporal Dynamics and Chaos in Semiconductors* deals with complex nonlinear dynamics, pattern formation, and chaotic behavior in such systems. In doing so it bridges the gap between two well-established fields: the theory of dynamic systems, and nonlinear charge transport in semiconductors. This unified approach is used to consider important electronic transport instabilities. The initial chapters lay a general framework for the theoretical description of nonlinear self-organized spatio-temporal patterns, such as current filaments, field domains, and fronts, and for analysis of their stability. Later chapters consider important model systems in detail: impact-ionization-induced impurity breakdown, Hall instabilities, superlattices, and low-dimensional structures. State-of-the-art results include chaos control, spatio-temporal chaos, multistability, pattern selection, activator–inhibitor kinetics, and global coupling, linking fundamental issues to electronic-device applications. This book will be of great value to semiconductor physicists and nonlinear scientists alike.

ECKEHARD SCHÖLL was born on 6 February 1951 in Stuttgart, Germany. He received his Diplom degree in physics (MSc) from the University of Tübingen, Germany, in 1976, his PhD in applied mathematics from the University of Southampton, UK, in 1978, and the Dr rer. nat. degree and the *venia legendi* from Aachen University of Technology (RWTH Aachen), Germany, in 1981 and 1986 respectively. During 1983–1984 he was a visiting assistant professor in the Department of Electrical and Computer Engineering, Wayne State University Detroit, Michigan, USA. Since 1989 he has been a Professor of Theoretical Physics at the Technical University of Berlin, Germany. His research interests include the theory of nonlinear charge transport and current instabilities in semiconductors, in particular, low-dimensional structures; nonlinear spatio-temporal dynamics, chaos, and pattern formation; and electro-optical nonlinearities. Dr Schöll is the author of the book *Nonequilibrium Phase Transitions in Semiconductors* (1987), translated into Russian in 1991, the coauthor of *The Physics of Instabilities in Solid State Electron Devices* (1992), and has recently edited a monograph on the *Theory of Transport Properties of Semiconductor Nanostructures* (1998). He has published almost 200 articles in international journals, including *Physical Review Letters*, *Physical Review*, *Semiconductor Science and Technology*, *Applied Physics Letters*, the *Journal of Applied Physics*, and many others. In 1997 he was awarded a prize as a Champion of Teaching by the Technical University of Berlin.

Cambridge Nonlinear Science Series 10

Editors

Professor Boris Chirikov
Budker Institute of Nuclear Physics, Novosibirsk

Professor Predrag Cvitanović
Niels Bohr Institute, Copenhagen

Professor Frank Moss
University of Missouri, St Louis

Professor Harry Swinney
Center for Nonlinear Dynamics,
The University of Texas at Austin

Titles in print in this series

Nonlinear Spatio-Temporal Dynamics and Chaos in Semiconductors

Eckehard Schöll
Technische Universität Berlin

CAMBRIDGE
UNIVERSITY PRESS

PUBLISHED BY THE PRESS SYNDICATE OF THE UNIVERSITY OF CAMBRIDGE
The Pitt Building, Trumpington Street, Cambridge, United Kingdom

CAMBRIDGE UNIVERSITY PRESS
The Edinburgh Building, Cambridge CB2 2RU, UK
40 West 20th Street, New York, NY 10011-4211, USA
10 Stamford Road, Oakleigh, VIC 3166, Australia
Ruiz de Alarcón 13, 28014, Madrid, Spain
Dock House, The Waterfront, Cape Town 8001, South Africa

http://www.cambridge.org

First published 2001

Printed in the United Kingdom at the University Press, Cambridge

Typeface Times 10.25/13.25pt.   *System* LATEX 2_ε [DBD]

A catalogue record of this book is available from the British Library

ISBN 0 521 45186 8   hardback

Contents

Preface

More than a dozen years after my book on *Nonequilibrium Phase Transitions in Semiconductors – Self-organization Induced by Generation and Recombination Processes* appeared, the subject of nonlinear dynamics and pattern formation in semiconductors has become a mature field. The aim of that book had been to link two hitherto separate disciplines, semiconductor physics and nonlinear dynamics, and advance the view of a semiconductor driven far from thermodynamic equilibrium as a nonlinear dynamic system. It focussed on one particular class of instabilities related to nonlinear processes of generation and recombination of carriers in bulk semiconductors, and was essentially restricted to either purely *temporal* nonlinear dynamics, or nonlinear *stationary spatial* patterns. Within the past decade extensive research, both theoretical and experimental, has elaborated a great wealth of complex self-organized spatio-temporal patterns in various semiconductor structures and material systems. Thus semiconductors have been established as a model system with several advantages over the classical systems in which self-organization and nonlinear dynamics have been studied, viz. hydrodynamic, optical, and chemically reacting systems. First, semiconductor structures nowadays can be designed and fabricated by modern epitaxial growth technologies with almost unlimited flexibility. By controlling the vertical and lateral dimensions of those structures on an atomic length scale, systems with specific electric and optical properties can be tailored. Second, the dynamic variables describing nonlinear charge-transport properties are directly and easily accessible to measurement as electric quantities. Elaborate experimental techniques for detecting spatio-temporal patterns with high spatial and temporal resolution have recently been developed. Third, semiconductors may provide particularly useful applications of fundamental concepts and results of nonlinear dynamics. Many modern microelectronic devices operate in the regime of controlled electric instabilities. Semiconductor nanostructures, due to the occurrence of large electric fields, inherently possess strongly nonlinear transport properties that may be applied in electronic oscillating or switching devices.

This book provides a theoretical framework for complex nonlinear spatio-temporal dynamics, pattern formation, and chaotic behavior in semiconductors, illustrated by computer simulations and experimental results. It expands the scope of my earlier book substantially in a three-fold way. First, the emphasis is on nonlinear *spatially and temporally* modulated self-organized patterns. While purely temporal or spatial structures generally occur as first bifurcations of a dynamic system on going away from thermodynamic equilibrium, within the past decade research in the field of nonlinear dynamics has focussed more and more on higher bifurcations leading to complex structures such as breathing or spiking modes and spatio-temporal chaos. Second, the analysis of semiconductor instabilities is extended to a wide variety of transport mechanisms. Since the interest in semiconductor physics has generally shifted away from bulk materials, toward low-dimensional structures, this book features in particular modulated semiconductor structures, including heterojunctions between layers of different materials, such as quantum wells, superlattices, and resonant tunneling structures. Third, the book aims to introduce modern concepts and methods of dynamic systems into the field of semiconductors and thereby provide a unified and coherent approach to diverse models and results of electronic instabilities. Therefore the systematic methodology is stressed and illustrated by simple models offering insight, rather than by analyzing sophisticated, highly specialized transport models, even though those might provide a better quantitative description of specific cases.

The aim of this book is to build a bridge between two well-established fields: the theory of dynamic systems, and nonlinear charge transport in semiconductors. A unified approach toward various scattered results on electronic transport instabilities is adopted. The first three chapters lay the general framework for the theoretical description of nonlinear self-organized spatio-temporal patterns, such as current filaments, field domains, and fronts, and the analysis of their stability, while in the last four chapters some important model systems are treated in detail: impact-ionization-induced low-temperature impurity breakdown; Hall instabilities in crossed electric and magnetic fields; vertical high-field transport in superlattices; and spatio-temporal chaos in layered and low-dimensional semiconductor structures. These model systems are not exhaustive; rather they should be viewed as examples and give guidance for the application of the general concepts to other semiconductor structures. Particular emphasis is placed on state-of-the-art results such as chaos control, spatio-temporal chaos, multistability, pattern selection, activator–inhibitor kinetics, and global couplings due to the circuit. Wherever appropriate, I have tried to link fundamental issues with aspects of applications in electronic devices.

The book is aimed at physicists, electronic engineers, applied mathematicians, and materials scientists. It should be of interest to graduate students and researchers who want to familiarize themselves with this new field. A brief summary of nonlinear dynamics and chaos theory, and of semiconductor transport theory is provided in Chapters 1 and 2, respectively, with further references to more detailed treatises. Although this book can not substitute for an introduction to dynamic systems or semiconductors on a textbook level, it nevertheless introduces all of the basic notions and concepts needed. It has been written with the intention of providing the reader with the tools to apply the methods of nonlinear dynamics to further models of his specific interest.

This book would not have been possible without the interaction with many of my colleagues and students. I am grateful for valuable discussions with K. Aoki, M. Asche, N. Balkan, L. L. Bonilla, L. Eaves, H. Engel, H. Gajewski, H. P. Herzel, W. Just, H. Kostial, K. Kunihiro, A. S. Mikhailov, F. J. Niedernostheide, V. Novak, J. Parisi, J. Peinke, W. Prettl, H. G. Purwins, B. K. Ridley, P. Rodin, L. Schimansky-Geier, J. Socolar, S. W. Teitsworth, P. Vogl, and S. M. Zoldi. I am indebted to my collaborators and students who have over many years contributed substantially to the results presented in this book. In particular, special thanks are due to A. Amann, S. Bose, M. Meixner, P. Rodin, G. Schwarz, and A. Wacker for helpful comments and for critically reading parts of the manuscript. I am deeply indebted to my academic teachers P. T. Landsberg and F. Schlögl, who introduced me to this field. Last but not least, I want to thank my family for their patience and encouragement.

Much of this work was supported by various grants from the Deutsche Forschungsgemeinschaft and the Alexander-von-Humboldt-Stiftung.

Berlin, January 2000
Eckehard Schöll

Chapter 1

Semiconductors as continuous nonlinear dynamic systems

This book deals with complex nonlinear spatio-temporal dynamics, pattern formation, and chaotic behavior in semiconductors. Its aim is to build a bridge between two well-established fields: The theory of dynamic systems, and nonlinear charge transport in semiconductors. In this introductory chapter the foundations on which the theory of semiconductor instabilities can be developed in later chapters will be laid. We will thus introduce the basic notions and concepts of continuous nonlinear dynamic systems. After a brief introduction to the subject, highlighting dissipative structures and negative differential conductivity in semiconductors, the most common bifurcations in dynamic systems will be reviewed. The notion of deterministic chaos, some common scenarios, and the particularly challenging topic of chaos control are introduced. Activator–inhibitor kinetics in spatially extended dynamic systems is discussed with specific reference to semiconductors. The role of global couplings is illuminated and related to the external circuits in which semiconductor elements are operated.

1.1 Introduction

Semiconductors are complex many-body systems whose physical, e.g. electric or optical, properties are governed by a variety of nonlinear dynamic processes. In particular, modern semiconductor structures whose structural and electronic properties vary on a nanometer scale provide an abundance of examples of nonlinear transport processes. In these structures nonlinear transport mechanisms are given, for instance, by quantum mechanical tunneling through potential barriers, or by thermionic emission of *hot electrons* that have enough kinetic energy to overcome the barrier. A further important feature connected with potential barriers and quantum wells in such semiconductor structures is the ubiquitous presence of space charge. This, according to Poisson's equation, induces a further feedback between the charge-carrier distribution and the electric potential distribution governing the transport. This mutual nonlinear interdependence is particularly pronounced in the cases of semiconductor heterostructures and low-dimensional structures, in which abrupt junctions between different materials on an atomic length scale cause conduction band discontinuities resulting in potential barriers and wells. The local charge accumulation in these potential wells and nonlinear processes for transport of charge across the barriers have been found to provide a number of nonlinearities.

Another important class of nonlinear processes that influences, in particular, the electric transport properties, but also optical phenomena, is generation and recombination processes of nonequilibrium charge carriers (Schöll 1987). These are generally described by rate equations for the change in time of the carrier concentrations, which are nonlinear functions of these concentrations and of the electric field. Other nonlinearities are exhibited by nonlinear scattering processes of hot carriers, which may lead to a field-dependent mobility, and thus to a current density that is a nonlinear function of the local electric field.

1.1.1   Dissipative structures

Although the features described above have been known for a long time, the view of a semiconductor as a nonlinear dynamic system is a fairly recent development. Such nonlinear dynamic systems can exhibit a variety of complex behaviors such as bifurcations, phase transitions, spatio-temporal pattern formation, self-sustained oscillations, and deterministic chaos. Semiconductors are *dissipative* dynamic systems, i.e. a steady state can be maintained only by a continuous flux of energy, and possibly matter, through them. Mathematically, this is described by the feature that volume elements in a suitable space of dynamic variables – the *phase space* – shrink with increasing time. In the language of thermodynamics, this represents an *open* system that is driven by external fluxes and forces so far from thermodynamic equilibrium that linear dynamic laws no longer hold. Owing to the driving forces and the inherent nonlinearities of these systems, they may spontaneously evolve into a state of highly ordered spatial or temporal structures, so-called *dissipative structures*. Unlike an isolated, closed system, which after a perturbation always returns to a thermodynamic equilibrium state characterized by maximum entropy, an open dissipative nonlinear system may exhibit a process of *self-organization*, in which the entropy is locally decreased. Such processes usually involve qualitative changes in the state of the system, similar to phase transitions. Nonequilibrium phase transitions and dissipative structures have been noted in a great number of very different dissipative systems occurring in physics, chemistry, biology, ecology

(Glansdorff and Prigogine 1971, Nicolis and Prigogine 1977, Haken 1983, Bergé *et al.* 1987, Feistel and Ebeling 1989, Manneville 1990, Murray 1993, Cross and Hohenberg 1993, Mikhailov 1994, Walgraef 1997, Mori and Kuramoto 1998, Busse and Müller 1998), and even economics and social sciences (Weidlich and Haag 1984, Mantegna and Stanley 1999, Moss de Oliveira *et al.* 1999), but the phenomena observed are similar. Famous examples are the laser (Graham and Haken 1968), the Bénard and Taylor instabilities in hydrodynamics (Swinney and Gollub 1984), and chemical reaction systems (Kuramoto 1988). In the field of semiconductor physics, nonlinear generation–recombination processes (Schöll 1987) and nonlinear optical effects (Haug 1988) may give rise to nonequilibrium phase transitions. In the first case they manifest themselves as electric instabilities such as current runaway, threshold switching between a nonconducting and a conducting state, spontaneous oscillations of the current or voltage, and nucleation and growth of current filaments or high-field domains, if sufficiently high electric or magnetic fields, injected currents, or optical or microwave irradiations are applied. The study of such nonlinear effects in semiconductors is now established as a mature part of the interdisciplinary field of *synergetics* which was pioneered by Haken (1983, 1987).

Though these matters have become an active field of semiconductor research only recently, there was some singular early work, for example on the bifurcation of current filaments in connection with dielectric breakdown of solids (Lueder *et al.* 1936), and phase-portrait analysis of field domains in CdS crystals (Böer and Quinn 1966). The analogy of an overheating instability of the electron gas with an equilibrium phase transition was pointed out by Volkov and Kogan (1969), and Pytte and Thomas (1969) drew this analogy in the case of the Gunn instability of the electron drift velocity at about the same time. The early theory of domain instabilities in semiconductors was reviewed by Bonch-Bruevich *et al.* (1975). Generation–recombination-induced phase transitions in semiconductors were first noted by Landsberg and Pimpale (1976), stimulated by the similarity with Schlögl's chemical reaction models for nonequilibrium phase transitions (Schlögl 1972). Impact ionization of electrons or holes by hot carriers across the bandgap or from localized levels was recognized as the main autocatalytic process which is necessary for phase transitions, and low-temperature impurity breakdown was studied as a prominent example (Landsberg *et al.* 1978, Schöll and Landsberg 1979, Robbins *et al.* 1981). Current filamentation was treated as a process of self-organization in a system far from equilibrium (Schöll 1982*b*). The interest in this field was greatly increased by the experimental discovery of deterministic chaos in semiconductors (Aoki *et al.* 1981, Teitsworth *et al.* 1983). The introduction of concepts and methods from nonlinear dynamics subsequently stimulated a large amount of experimental and theoretical work on chaos and spatio-temporal self-organized pattern formation in a variety of semiconducting materials, e.g. p-Ge and n-GaAs in the regime of low-temperature impurity breakdown, and in layered semiconductor structures such as p–i–n and p–n–p–n diodes, double-barrier resonant-tunneling structures, heterostructure hot-electron diodes, and semiconductor superlattices (Schöll 1987, 1998b, Shaw et al. 1992, Thomas 1992, Peinke et al. 1992, Kerner and Osipov 1994, Niedernostheide 1995, Aoki 2000).

Progress has recently been made by using elaborate experimental techniques such as scanning electron microscopy (Mayer et al. 1988, Wierschem et al. 1995), scanning laser microscopy (Brandl et al. 1989, Spangler et al. 1994, Kukuk et al. 1996), potential probe measurements (Baumann et al. 1987, Niedernostheide et al. 1992a), and quenched photoluminescence (Eberle et al. 1996, Belkov et al. 1999) to detect spatially and temporally resolved structures. Detailed computer simulations in one and two spatial dimensions have allowed a quantitative comparison between theory and experiment. Whereas in earlier work irregular and chaotic *temporal* behavior was the center of interest (Abe 1989, Schöll 1992), the focus has now shifted toward more complex *spatio-temporal* dynamics including the dynamics of solitary filaments and multifilamentary states, spatio-temporal chaos, higher bifurcations of the elementary dissipative structures, interaction of defects and structural imperfections with pattern formation, and complex two-dimensional sample and contact geometries.

1.1.2   Negative differential conductivity

The electric transport properties of a semiconductor show up most directly in its current–voltage characteristic under time-independent bias conditions (dc, direct current). It is determined in a complex way by the microscopic properties of the material, which specify the current density j as a function of the local electric field $\mathcal{E}$, and by the contacts. A local, static, scalar $j(\mathcal{E})$ relation need not always exist, but in fact does in many cases.

Close to thermodynamic equilibrium, i.e. at sufficiently low bias voltage, the $j(\mathcal{E})$ relation is linear (*Ohm's Law*), but under practical operating conditions it will generally become nonlinear and may even display a regime of *negative differential conductivity*

$$\sigma_{\text{diff}} = \frac{dj}{d\mathcal{E}} < 0. \tag{1.1}$$

Thus the current density decreases with increasing field, and vice versa, which in general corresponds to an unstable situation. The actual electric response depends, for instance, upon the contact conditions and the attached circuit, which in general contains – even in the absence of external load resistors – unavoidable resistive and reactive components such as lead resistances, lead inductances, package inductances, and package capacitances.

Two important cases of negative differential conductivity (NDC) are described by an N-shaped or an S-shaped $j(\mathcal{E})$ characteristic, and denoted by NNDC and SNDC, respectively (Fig. 1.1). However, more complicated forms such as Z-shaped, loop-shaped, or disconnected characteristics are also possible (Wacker and Schöll 1995).

NNDC and SNDC are associated with voltage- and current-controlled instabilities, respectively. In the NNDC case the current density is a single-valued function of the field, but the field is multivalued: The $\mathcal{E}(j)$ relation has three branches in a certain range of j. The SNDC case is complementary in the sense that $\mathcal{E}$ and j are interchanged. This duality is in fact far-reaching, and will be elaborated upon subsequently.

The *global* current–voltage characteristic $I(U)$ of a semiconductor can in principle be calculated from the *local* $j(\mathcal{E})$ relation by integrating the current density j over the cross-section A of the current flow

$$I = \int_A j \, df \tag{1.2}$$

and the electric field $\mathcal{E}$ over the length L of the sample

$$U = \int_0^L \mathcal{E} \, dz. \tag{1.3}$$

Unlike the $j(\mathcal{E})$ relation, the $I(U)$ characteristic is not only a property of the semiconductor material, but also depends on the geometry, the boundary conditions, and the contacts of the sample. Only for the idealized case of spatially homogeneous states are the $j(\mathcal{E})$ and the $I(U)$ characteristics identical, up to re-scaling. The $I(U)$ relation is said to display *negative differential conductance* if

$$\frac{dI}{dU} < 0. \tag{1.4}$$

In case of NNDC, the NDC branch is often but not always – depending upon external circuit and boundary conditions – unstable against the formation of electric field domains, whereas in the SNDC case current filamentation generally occurs (Ridley 1963), as we shall discuss in detail in Chapter 3. These primary self-organized spatial patterns may themselves become unstable in secondary bifurcation, leading to periodically or chaotically breathing, rocking, moving, or spiking filaments or domains, or even solid-state turbulence and spatio-temporal chaos.

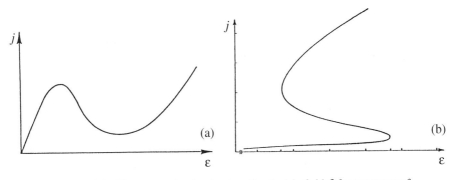

Figure 1.1. The current density j versus the electric field $\mathcal{E}$ for two types of negative differential conductivity (NDC): (a) NNDC and (b) SNDC (schematic).

1.2   Bifurcations in dynamic systems

1.2.1   Basic notions

Generally, a dynamic system is specified by

(i) a set of dynamic variables $q \equiv (q_1, \ldots, q_N) \in \Gamma$ that depends upon time t and spatial coordinates r, and forms the *phase space* Γ;

(ii) a law determining the temporal evolution $q(t)$, i.e. the phase trajectories or orbits, for given initial conditions q_0; and

(iii) a set of *control parameters* $k \equiv (k_1, \ldots, k_M)$ by which the dynamic law can be changed.

For a semiconductor the dynamic variables may represent, for instance, the local electric field $\mathcal{E}$, the carrier densities in the conduction and valence bands, and in impurity levels, the mean carrier energy and momentum, or the radius of a current filament. The control parameters are given by external parameters such as applied currents, electric or magnetic fields, and optical irradiation intensities.

The modeling of a spatially extended nonlinear dynamic system, such as a semiconductor, may be cast into one of the following common forms, depending on whether space, time, and the dynamic variables are chosen as continuous or discrete:

(i) a set of partial differential equations
    (space, time, and dynamic variables all continuous);

(ii) a set of ordinary differential equations
    (space discrete, time and dynamic variables continuous);

(iii) iterated maps
    (space and time discrete, dynamic variables continuous); and

(iv) Cellular automata
    (space, time, and dynamic variables all discrete).

Semiconductor transport theory and generation–recombination kinetics are usually formulated in terms of (i) or (ii), as will be discussed in Chapter 2. Only in a few exceptional cases have models of the discrete type (iii) been used, e.g. to describe superlattices (Bonilla *et al.* 1994*b*) or the onset of chaos due to generation–recombination processes (Landsberg *et al.* 1988). High-field transport has also been modeled in terms of cellular automata (Kometer *et al.* 1992, Schöll 1998*b*).

In the following we shall briefly review some basic aspects of stability and bifurcation (Haken 1983), confining our attention to the spatially homogeneous case (ii). The dynamic system is given by

$$\frac{dq}{dt} = F(q, k), \tag{1.5}$$

where $\boldsymbol{F} \equiv (F_1, \ldots, F_N)$ is a set of nonlinear functions of $\boldsymbol{q}$. More precisely, eq. (1.5) is called an *autonomous* dynamic system if the functions $\boldsymbol{F}$ do not depend explicitly upon time, i.e. if all the time dependency is through $\boldsymbol{q}$. For every phase point $\boldsymbol{q}$, except for the *fixed points* $\boldsymbol{q}^*$ where $\boldsymbol{F}(\boldsymbol{q}^*, \boldsymbol{k}) = 0$, the differential equations (1.5) give a unique direction of the phase flow $\boldsymbol{q}(t)$. Thus one can construct a phase portrait of the *trajectories*, or *orbits*, similar to the streamlines of a fluid. The fixed points represent time-independent steady states, which are in general states far from thermodynamic equilibrium. Any solution $\boldsymbol{q}(t)$, which may be regarded as the path of a particle in $\boldsymbol{q}$-space, is uniquely determined by its initial value $\boldsymbol{q}_0$ at the initial time t_0, and will be denoted by $q \equiv \boldsymbol{u}_{q_0}(t)$. Since the initial values are normally subject to perturbations, the question of interest becomes this: What happens to the path of the particle if the initial condition is slightly different from the original choice? A trajectory is said to be *locally stable* if other trajectories, obtained from other initial conditions, different from, but close to, $\boldsymbol{q}_0$, remain close to the original trajectory for all later times. It is *locally asymptotically stable* if, for all trajectories $\boldsymbol{v}_{q'_0}$, close to $\boldsymbol{u}$, whose phase points satisfy the stability criterion, the condition

$$|\boldsymbol{u}_{q_0}(t) - \boldsymbol{v}_{q'_0}(t)| \to 0 \quad \text{as} \quad t \to \infty \tag{1.6}$$

also holds. A bounded, undecomposable, invariant, and locally asymptotically stable subset of the phase space is called an *attractor*. An attractor is *globally stable* if its basin of attraction includes the whole phase space. Note that attractors can occur only in *dissipative* dynamic systems, in which phase volumes contract. These definitions include the special case that $\boldsymbol{u}$ consists of a single fixed point: A point attractor. If the phase space is at least two-dimensional, periodic attractors (*limit cycles*) are also possible.

Three dynamic degrees of freedom are required for a qualitatively new type of attractor: A *strange* or *chaotic* attractor. It consists in an attracting, bounded set of phase points with the property that trajectories within the attractor that are initially very close to each other become separated exponentially fast with time. Hence, this implies that there is a sensitive dependency upon initial conditions, and an unstable motion *within* the attractor.

1.2.2   Linear stability analysis

An important method for determining the local stability of fixed points is the linear mode or linear stability analysis. By way of example, we shall examine a two-variable autonomous dynamic system of the form (1.5). In fact, this treatment holds even more generally since higher-dimensional systems can generically be reduced to a one- or two-dimensional *center manifold* and to a normal form of the resulting effective dynamic equations. In the vicinity of a fixed point (q_1^*, q_2^*) the dynamic system (1.5) may be linearized around (q_1^*, q_2^*). In this approximation the trajectories near a fixed point can be computed explicitly. On setting

$$\delta q_1 \equiv q_1 - q_1^*,$$

$$\delta q_2 \equiv q_2 - q_2^*, \tag{1.7}$$

we obtain from (1.5)

$$\frac{d}{dt}\delta\boldsymbol{q} = \mathcal{A}(q_1^*, q_2^*)\,\delta\boldsymbol{q}, \tag{1.8}$$

where the Jacobian matrix $\mathcal{A}(q_1^*, q_2^*)$ is defined by its matrix elements

$$\mathcal{A}_{ij} \equiv \frac{\partial F_i}{\partial q_j}(\boldsymbol{q}^*) \qquad i, j = 1, 2. \tag{1.9}$$

The general solution of the system of linear differential equations (1.8) in the generic case $(\lambda_1 \neq \lambda_2)$ is

$$\delta\boldsymbol{q}(t) = c_1 \boldsymbol{\eta}_1 e^{\lambda_1 t} + c_2 \boldsymbol{\eta}_2 e^{\lambda_2 t}, \tag{1.10}$$

where c_1 and c_2 are constants, $\boldsymbol{\eta}_1$ and $\boldsymbol{\eta}_2$ are eigenvectors of $\mathcal{A}$, and λ_1 and λ_2 are the eigenvalues of $\mathcal{A}$, given by the characteristic equation

$$\lambda^2 - T\lambda + D = 0 \tag{1.11}$$

with the solution

$$\lambda_{1,2} = \tfrac{1}{2}\big[T \pm (T^2 - 4D)^{1/2}\big]. \tag{1.12}$$

Here

$$T \equiv \mathcal{A}_{11} + \mathcal{A}_{22}, \tag{1.13}$$

$$D \equiv \mathcal{A}_{11}\mathcal{A}_{22} - \mathcal{A}_{12}\mathcal{A}_{21} \tag{1.14}$$

are the trace and the determinant of the matrix $\mathcal{A}$, respectively. For an asymptotically stable fixed point all trajectories (1.10) are required to approach the fixed point as $t \to \infty$, hence the real parts of both λ_1 and λ_2 must be negative.

The five qualitatively different regimes of solutions (1.10) are shown in Fig. 1.2.

(a) *Stable focus*: $D > 0$, $T < 0$, and $T^2 < 4D$. The eigenvalues are conjugate complex and have negative real parts. The solutions describe damped oscillations in the phase plane with angular frequency $\omega = (D - T^2/4)^{1/2}$ and damping constant $T/2$. The general form of the trajectory is an elliptical spiral.

(b) *Unstable focus*: $D > 0$, $T > 0$, and $T^2 < 4D$. The eigenvalues are conjugate complex and have positive real parts; here the elliptical spiral diverges.

(c) *Stable node*: $D > 0$, $T < 0$, and $T^2 > 4D$. Both eigenvalues are negative. The solutions can be chosen to be real, and decay exponentially.

(d) *Unstable node*: $D > 0$, $T > 0$, and $T^2 > 4D$. Both eigenvalues are positive. The solutions can be chosen to be real, and diverge exponentially.

(e) *Saddle-point*: $D < 0$. The eigenvalues are real, with one eigenvalue positive and one negative. The positive eigenvalue causes an instability in the direction of the corresponding eigenvector (an *unstable manifold*), whereas the other eigenvector is attractive (a *stable manifold*).

The borderlines between these five regimes represent structurally unstable cases. They require more detailed investigations involving higher orders in the Taylor expansion of the system around the fixed point. In particular, for $T = 0$ and $D > 0$ the trajectories near the fixed point are either ellipses, or slowly converging or diverging spirals, depending upon the nonlinear terms of the expansion. In the first

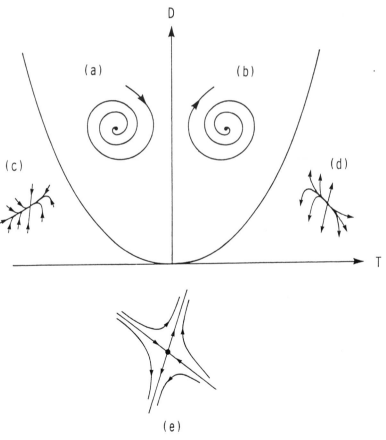

Figure 1.2. Regimes of stability of the fixed points of a two-variable autonomous dynamic system. T and D denote the trace and the determinant, respectively, of the Jacobian matrix. The insets show the qualitative natures of the phase portraits near the respective fixed points for the five generic cases: (a) a stable focus, (b) an unstable focus, (c) a stable node, (d) an unstable node, and (e) a saddle-point.

case the fixed point is called a *center*, in the second case a *weakly* stable or unstable focus. The borderlines between regimes (a) and (b), and between (c) and (e) are associated with various bifurcations that we shall study in the next subsection.

1.2.3   Classification of bifurcations

Bifurcation is a phenomenon peculiar to *nonlinear* dynamic systems and is closely related to the loss of stability (Andronov *et al.* 1971, 1973, Sattinger 1973, Guckenheimer and Holmes 1983, Thompson and Stewart 1986, Schuster 1988, Troger and Steindl 1991, Glendinning 1994). It describes the branching of solutions (fixed points or periodic orbits or even chaotic attractors) as a control parameter k is varied. If different solution branches intersect or coalesce, they usually change their stability character. In case of fixed points this means that at least one of the eigenvalues of the Jacobian $\mathcal{A}$ has a vanishing real part. At the bifurcation point the system undergoes a structural change. If it is sufficient to vary one control parameter, the bifurcation is of *codimension one*. Codimension-two bifurcations require the simultaneous adjustment of two control parameters.

In the following we give a classification of the most common types of codimension-one bifurcations for the example of a two-variable autonomous dynamic system (with the exception of cases (C2) and (C3) below, which require a three-variable system).

(A) Zero-eigenvalue bifurcations

The simplest *local* bifurcation occurs if a single eigenvalue λ of $\mathcal{A}$ turns from positive to negative values upon variation of the control parameter k. At the bifurcation point k_0 the eigenvalue is zero. Thus the center manifold, which is the invariant manifold associated with the eigenspace of the zero eigenvalue, is one-dimensional, and the bifurcation can already arise in a one-variable system. The condition for the Jacobian matrix $\mathcal{A}$ is $D = 0$ while $T \neq 0$. The following subclasses may occur (see Fig. 1.3):

(A1) Saddle-node bifurcation or fold
A saddle-point (sa) and a stable node (sn) coalesce and disappear. This bifurcation occurs, e.g. at the two turning points of an SNDC current–voltage characteristic. The normal form is given by the one-variable system $\dot{q} = k - q^2$, where the dot denotes the time derivative.

(A2) Transcritical bifurcation
A saddle-point and a stable node cross and exchange their stability characters. Often the unstable branch, upon further increase of the control parameter, turns back in a saddle-node bifurcation of type (A1). The normal form is $\dot{q} = kq - q^2$.

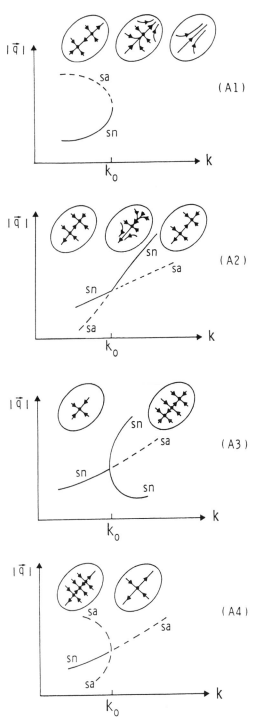

Figure 1.3. Bifurcation diagrams for a zero-eigenvalue bifurcation. The coordinate q of the fixed point is plotted versus the control parameter k. Full lines represent stable states, broken lines are unstable states. Saddle-points and stable nodes are denoted by sa and sn, respectively. The insets present schematic phase portraits corresponding to values of $k < k_0$, $k = k_0$, and $k > k_0$, where k_0 is the bifurcation point.

(A3) Supercritical pitchfork bifurcation

A saddle-point and two stable nodes coalesce into a single stable node. This is typical of a symmetry-breaking bifurcation, such as occurs at a critical point of a phase transition. The normal form is $\dot{q} = kq - q^3$.

(A4) Subcritical pitchfork bifurcation

Two saddle-points and a stable node coalesce into a single unstable fixed point (a saddle-point). The normal form is $\dot{q} = -kq + q^3$.

(B) Hopf bifurcation

If a pair of complex-conjugate eigenvalues of $\mathcal{A}$ crosses the imaginary axis transversely, the fixed point changes from stable to unstable focus, and a time-dependent periodic solution bifurcates, as shown in Fig. 1.4. This periodic solution is asymptotically orbitally stable, i.e. all trajectories starting in a neighborhood (inside and outside) of this closed orbit in the phase plane will asymptotically approach it. It is therefore called a *periodic attractor* or *limit cycle*. The bifurcation depicted in Fig. 1.4 is *supercritical* (a stable focus bifurcates into an unstable focus and a stable limit cycle), but *subcritical Hopf bifurcation* (an unstable focus bifurcates into a stable focus and an unstable limit cycle) also exists. The bifurcation point is given by the condition $T = 0$, $D > 0$. The limit cycle bifurcates at $k = k_0$ with angular frequency $\omega = \sqrt{D} \neq 0$, and the amplitude can be shown to increase as $\sqrt{k - k_0}$. Since the frequency does not tend to zero, i.e. does not "soften", the bifurcation is

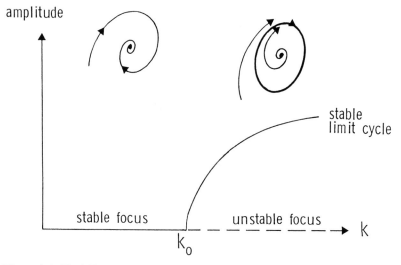

Figure 1.4. The bifurcation diagram for a supercritical Hopf bifurcation of a limit cycle. The amplitude of the limit cycle is plotted versus the control parameter k. Full lines represent stable states, broken lines are unstable states. The insets represent schematic phase portraits for $k < k_0$ (stable focus) and $k > k_0$ (unstable focus and stable limit cycle).

sometimes called a *hard mode instability*. This bifurcation has a two-dimensional center manifold, and hence requires at least two dynamic variables.

(C) Local bifurcations of limit cycles

Bifurcations may not only transform fixed points but also start from periodic attractors. Here we shall discuss three local bifurcations that a limit cycle may undergo.

(C1) Limit-cycle bifurcation by condensation of paths

A stable and an unstable limit cycle around the same fixed point can coalesce and disappear (Fig. 1.5). At the bifurcation point both the amplitude and the frequency are nonzero, as opposed to the Hopf bifurcation. This bifurcation is also called a *cyclic fold* since it is similar to the fold bifurcation discussed in (A1) above if one considers a transverse intersection of the orbits with a plane in phase space (a *Poincaré section*) that projects the two limit cycles onto two fixed points in the Poincaré plane. Therefore it is also regarded as a local bifurcation.

(C2) Period-doubling bifurcation of a limit cycle

Another local bifurcation of a limit cycle is the period-doubling or *flip instability* (Fig. 1.6). In this *subharmonic bifurcation* a limit cycle loses its stability, and another cycle whose period is twice the period of the original cycle, is born. The

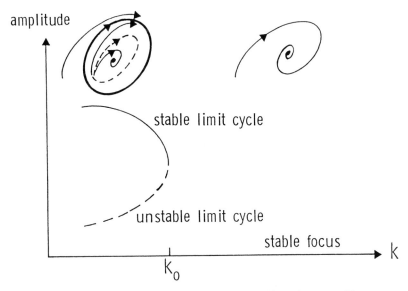

Figure 1.5. The bifurcation diagram for a limit-cycle bifurcation created by condensation of paths. Full lines represent stable states, broken lines are unstable states. The insets show a stable and an unstable limit cycle around a stable focus for $k < k_0$, and a stable focus for $k > k_0$.

new stable period-two orbit lies on the edge of a Möbius strip. The width of the Möbius strip increases continuously from zero as k passes the bifurcation point. It is associated with a torsion of orbits in the vicinity of the original period-one limit cycle, which leads to the bifurcation of the period-two orbit when the phase flip of nearby trajectories during one cycle becomes π. This bifurcation requires at least a three-dimensional phase space. In many cases a sequence of n period-doubling bifurcations is observed, in which limit cycles of period $2^n T$ are successively created. If the sequence is infinite, with a finite accumulation point, the resulting motion beyond that point is chaotic. On the other hand, it is also possible to observe a finite number of period doublings, followed, e.g. by the same number of reverse bifurcations.

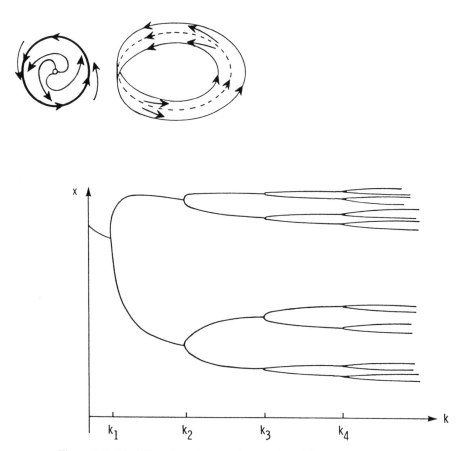

Figure 1.6. The bifurcation diagram of a period-doubling bifurcation of a limit cycle. The inset shows a stable limit cycle for $k < k_1$, and a stable period-two cycle and an unstable period-one cycle for $k > k_1$ (after Thompson and Stewart (1986)).

(C3) Secondary Hopf bifurcation of a limit cycle

A periodic orbit may itself undergo a second Hopf bifurcation (*Neimark bifurcation*) that generates a second incommensurate frequency ω_2 (i.e. the frequency ratio ω_1/ω_2 is irrational). It is convenient to parametrize the orbit in terms of the two angular coordinates $\phi_1 = \omega_1 t$ and $\phi_2 = \omega_2 t$ (Fig. 1.7). The doubly periodic motion can thus be visualized as motion on a two-dimensional torus (a *two-torus*). As the old limit cycle loses stability, a two-torus is born (Fig. 1.8). The motion is called *quasiperiodic* and the orbit fills densely the entire torus. Also this bifurcation requires at least a three-dimensional phase space. In some special cases the limit cycle can bifurcate into another periodic orbit whose period is an integer multiple of the first (*mode locking*). The resulting motion lies on a two-dimensional torus, but does not fill the whole surface.

(D) Global bifurcations of limit cycles

These bifurcations are called *global* bifurcations, since they involve not merely local changes of the phase portrait but rather global changes of the topological configuration, e.g. the invariant manifolds of saddle-points. Two possible topologies are shown in Fig. 1.9. In both cases the limit cycle disappears upon collision with a saddle-point. One may also visualize these bifurcations as *bifurcations of limit cycles from a separatrix*. A separatrix is an invariant manifold that separates two attractor basins or topologically different regions of phase space; specifically, in

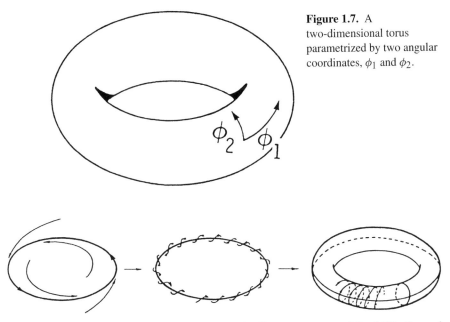

Figure 1.7. A two-dimensional torus parametrized by two angular coordinates, ϕ_1 and ϕ_2.

Figure 1.8. Secondary Hopf bifurcation leading to a two-torus (after Mikhailov and Loskutov (1996)).

(D1) below the separatrix is a closed loop consisting of the unstable manifold of the saddle-point with the two branches joined together at a stable fixed point (node); in (D2) below the separatrix represents a closed-loop orbit that is formed if one branch of the unstable manifold and one branch of the stable manifold of the saddle-point

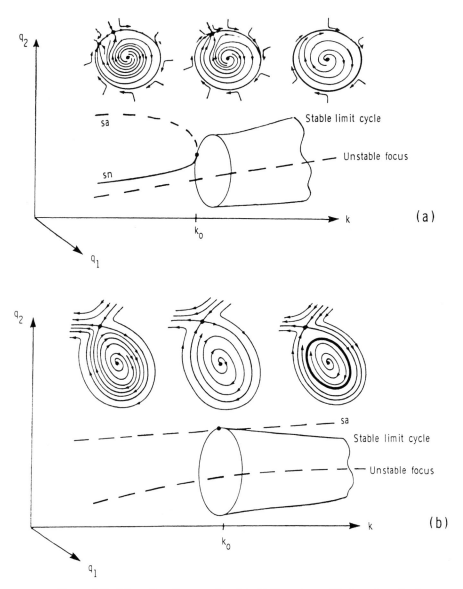

Figure 1.9. Bifurcation diagrams for global bifurcations of limit cycles. The limit cycle is represented by a cylindrical surface in the three-dimensional space of dynamic variables (q_1 and q_2) and the control parameter k. (a) Omega explosion (saddle-node bifurcation on a limit cycle). (b) Homoclinic bifurcation (the blue-sky catastrophe).

coincide in what is called a *homoclinic orbit*. In both cases the limit cycle appears with a nonzero amplitude but with zero frequency. Thus the bifurcation is associated with a critical slowing down of the oscillation frequency according to a universal scaling law. Such bifurcations have indeed been observed experimentally (Peinke *et al.* 1989) and modeled theoretically (Hüpper *et al.* 1989, Döttling and Schöll 1992, Patra *et al.* 1998) in semiconductors in the regime of low-temperature impurity breakdown, in real-space transfer heterostructures, and in superlattices.

(D1) Omega explosion

In Fig. 1.9(a), for $k < k_0$, a saddle-point and a stable node lie on a closed loop that is formed by the two branches of the unstable manifold of the saddle-point. At $k = k_0$ they coalesce and disappear in a saddle-node bifurcation, whereby a stable limit cycle bifurcates from this separatrix. Thus an abrupt transition from a point attractor to a stable limit cycle occurs.

(D2) Homoclinic bifurcation

In Fig. 1.9(b), as k increases, a homoclinic orbit (a saddle-to-saddle loop consisting in two coinciding branches of the stable and the unstable manifolds of the saddle-point) is formed at $k = k_0$. From this homoclinic connection a stable limit cycle bifurcates. Here the saddle-point does not disappear. Since the limit-cycle attractor appears "out of the blue sky", this bifurcation has also been termed the *blue-sky catastrophe*. Often the homoclinic bifurcation is associated with bistability between the limit cycle and a further point attractor that is born by a fold bifurcation from the saddle-point. This behavior typically occurs in semiconductors with S-shaped current–voltage characteristics and is associated with hysteresis between self-sustained current oscillations and a steady state upon up-sweep and down-sweep of the voltage (Döttling and Schöll 1992).

(E) Bifurcations of spatial patterns

So far we have considered low-dimensional dynamic systems described by ordinary differential equations. In spatially extended systems there exists a rich menagery of more complex bifurcations, which leads to the formation of nonuniform spatial or spatio-temporal dissipative structures, as a result of the occurrence of spatial derivatives describing diffusion and drift (Cross and Hohenberg 1993). A prototype example is the nonlinear reaction–diffusion system

$$\frac{\partial}{\partial t} \boldsymbol{q}(x, t) = \boldsymbol{F}(\boldsymbol{q}, k) + D \,\Delta \boldsymbol{q}, \tag{1.15}$$

where D is a diffusion matrix. The linear stability analysis proceeds as in Section 1.2.2, if the spatial dependency is expressed in terms of a suitable set of modes satisfying appropriate boundary conditions, e.g. by Fourier modes $\exp(i\boldsymbol{K}x)$, where $\boldsymbol{K}$ is a wave vector and x is the spatial coordinate. In case of the reaction–diffusion system (1.15) the stability of the uniform steady state is determined by the linear evolution matrix

$$\mathcal{A}_{ij} \equiv \frac{\partial F_i}{\partial q_j}(\boldsymbol{q}^*) - D_{ij}\boldsymbol{K}^2. \tag{1.16}$$

The eigenvalues of $\mathcal{A}$ are thus given by the dispersion relation $\lambda(\boldsymbol{K})$. The uniform steady state is asymptotically stable if all eigenvalues $\lambda(\boldsymbol{K})$ have negative real parts; then small periodic spatial fluctuations are damped out. If $\operatorname{Re}\lambda(\boldsymbol{K})$ changes sign at some value of $\boldsymbol{K}$, a spatially periodic solution with the wave vector $\boldsymbol{K}$ bifurcates from the uniform steady state. Upon variation of the control parameters the dispersion relation $\lambda(\boldsymbol{K})$ is changed, and a whole family of spatially nonuniform modes $\boldsymbol{K}$ becomes unstable and their coupling may lead to complex instabilities. Here we consider only the following simple cases.

(E1) Turing instability

If, at the bifurcation point, $\mathcal{A}$ has a zero eigenvalue $\lambda(\boldsymbol{K}) = 0$ at $|\boldsymbol{K}| = K_c \neq 0$, the bifurcation is a pattern-forming instability leading to stationary spatially periodic structures with an intrinsic wave vector K_c. In particular, such *Turing structures* occur in two-variable reaction–diffusion systems with a slowly diffusing activator variable and a fast-moving inhibitor variable, as we shall see in Section 1.4. The generic dispersion relation near the instability is given by

$$\lambda(K) = \epsilon - b(K^2 - K_c^2)^2, \tag{1.17}$$

where $\epsilon = (k - k_0)/k_0$ measures the distance from the bifurcation point given by the control parameter $k = k_0$. Note that, in systems with two or more spatial dimensions, the problem of orientational degeneracy and of the selection of patterns of different symmetries needs further analysis (Walgraef 1997).

(E2) Standing and traveling waves

If, at the bifurcation point, $\mathcal{A}$ has two complex-conjugate eigenvalues with vanishing real parts $\lambda(\boldsymbol{K}) = \pm i\omega$ at $|\boldsymbol{K}| = K_c \neq 0$, the instability leads to oscillatory spatially periodic structures, i.e. traveling or standing waves with wave vector K_c, corresponding to the generic dispersion relation near the instability

$$\lambda(K) = \epsilon - b(K^2 - K_c^2)^2 \pm i\big(\omega + \mathcal{O}(\epsilon, (K^2 - K_c^2))\big). \tag{1.18}$$

(E3) Hopf bifurcation

If, at the bifurcation point, $\mathcal{A}$ has two complex-conjugate eigenvalues with vanishing real parts at $|\boldsymbol{K}| = K_c = 0$, the dispersion relation near the instability is of the form

$$\lambda(K) = \epsilon - b(K^2)^2 \pm i\big(\omega + \mathcal{O}(\epsilon, K^2)\big) \tag{1.19}$$

and we have a Hopf bifurcation leading to spatially uniform limit-cycle oscillations as in (B) above. The nonlinear coupling with spatial modes may, however, induce more complex oscillatory spatio-temporal patterns, e.g. spiral waves (Mikhailov 1994).

Of course the zero eigenvalue bifurcation (A) of uniform fixed points is also possible, viz. if $\lambda(K) = 0$ at $|K| = 0$.

1.3   Deterministic chaos

In two-variable autonomous purely temporal dynamic systems, such as we have mainly discussed so far, the field of directions of the phase flow is uniquely defined, and therefore trajectories can not cross. All trajectories in a bounded dissipative system must tend asymptotically to either a fixed point or a limit cycle, i.e. a zero- or one-dimensional attractor. Now we shall consider autonomous systems involving at least three dynamic variables. This third degree of freedom adds a qualitatively new type of behavior, which is associated with an irregular and unpredictable temporal evolution of the system. If such irregular but bounded motion depends sensitively upon the initial conditions it is called *chaotic*. It is distinct from the irregularities caused by stochastic fluctuations in systems with many microscopic degrees of freedom since here we are considering a deterministic system with few degrees of freedom. The sensitivity to small variations of the initial conditions is characterized by an exponential divergence of neighboring trajectories. If two initial points are separated by a small distance $\delta x(0)$, it may grow as $\delta x(t) = \delta x(0) \exp(\lambda t)$, where $\lambda > 0$ measures the mean rate of divergence of the orbits.

An autonomous system of three ordinary differential equations of first order can already produce such deterministic chaos. In a three-dimensional phase space the trajectories can be wrapped or twisted around each other in quite intricate ways. Simple examples are provided by the Lorenz equations (Lorenz 1963) and the Rössler equations (Rössler 1976). In appropriate parameter ranges these systems possess a *strange attractor*. It consists in a set of phase points embedded in a bounded region of phase space. Trajectories outside that region but close enough to it are attracted to that region. Trajectories within that strange attractor that are initially very close to each other become separated exponentially fast with time. Intuitively, this implies that there is unstable motion *within* the attractor. Since the attractor is bounded, the trajectories are folded back at some point. This stretching and folding of trajectories is characteristic of chaotic motion. The "strangeness" of the attractor consists in the fact that it is not a fixed point (dimension zero), a limit cycle (dimension one) or a torus (dimension two), but has in general a noninteger *fractal dimension*. For any bounded set of points in $\mathbb{R}^p$, representing the strange attractor, its fractal dimension D_0 (the Hausdorff dimension) can be defined as

$$D_0 = \lim_{\epsilon \to 0} \frac{\log N(\epsilon)}{\log(1/\epsilon)}, \tag{1.20}$$

where $N(\epsilon)$ is the minimum number of p-dimensional cubes of side length ϵ needed to cover the set. Thus

$$N(\epsilon) \sim \epsilon^{-D_0} \quad \text{for} \quad \epsilon \to 0, \tag{1.21}$$

which gives $D_0 = 0$, 1, and 2 for the trivial cases of a point, a line, and a surface, as expected. It is important to stress that the fractal dimension D_0 calculated from the above scaling law is meaningful only if it is independent from the embedding dimension p of the embedding space, which requires at least $p > 2D_0 + 1$.

In practical experiments and numerical simulations, the computation of D_0 is often made difficult because a very large number of points on the attractor is needed in order to ensure the validity of the asymptotic scaling law. A more rapidly converging algorithm has been proposed by Grassberger and Procaccia (1983a, 1983b), which generally gives a very good lower bound $d \leq D_0$. Here the *correlation dimension* d is defined via the scaling law for $\epsilon \to 0$:

$$C(\epsilon) \sim \epsilon^d \quad \text{or} \quad d = \lim_{\epsilon \to 0} \frac{\log C(\epsilon)}{\log \epsilon}, \tag{1.22}$$

where the correlation integral

$$C(\epsilon) = \lim_{N \to \infty} \frac{1}{N^2} \sum_{i,j=1}^{N} \Theta(\epsilon - |\boldsymbol{q}_i - \boldsymbol{q}_j|) \tag{1.23}$$

measures the numbers of pairs of points $\boldsymbol{q}_n \equiv (q_n, \ldots, q_{n+p-1})$ whose distance is less than ϵ in a p-dimensional embedding space with sufficiently large p. Θ is the Heaviside function. More generalized notions such as the multifractal spectrum of Renyi dimensions D_q ($q = 0, 1, 2, \ldots$) have also been discussed (Beck and Schlögl 1993).

Another important quantity used to characterize chaos is the spectrum of *Lyapunov exponents*. These are generalizations of the eigenvalues of the Jacobian matrix determining the linear stability of fixed points:

$$\lambda = \lim_{t \to \infty} \sup \left(\frac{1}{t} \ln |\boldsymbol{q}(t) - \boldsymbol{q}^*(t)| \right), \tag{1.24}$$

where $\boldsymbol{q}(t)$ are trajectories in the neighborhood of a reference trajectory $\boldsymbol{q}^*(t)$. The sign of the Lyapunov exponent indicates whether or not $\boldsymbol{q}^*(t)$ is asymptotically approached by other trajectories $\boldsymbol{q}(t)$. A positive Lyapunov exponent describes the exponential separation of nearby trajectories which is the defining property of deterministic chaos. It characterizes the expansion of a volume element in phase space in some direction. An invariant set always (with the exception of fixed points) has one vanishing Lyapunov exponent associated with the direction of motion. It reflects the fact that a volume element is neither expanded nor contracted along the direction of the orbit. Finally, an attractor has at least one negative Lyapunov exponent associated with the attracting direction; it describes the contraction of volume elements in dissipative dynamic systems. Thus, in case of a chaotic attractor in a three-variable dynamic system, one of the Lyapunov exponents is positive, one is zero, and one is negative. This may be symbolically expressed by the signature $(+, 0, -)$. Likewise, a fixed point attractor has the signature $(-, -, -)$, a limit cycle has $(0, -, -)$, and a torus attractor has $(0, 0, -)$.

A further characteristic feature of chaotic motion shows up in the correlations of the variables at different times: The *autocorrelation function*

$$\langle q(t)q(t + \tau)\rangle \equiv \lim_{T\to\infty} \frac{1}{2T} \int_{-T}^{T} q(t)q(t + \tau)\, dt \qquad (1.25)$$

tends to zero for $\tau \to \infty$. Such behavior is different from simply and even multiply periodic motion. For periodic motion one always obtains an undamped oscillatory correlation function, e.g. $\langle q(t)q(t + \tau)\rangle = \frac{1}{2}\cos(\omega\tau)$ for $q(t) = \sin(\omega t)$. In case of stochastic motion the autocorrelation function is often zero or negligible for τ greater than some finite autocorrelation time τ_c.

Finally, a characterization of chaos involves also the Fourier spectrum of the temporal evolution. Defining the spectral power density or *power spectrum* by

$$S(\omega) \equiv \lim_{T\to\infty} \frac{\pi}{T} |\hat{q}(\omega; T)|^2, \qquad (1.26)$$

where

$$\hat{q}(\omega; T) \equiv \frac{1}{2\pi} \int_{-T}^{T} q(t)e^{i\omega t}\, dt \qquad (1.27)$$

is the Fourier transform of $q(t)$ in a finite but sufficiently long observation interval $[-T, T]$, we find, using eq. (1.26),

$$S(\omega) = \frac{1}{2\pi} \int_{-\infty}^{\infty} \langle q(t)q(t + \tau)\rangle e^{i\omega\tau}\, d\tau. \qquad (1.28)$$

This means that the power spectrum is the Fourier transform of the autocorrelation function (Wiener–Khinchine theorem). For a simply or multiply periodic motion the power spectrum has discrete lines at the fundamental frequencies. Chaotic motion, however, is characterized by a continuous broad band of frequencies.

In spatially extended systems, *spatio-temporal chaos* may arise, which involves not only temporal degrees of freedom but also spatio-temporal dynamics (Mikhailov and Loskutov 1996). This usually goes along with the break-up of spatial coherence and may eventually lead to fully developed turbulence. Phase and amplitude instabilities, defect-mediated turbulence, and spatio-temporal intermittency, where laminar and turbulent regions coexist in space, are among the scenarios of spatio-temporal chaos (Manneville 1990, Mori and Kuramoto 1998). The degree of spatio-temporal chaos may be characterized by a larger number of positive Lyapunov exponents.

Here we can give only a brief survey of deterministic chaos. For a more thorough introduction the reader is referred to a number of textbooks and monographs (Bergé *et al.* 1987, Holden 1986, Schuster 1988, Ruelle 1989, Baker and Gollub 1990, Beck and Schlögl 1993, Ott 1993, Mikhailov and Loskutov 1996, Kantz and Schreiber 1997, Mori and Kuramoto 1998).

1.3.1   Routes to chaos

The onset of temporal chaos is usually preceded by other bifurcations, such as those discussed above. For example, with increasing value of the control parameter, the system may first exhibit a Hopf bifurcation from a stable fixed point to a limit cycle, and then, at still higher values of the control parameter, a transition to chaos. Various routes to chaos have been found. In the following we discuss three common scenarios that have universal features and appear in many systems of quite different physical origin.

- First, let us consider the transition to chaos via *period doubling* (the Feigenbaum scenario). When a limit cycle embedded in a three-dimensional phase space becomes unstable via a flip instability (see (C2) above), it bifurcates into another limit cycle with doubled period, i.e. half the frequency (*subharmonic bifurcation*). Upon further increase of the control parameter, a sequence of period-doubling bifurcations may occur at successive values of the control parameter k_n with $n = 1, 2, 3, \ldots$ with periods $2^n T$. For $n \to \infty$ the k_n converge to a critical value k_∞, at which the motion ceases to be periodic, and becomes chaotic.

  A convenient description of period doubling and other nonlinear dynamic effects is in terms of discrete iterated maps. These arise if the set of variables $q(t)$ is sampled at discrete times t_m. The sequence $q_m = q(t_m)$ can be constructed, for example, by taking the transverse intersections of the trajectories in N-dimensional phase space with an $(N-1)$-dimensional hypersurface (a *Poincaré section*). Note that the phase flow through the Poincaré surface induces an $(N-1)$-dimensional map $q_m \to q_{m+1}$, the *Poincaré map*, whereby the differential equations are replaced by difference equations. In particular, as the Poincaré section of a two-variable system with a limit cycle around the origin, we may choose the intersections with one of the coordinate half-axes, x_m. Each intersection point x_{m+1} is determined, once the previous point x_m is fixed. Therefore, a one-dimensional Poincaré map (or *first-return map*) $x_{m+1} = f(x_m)$ exists. A limit cycle is then represented by a single point of intersection x_∞ to which the intersection points of all neighboring orbits converge for $m \to \infty$. The period-doubling bifurcation sequence corresponds to a cascade of pitchfork bifurcations of the

Poincaré map (Fig. 1.6). The cascade of bifurcation values k_n has been found to obey the simple law (Grossmann and Thomae 1977, Feigenbaum 1978)

$$\lim_{n \to \infty} \frac{k_n - k_{n-1}}{k_{n+1} - k_n} = \delta \equiv 4.669\,2016\ldots \tag{1.29}$$

for a large class of one-dimensional maps with quadratic maxima. The prototype example is the logistic map $f(x) = kx(1 - x)$. Beyond k_∞, the chaotic regime may be interrupted by periodic windows, and further period-doubling cascades.

■ A second route to chaos is via *quasiperiodicity* (the Ruelle–Takens scenario). A limit cycle with frequency ω_1 undergoes a second Hopf bifurcation (see (C3) above) leading to a quasiperiodic motion on a two-torus with incommensurate frequencies ω_1 and ω_2. A subsequent bifurcation may either take the system back to a limit cycle with a single frequency (which is called *frequency-locking* or *mode-locking*), or add a third fundamental frequency ω (which transforms the two-torus into a three-torus). A condition for the latter is that the frequencies must be sufficiently irrational with respect to each other (the KAM condition). However, this is a structurally unstable situation, and small structural perturbations will destroy the three-torus and lead to a strange attractor. The observed scenario again obeys a universality law, which can be most easily studied in terms of an iterated map, viz. the *circle map* (Schuster 1988):

$$\Theta_{n+1} = \Theta_n + \Omega - \frac{K}{2\pi} \sin(2\pi \Theta_n), \tag{1.30}$$

where Θ is the iterated angular variable on the circle (modulo 1), and the parameters Ω and K denote the frequency ratio and the coupling strength of the two competing oscillatory modes, respectively. Increasing the strength of the nonlinear coupling between the oscillators develops an increasing tendency to lock into incommensurate motion for which the ratio of the oscillation frequencies is rational. The resonance-frequency-locked states form horn-like structures within the (Ω, K) control parameter plane known as *Arnold tongues*. They are ordered hierarchically according to the *Farey tree*, which orders all rationals in the interval [0, 1] with increasing denominator according to the rule that the rational with the smallest denominator between p/q and p'/q' is $(p + p')/(q + q')$, where $p, p', q,$ and q' are integers. In the field of semiconductor instabilities, the Farey tree has been observed, e.g. experimentally in Si p–i–n diodes (Coon *et al.* 1987), in p-Ge impurity breakdown (Peinke *et al.* 1992), and in driven superlattices (Luo *et al.* 1998*b*), and also in semiconductor models (Naber and Schöll 1990). At the critical line $K = 1$, the mode-locking intervals form a characteristic self-similar structure (a *devil's staircase*), i.e. it looks the same at any scale of magnification, and the Arnold tongues cover the entire Ω interval. For $K > 1$ the map ceases to be invertible, and chaos may set in, especially if Ω is chosen equal to the Golden Mean $(\sqrt{5} - 1)/2$, which is in some sense the most irrational number.

▪ A third common route to chaos is the *intermittency* scenario (Pomeau–Manneville). The intermittent motion of a dynamic system is characterized by the alternation of bursts of apparently chaotic behavior and intervals of almost periodic oscillations of length l (laminar flow). The associated Poincaré sections exhibit saddle-node bifurcations, in which a stable node and a saddle-point collide and annihilate. The repetition frequency of the chaotic bursts depends upon the external control parameter. There exist three different types of intermittency distinguished by different scaling laws of the mean laminar length $\langle l \rangle$ as a function of the normalized control parameter $\epsilon = (k - k_c)/k_c$ which measures the distance from the critical point where periodic motion sets on. Type-I intermittency is characterized by the scaling behavior

$$\langle l \rangle \sim |\epsilon|^{-1/2}. \tag{1.31}$$

The distribution of the laminar lengths $P(l)$ behaves as $P(l) \sim l^{-1/2}$ for small l and diverges also at a finite cut-off value $l_{max} = \epsilon^{-1/2}$. In the logistic map, for instance, type-I intermittencies appears at the onset of the period-three window at $k_c = 1 + \sqrt{8}$. For $k < k_c$ there appear laminar (periodic) intervals interrupted by chaos, and their length increases according to (1.32) as $k \to k_c$, and for $k > k_c$ there is a cycle of period three. Type-II and type-III intermittencies are characterized by the scaling law

$$\langle l \rangle \sim |\epsilon|^{-1} \tag{1.32}$$

and the distributions of the laminar lengths scale as $P(l) \sim l^{-2}$ and $P(l) \sim l^{-3/2}$ for small l for intermittencies of type II and type III, respectively, and tend to zero exponentially for $l \to \infty$ (Schuster 1988).

It should be noted that there exist other, more complex scenarios, e.g. homoclinic explosions and Shilnikov chaos, but these are beyond the present survey.

1.3.2   Chaos control

Chaotic behavior of dynamic systems, though it is widespread, is often undesired. For instance, in electric circuits and in lasers, the occurrence of chaos can sensitively affect the performance and in some cases even destroy the device. Likewise, in biological systems chaotic behavior, e.g. cardiac arhythmia and voice pathology, is highly undesirable and should be suppressed, as in mechanical engineering, chemical engineering, and financial markets. This raises the question of how to control chaotic dynamic systems in such a way as to convert chaotic motion into periodic motion by imposing deliberate perturbations. Since the seminal work by Ott, Grebogi, and Yorke (1990) in which they demonstrated that small time-dependent changes in the control parameter of the system can turn a previously chaotic trajectory into a stable periodic motion, there has been a lot of interest in this new field (Schuster 1999).

The key to controlling chaos is the observation that small perturbations by external forces (Ott *et al.* 1990) or time-delayed feedback (Pyragas 1992) can eliminate chaotic motion by stabilizing one of the unstable periodic orbits (UPO) which are embedded in any chaotic attractor. Analytic insight into the mechanism of delayed feedback control and the optimization of parameters (Bleich and Socolar 1996, Nakajima 1997, Just *et al.* 1997, 1998, 1999, Just 1999) has only recently been achieved. Although the control of electronic circuits (Ditto *et al.* 1990, Pyragas and Tamaševičius 1993, Sukow *et al.* 1997), plasmas (Pierre *et al.* 1996), and optical (Bielawski *et al.* 1994, Simmendinger and Hess 1996, Boccaletti *et al.* 1997), chemical (Zykov *et al.* 1997), and biological (Lourenço and Babloyantz 1994, Hall *et al.* 1997) systems, for instance, has been studied widely, little work has been devoted to controlling current instabilities in semiconductors, despite the fact that these systems would offer particularly useful applications if chaotic dynamics could be converted into a stable, tunable high-frequency electronic oscillator (Schöll and Pyragas 1993, Reznik and Schöll 1993, Cooper and Schöll 1995, Franceschini *et al.* 1999). In the following chapters we shall present a few examples that are designed to fill this gap by demonstrating that chaotic current oscillations in semiconductor devices can be very effectively controlled by a time-delay autosynchronization method.

The various control methods may be grouped into those which do not use feedback (*open loop*) and those which do (*closed loop*). In algorithms without feedback, generally a control parameter is modulated by a weak periodic signal in order to synchronize the system to periodic behavior and suppress chaos. Thus, by resonance stimulation, new dynamics is imposed upon the system. Examples have been given, e.g., by Hübler and Lüscher (1989) and by Braiman and Goldhirsch (1991).

Feedback methods, on the other hand, stabilize UPOs that are already contained in the uncontrolled dynamics. First we shall briefly review the method proposed by Ott, Grebogi, and Yorke (OGY) (Ott *et al.* 1990). They noted that a chaotic attractor has embedded within it a large number of low-period UPOs. In addition, because of ergodicity, the trajectory visits the neighborhood of each one of these. Some of these UPOs may correspond to the desired performance of a system. Chaos, while signifying that there is a sensitive dependency on small changes to the current state, also implies that the system's behavior can be altered by using small perturbations. Specifically, the OGY approach proceeds as follows: One first determines some of the low-period UPOs that are embedded in the chaotic attractor. One then chooses one that yields the desired performance of the system. Finally, as soon as the trajectory comes close to this orbit one applies a small feedback control to stabilize this desired UPO. For this purpose one constructs the $(N-1)$-dimensional Poincaré section of the dynamic system:

$$\xi_{n+1} = f(\xi_n, p), \tag{1.33}$$

where p is a control parameter. A period-one UPO of the continuous dynamic system corresponds to a fixed point (saddle-point) ξ^* of the discrete Poincaré map. Assume the $\xi^* = 0$ for $p = p_0$ and let λ_s, λ_u, e_s, and e_u denote the stable and unstable eigenvalues and corresponding eigenvectors at the fixed point; f_s and f_u are the contravariant basis vectors corresponding to e_s and e_u, respectively, and $g \equiv \partial \xi^*/\partial p$ is the shift of the fixed point upon variation of p. If, after n iterations, ξ_n happens to be sufficiently close to the fixed point ξ^* (Fig. 1.10(a)), a perturbation δp of p_0 is applied such that the phase point ξ_n is thrown in the next iteration onto the stable manifold of the saddle-point ξ_{n+1} (Fig. 1.10(b)), from where it is attracted to the saddle-point $\xi^*(p_0)$ corresponding to the UPO. An analysis shows that, to ensure that this happens, the control parameter p has to be changed approximately to the new value

$$p_n = \frac{\lambda_u}{\lambda_u - 1} \frac{\xi_n \cdot f_u}{g \cdot f_u}. \tag{1.34}$$

By selecting small appropriate perturbations at each step of the iteration, it can be ensured that the phase point always remains near the stable manifold and thus the

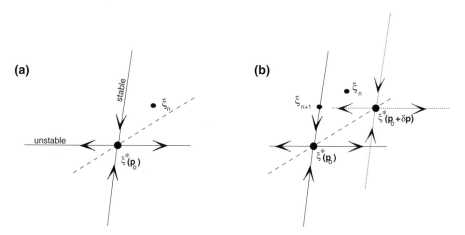

Figure 1.10. The OGY scheme of chaos control for an unstable periodic orbit (UPO) of period one. The Poincaré section of the UPO (saddle-point $\xi^*(p_0)$) is shown, and the stable and unstable manifolds are schematically indicated by arrows. (a) The nth iterate of the Poincaré map ξ_n falls in the neighborhood of the uncontrolled fixed point. (b) The control parameter p_0 is varied by a small amount δp such that the fixed point is shifted to $\xi^*(p_0 + \delta p)$ (dotted), and the next iterate ξ_{n+1} falls onto the stable manifold of $\xi^*(p_0)$.

UPO is stabilized. Several variants of this method exist and have been reviewed (Ditto and Showalter 1997, Schuster 1999).

The OGY method uses feedback control only at discrete times. Its disadvantage is, besides its large computational effort, that UPOs whose largest Lyapunov exponents are of the order of the inverse period become very sensitive to noise and bursts of de-synchronization. A simple method of *time-continuous* feedback control has been suggested by Pyragas (1992). In order to stabilize UPOs of the chaotic dynamic system (1.5)

$$\frac{d\boldsymbol{q}}{dt} = \boldsymbol{F}(\boldsymbol{q}, \boldsymbol{k}) \tag{1.35}$$

one may employ continuous feedback to synchronize the current state of the system with a time-delayed version of itself ("time-delay autosynchronization", TDAS), as shown schematically in Fig. 1.11. Usually it is sufficient to use one of the dynamic variables, say $q_1(t)$, for the construction of the control signal $\epsilon(t)$ and feed this signal back into the same variable, i.e. eq. (1.5) is amended to

$$\frac{dq_1}{dt} = F_1(\boldsymbol{q}, \boldsymbol{k}) + \epsilon(t),$$
$$\epsilon(t) \equiv K\big(q_1(t - \tau) - q_1(t)\big), \tag{1.36}$$

where K is the control amplitude, and τ is the delay time. Control can be achieved if the parameters K and τ are chosen appropriately. If the delay time τ coincides with the UPO period T, the control signal ϵ vanishes, indicating that the stabilization of a periodic orbit of the uncontrolled system has occurred.

A natural extension of the Pyragas scheme is given by a feedback that takes into account several previous states and for which the term "extended time delay autosynchronization" (ETDAS) was proposed (Socolar *et al.* 1994). Using this extension one obtains the control signal

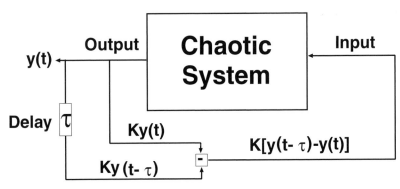

Figure 1.11. The scheme of chaos control by time-delay autosynchronization. The system variable $y(t)$ is used as feedback, K is the control amplitude, and τ is the delay time.

$$\epsilon(t) = K \sum_{m=0}^{\infty} R^m \{q_1[t - (m + 1)\tau] - q_1(t - m\tau)\} \tag{1.37}$$

$$= K\left((1 - R) \sum_{m=1}^{\infty} R^{m-1} q_1(t - m\tau) - q_1(t)\right)$$

$$= K[q_1(t - \tau) - q_1(t)] + R\epsilon(t - \tau), \qquad 0 \leq R < 1,$$

where R controls the contributions of previous states; small values of R indicate a short time memory, whereas larger values give a stronger weight to all previous states. In general, this method works more efficiently than does the simple TDAS scheme which corresponds to the case $R = 0$. With increasing R orbits of higher order and orbits with larger Lyapunov exponents can be stabilized (Bleich and Socolar 1996, Just *et al.* 1999, Franceschini *et al.* 1999). Finally we note that control schemes combining the time-discrete OGY approach with a time delay like that in the Pyragas method have also been suggested (Claussen *et al.* 1998).

1.4   Activator–inhibitor kinetics

1.4.1   Active media

Semiconductors are dissipative dynamic systems due to the dissipation of electric energy in charge transport. They are nonlinear due to the inherent nonlinearities in the charge-transport processes at high fields. Thus they can be considered active media (Mikhailov 1994), a concept that has successfully been applied to a variety of dissipative systems in physics, chemistry, and even biology and ecology. Although the microscopic mechanisms underlying nonlinear spatio-temporal dynamics in semiconductors and in other nonequilibrium systems such as gas discharges and chemical reaction–diffusion systems are different, the macroscopic phenomena of self-organized pattern formation have many common features (Engel *et al.* 1996, Busse and Müller 1998). An essential ingredient of the nonlinearity is the kinetics of particles, which may be neutral or charged, as in the case of transport in a semiconductor, gas discharges or electrochemistry. The analogy between processes of generation and recombination of carriers in semiconductors (Schöll 1987) and chemical reaction kinetics was noted long ago (Landsberg and Pimpale 1976, Schöll and Landsberg 1979). For nondegenerate or diluted systems the reaction rates follow the law of mass action and are sums of products of the various concentrations; therefore, semiconductors are nonlinear systems in general. Finally they must be considered continuous media because of spatial coupling due to diffusion. From these properties various spatial, temporal, and spatio-temporal patterns are expected in semiconductors (Schöll 1987, Shaw *et al.* 1992, Thomas 1992, Peinke *et al.* 1992, Kerner and Osipov 1994, Niedernostheide 1995, Schöll *et al.* 1998a). These patterns and their descriptions are expected to be similar to those for other dissipative systems, e.g. surface reactions, or autocatalytic chemical reaction systems. In the case of charged particles, the coupling with electric fields provides an important nonlinear feedback mechanism. Systems of this type are often adequately described in terms of activator–inhibitor kinetics. This concept was introduced and successfully applied to the self-organization of living organisms by various authors (Gierer and Meinhardt 1972, Fife 1983, Smoller 1983, Murray 1993). An activator–inhibitor system is most simply modeled as a system of two nonlinear reaction–diffusion equations (Mikhailov 1994)

$$\tau_a \frac{\partial a(x,t)}{\partial t} = f(a,u) + l^2 \frac{\partial^2 a}{\partial x^2}, \tag{1.38}$$

$$\tau_u \frac{\partial u(x,t)}{\partial t} = g(a,u) + L^2 \frac{\partial^2 u}{\partial x^2}, \tag{1.39}$$

where a and u denote the local activator and the inhibitor, respectively, τ_a and τ_u are time scales, l and L are characteristic diffusion lengths, and f and g are appropriate functions of a and u. The production of the inhibitor u increases with activator concentration a and saturates with increasing u. The activator a is increased by autocatalytic reproduction, and is inhibited by u. In the generic case, $f(a,u)$ is thus a nonmonotonic function of a, increasing in a certain interval of a, and is decreasing with u, while g may in the simplest case be of the form $g = a - u$. Depending upon the intersection between the null-clines $f(a,u) = 0$ and $g(a,u) = 0$, bistable, excitable, or oscillatory (Turing–Hopf) behavior is found (Fig. 1.12). These intersection points determine the spatially uniform fixed points of the system.

■ **A bistable medium:** There exist two stable fixed points separated by a saddle-point (Fig. 1.12(a)). The two stable fixed points lie on the upper and lower branches of the nonmonotonic null-cline $f(a,u) = 0$. Transitions

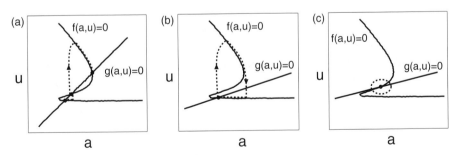

Figure 1.12. Classification of activator–inhibitor systems. The null-clines $f(a,u) = 0$ and $g(a,u) = 0$ of the activator a and the inhibitor u, respectively, are schematically plotted in the (a,u) phase plane for three different cases: (a) bistable, (b) excitable, and (c) oscillatory (Turing–Hopf). The intersection points of the null-clines determine the spatially uniform fixed points of the system. The dotted curves represent typical phase trajectories starting near the fixed points.

between the two fixed points may be induced by sufficiently large fluctuations (*trigger waves*, fronts). Such fronts generally propagate through the medium with a constant profile due to the diffusive coupling.

- **An excitable medium:** There exists only one stable fixed point, but, if a sufficiently large initial perturbation takes the system across the middle branch of the null-cline $f(a, u) = 0$, it returns to the stable fixed point only after a large excursion in phase space (the dotted curve in Fig. 1.12(b)). Thus a pulse is excited in the medium. In two-dimensional spatial domains concentric circular waves (*target patterns*) or spiral waves may be generated.

- **An oscillatory medium:** There exists a single fixed point on the middle branch of the null-cline $f(a, u) = 0$ (Fig. 1.12(c)). If the slope of the null-cline $g(a, u) = 0$ is less than that of $f(a, u) = 0$, the behavior depends upon the ratio of the time scales, τ_a/τ_u. For $\tau_a \ll \tau_u$ the only stationary state of such a system is an unstable focus, and temporal oscillations (limit cycles created by Hopf bifurcations) are expected. For larger values of $\alpha \equiv \tau_a/\tau_u$ stable stationary structures are possible. Long-range inhibition ($L \gg l$) favors the formation of static spatial structures, e.g. spatially periodic Turing patterns, whereas for smaller values of L/l, i.e. a slowly diffusing inhibitor, spiral waves and complex spatio-temporal behavior are found.

Although a description of transport in terms of the simple reaction–diffusion model (1.38) and (1.39) is possible for various semiconductor structures such as p–n–p–n diodes, p–i–n diodes, heterostructure hot-electron diodes, double-barrier resonant-tunneling structures, and other layered structures (Jäger *et al.* 1986, Radehaus *et al.* 1987, Baumann *et al.* 1987, Purwins *et al.* 1988, Vashchenko *et al.* 1990, Niedernostheide *et al.* 1992*b*, Gorbatyuk and Rodin 1992*c*, Wacker and Schöll 1994*b*, Glavin *et al.* 1997, Mel'nikov and Podlivaev 1998, Meixner *et al.* 2000*a*), some semiconductor systems may require refined modeling in terms of more sophisticated transport models, e.g. drift–diffusion equations combined with Poisson's equation for the electric potential, hydrodynamic balance equations for the carrier densities, mean carrier momenta, and mean carrier energies, or even a kinetic approach in terms of the semiclassical Boltzmann equation or a full quantum transport theory (Schöll 1998*b*). Often a combination of these approaches is used.

1.4.2   The dynamics of the electric field

In applying the activator–inhibitor principle to the description of semiconductor transport, one first has to identify the activator and inhibitor variables. The activator is generally an internal dynamic variable that triggers the instability via an autocatalytic process. Examples are provided by the carrier (electron or hole) density in doped semiconductors. Impact ionization of electrons or holes either across the bandgap or from localized levels is an autocatalytic process that leads to a multiplication of the number of carriers. Schematically it may be written as one of the following reaction equations:

$$e \longrightarrow 2e + h \tag{1.40}$$

$$e + e_t \longrightarrow 2e + h_t \tag{1.41}$$

$$h \longrightarrow 2h + e \tag{1.42}$$

$$h + h_t \longrightarrow 2h + e_t \tag{1.43}$$

where e and h denote free electrons and holes, respectively, and e_t and h_t stand for electrons and holes trapped at impurities (donors, acceptors, or deep levels). In doped semiconductors, impurity-impact ionization is essential for explaining the SNDC observed in the regime of low-temperature impurity breakdown. This will be the subject of Chapters 4 and 5.

Another activator variable may be provided by the interface charge density ρ_s in semiconductor heterostructures and low-dimensional structures, in which abrupt junctions between different materials on an atomic length scale cause conduction band discontinuities resulting in potential barriers and quantum wells. The local charge accumulations in these wells, together with nonlinear processes of transport across the barriers, may induce the positive nonlinear feedback between the carrier distribution and the electric potential distribution that is characteristic of an activator. In other cases the activator is given by the p–n emitter junction voltage in p–n–p–n structures. The inhibitor in such layered structures is the voltage drop across the second p–n junction, corresponding to the injected minority carrier density p, or the gate potential in case of a thyristor structure. Models of this type are introduced in Section 2.4, and analyzed in Chapters 3 and 7.

The inhibitor is often associated with the electric field $\mathcal{E}$, or the corresponding voltage drop across the device. Dielectric relaxation of the field acts as an inhibiting process that reduces the internal space-charge field. This generally follows from Maxwell's equations

$$\nabla \times \mathcal{E} = -\frac{\partial \boldsymbol{B}}{\partial t}, \tag{1.44}$$

$$\nabla \cdot \boldsymbol{B} = 0, \tag{1.45}$$

$$\frac{1}{\mu_0} \nabla \times \boldsymbol{B} = \epsilon \frac{\partial \mathcal{E}}{\partial t} + j \equiv \boldsymbol{J} \tag{1.46}$$

$$\epsilon \nabla \cdot \mathcal{E} = \rho, \tag{1.47}$$

with local charge density $\rho = e(n_D^+ - n_A^-) - e(n - p)$, where n_D^+ and n_A^- are the ionized donor and acceptor concentrations, respectively, n and p are the electron and hole concentrations, respectively, and $\epsilon = \epsilon_0 \epsilon_r$ and μ_0 are the permittivity and permeability, respectively. Generally, in the cases considered in this book, the magnetic induction $\boldsymbol{B}$ does not depend on time, so we assume always a curl-free electric field: $\nabla \times \boldsymbol{\mathcal{E}} = 0$.

From Ampère's law eq. (1.46) it follows that

$$\nabla \cdot \boldsymbol{J} = 0 \tag{1.48}$$

with

$$\boldsymbol{J} = \epsilon \dot{\boldsymbol{\mathcal{E}}} + \boldsymbol{j}, \tag{1.49}$$

where $\boldsymbol{J}$ is the total current density composed of the displacement current density $\epsilon \dot{\boldsymbol{\mathcal{E}}}$ and the conduction current density $\boldsymbol{j}$. The overdot denotes the temporal derivative. For a one-dimensional homogeneous conductor this gives the current continuity equation

$$\epsilon \dot{\mathcal{E}} = J - en\mu \mathcal{E}, \tag{1.50}$$

where the external current density J is constant, and $j = en\mu \mathcal{E}$ is the drift current density with mobility μ. This is the simplest example of a dynamic inhibitor equation of the type (1.39) with $L = 0$. Dielectric relaxation describes the delayed response of the electric field upon the change of the current. Indeed, assuming for the moment a constant n, eq. (1.50) can be integrated to give

$$\mathcal{E}(t) = \frac{J}{en\mu} \left(1 - e^{-t/\tau_{\mathrm{M}}}\right), \tag{1.51}$$

which describes the build-up of the electric field $\mathcal{E}$ on the time scale of Maxwell's dielectric relaxation time $\tau_{\mathrm{M}} = \epsilon/(en\mu)$ if a constant external current density J is imposed. At the beginning the total current is carried by the displacement current only, and, as the electric field relaxes to its steady-state value, the drift current takes over, increasing from zero up to its steady-state value J. The finite dielectric relaxation time of the electric field inhibits the instantaneous rise of the drift current.

1.5   Global coupling

1.5.1   Global inhibitory constraints

In many cases, in addition to a local diffusive coupling, spatially extended nonlinear systems also experience a *global coupling*. Generally, global coupling is related to external constraints imposed upon the system's dynamics. In the presence of global coupling, some dynamic variables of the active media (e.g. the global excitation level) depend on the spatially averaged parameters of the self-organized pattern which the system exhibits. Global coupling has recently been recognized widely to be an important factor for spatio-temporal dynamics in spatially extended systems.

It has been studied in various models, e.g. Ginzburg–Landau amplitude equations (Falcke *et al.* 1995, Falcke and Neufeld 1997, Battogtokh and Mikhailov 1996), arrays of discrete oscillators (Pikovsky *et al.* 1996), reaction–diffusion systems (Ohta 1989, Ohta *et al.* 1990, Elmer 1990, 1992, Schimansky-Geier *et al.* 1991, 1992, Engel *et al.* 1996, Niedernostheide *et al.* 1997, Hempel *et al.* 1998), surface reactions (Bär *et al.* 1994, Falcke and Engel 1994) and electro-chemical systems (Mazouz *et al.* 1997, Wang *et al.* 2000). For instance, in chemical surface reactions, the pressure of the gas phase, which is the same over the whole surface, represents such a global constraint. In thermokinetic reactions on surfaces with a large heat conductivity the temperature becomes a global parameter that depends upon the size of the surface where the reaction takes place and where heat is produced or absorbed. In optically bistable systems, the absorption coefficient may depend upon the density of the excited particles and hence the intensity of the penetrating light becomes a function of the depth to which the excited pattern extends into the bistable element.

In terms of the activator–inhibitor kinetics (1.38) and (1.39) the global constraint may be given by an integral equation for the inhibitor u of the form

$$\tau_u \frac{\partial u}{\partial t} = g(\langle a \rangle, u) \tag{1.52}$$

where $\langle a \rangle = (1/L) \int_0^L a(x, t)\, dx$ represents the global coupling. More generally, $\langle a \rangle = (1/L) \int K(x-x') a(x', t)\, dx'$ may be defined with a kernel function $K(x-x')$ of finite range, which is sometimes referred to as *nonlocal* coupling.

In the context of transport in semiconductors, global coupling of the inhibitor arises naturally for spatially inhomogeneous current densities, e.g. current filaments. If the current density is modulated in the x- or y-direction perpendicular to the current flow, as in a current filament, but the electric field is constant in the whole sample (this may be a good approximation for long samples, for which the influence of the contacts is negligible), then the inhibitor equation (1.50) can be integrated over the current cross-section A, yielding

$$\epsilon \dot{\mathcal{E}} = I_0/A - e\langle n \rangle \mu \mathcal{E}, \tag{1.53}$$

where I_0 is the total current through the device and $\langle n \rangle = (1/A) \int_A n(x, y, t)\, dx\, dy$ represents the global coupling.

1.5.2 Operation in a load circuit

In *bistable semiconductor systems* the global coupling represents an inherent feature of the spatio-temporal dynamics of the current-density patterns (e.g. current filaments (Schöll 1987), and fronts and pulses (Kerner and Osipov 1994)). The mechanism of this coupling is as follows. For any evolution of the current-density pattern that is accompanied by the variation of the total current I through the device, the voltage drop at an external load and/or an internal series resistance changes. That causes a variation of the voltage u dropping across the device which usually characterizes the global excitation level of the bistable semiconductor system. This type of feedback is well known with respect to stationary (Volkov and Kogan 1969, Bass *et al.* 1970, Schöll 1987, Alekseev *et al.* 1998), breathing (Schöll and Drasdo 1990, Kunz and Schöll 1992), and spiking (Wacker and Schöll 1994*b*, Niedernostheide *et al.* 1996*b*) current filaments.

In the following we consider this important case in some more detail. The voltage u across the sample is not fixed if the device is operated in a circuit as shown in Fig. 1.13. The capacitance C parallel to the sample is given by the sum of the device capacitance, the external capacitance, and parasitic wire capacitances. Note that the resistance R usually is thought to be an external resistance, but it can also be the linear resistance of a part of the sample that is not associated with the nonlinear behavior (e.g. a contact resistance). Kirchhoff's laws lead to

$$U_0 = R I_0(t) + u(t), \tag{1.54}$$

$$I_0(t) = I(t) + C \frac{\partial u}{\partial t} \tag{1.55}$$

with the current through the device $I(t)$, the voltage drop across the device $u(t)$, the applied bias voltage U_0, and the total current through the resistor $I_0(t)$.

Hence the temporal behavior of the voltage $u(t) = \int_0^L \mathcal{E}\, dz$ is determined by the circuit equation

$$\frac{du(t)}{dt} = \frac{1}{C}\left(\frac{U_0 - u}{R} - I(a, u)\right), \tag{1.56}$$

where $I(a, u)$ is the current through the device. If the current density j is constant in the direction of transport (which defines the z-axis), I is simply given by the integral over the cross-section A of the current flow

$$I(a, u) = \int_A dx\, dy\, j(a(x, y), u). \tag{1.57}$$

The arguments of j indicate that the local current density in a cross-section of the device depends not only upon the voltage drop u but also upon an additional internal local variable $a(x, y)$ that distinguishes, e.g. different states on a

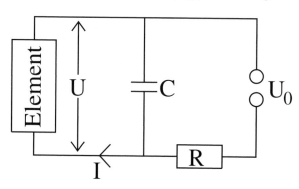

Figure 1.13. The bistable semiconductor element operated in a circuit with a load resistor of resistance R and a capacitor of capacitance C, and applied bias voltage U_0.

bistable current-density–field characteristic corresponding to the same voltage. In an activator–inhibitor system, as discussed above, a corresponds to the activator variable. Examples will be treated in Chapters 2, 3, and 7.

Equation (1.56) can be re-written as

$$\tau_u \frac{du}{dt} = U_0 - u - R \int_A dx\, dy\, j(a, u), \qquad \tau_u \equiv RC, \tag{1.58}$$

which is of the form of the globally coupled inhibitor equation (1.52).

Under steady-state conditions eq. (1.56) reduces to

$$I = \frac{U_0 - u}{R}, \tag{1.59}$$

which defines the dc *load line* in the (I, u)-plane. The stationary operating point of the system is determined by the intersection of the load line with the device characteristic $I(u)$.

If there is charge accumulation in the sample (such as, for instance, in the double-barrier resonant-tunneling diode or the heterostructure hot-electron diode, see Chapter 2) j is no longer constant in the direction of the current flow and I is given by a more complicated integral expression. Let the voltage be applied in the z-direction. We restrict ourselves to the important case of a layered structure, so that the material properties such as the dielectric constant $\epsilon(z)$ are only z-dependent. Ampère's law (1.46) gives $\mathrm{div}(\epsilon\, d\mathcal{E}/dt + j) = 0$. From this we obtain by integrating and applying Gauss' Theorem

$$\oint_{\partial V} df \cdot \left(\epsilon \frac{d\mathcal{E}}{dt} + j \right) = 0. \tag{1.60}$$

Now we choose the closed surface of integration ∂V shown in Fig. 1.14 by dotted lines, and find

$$\frac{C_{\mathrm{ext}}}{\epsilon(z)} \frac{du}{dt} + \int_A dx\, dy \left(\frac{\mathcal{E}_z}{dt} + \frac{dj_z}{\epsilon(z)} \right) - \frac{I_0}{\epsilon(z)} = 0, \tag{1.61}$$

where C_{ext} is the external capacitance, I_0 is the total current, and z is the position of the intersection of the surface with the sample of cross-section A. Here we have

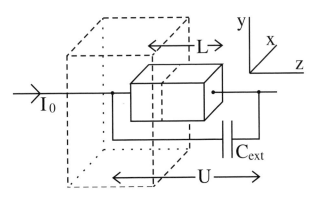

Figure 1.14. The surface of integration (dotted) used for determining the dynamics of the voltage across the device $u(t)$.

neglected the electric fields outside the sample, which is reasonable for structures whose diameters are much larger than the sample length L in the z-direction. Integrating over the z-direction we obtain

$$\left(A + C_{\text{ext}} \int_0^L dz\, \frac{1}{\epsilon(z)} \right) \frac{du}{dt} + \int_0^L dz \int_A dx\, dy\, \frac{j_z}{\epsilon(z)} - \int_0^L dz\, \frac{I_0}{\epsilon(z)} = 0 \qquad (1.62)$$

because $\int_0^L dz\, \mathcal{E}_z = u$ is not dependent on (x, y), if we assume that there are ideal metallic planar contacts at $z = 0$ and $z = L$ and neglect time-varying magnetic fields. Defining the intrinsic sample capacitance C_{int} by $C_{\text{int}}^{-1} = \int_0^L dz\, (A\epsilon(z))^{-1}$, we obtain

$$\frac{du}{dt} = \frac{1}{C_{\text{int}} + C_{\text{ext}}} \left(I_0 - \frac{C_{\text{int}}}{A} \int_A dx\, dy \int_0^L dz\, \frac{j_z}{\epsilon(z)} \right). \qquad (1.63)$$

From this expression we can easily identify the conduction current through the device

$$I = \frac{C_{\text{int}}}{A} \int_A dx\, dy \int_0^L dz\, \frac{j_z}{\epsilon(z)} \qquad (1.64)$$

and, with $C \equiv C_{\text{ext}} + C_{\text{int}}$ and $I_0 = (U_0 - u)/R$, the globally coupled dynamic equation (1.56) for the voltage $u(t)$ is recovered.

The global coupling leads to a feedback upon the spatio-temporal dynamics that can result in a variety of phenomena, which are to be discussed in later chapters. Important effects include accelerated, decelerated, and oscillatory current-density fronts; competition of current filaments in the form of a "winner-takes-all-dynamics" that is similar to Ostwald ripening of liquid droplets in thermodynamics; suppression of spiking multifilaments; and spatial filtering in chaos control – to mention but a few examples.

1.6   The scope of this book

In this book basic state-of-the-art theoretical methods, concepts, and results of the nonlinear dynamics of spatially extended systems will be presented and applied to a variety of electric transport phenomena in semiconductors that are of broad current interest. Some of these concepts, such as activator–inhibitor kinetics, global coupling, chaos control, pattern selection, and spatio-temporal chaos, have been reviewed in a general context in this introductory chapter. They represent modern developments in the interdisciplinary field of pattern formation in continuous nonlinear dynamic systems that are currently being analyzed intensively in classical fields such as hydrodynamics, optical systems, and chemical reaction systems. Nevertheless, they have received relatively little attention in the field of semiconductor transport. It is hoped that this book may stimulate future work in this direction by outlining the principles, which may be readily applied to a variety of semiconductor structures.

In our book these modern concepts will be elaborated and illustrated by using some model systems as examples. A common feature of these systems is that they exhibit generic nonlinear spatio-temporal structures irrespectively of their particular microscopic nature. Global couplings are important, and the activator–inhibitor concept can be fruitfully applied. The models represent bistable, excitable, and oscillatory systems. For the majority, a detailed comparison between theory and experiment has recently been achieved, or is at least within easy accessibility. Although the main focus of this book is on a theoretical analysis, comparisons with experiments are drawn whenever appropriate.

It should be noted that we shall not touch the wide field of optical phenomena in semiconductors, in which spatio-temporal instabilities and self-organized pattern formation are also abundant (Haug 1988, Weiss and Vilaseca 1991). Examples of recent interest include spatio-temporal complexity in the dynamic output of multi-stripe (Hess and Schöll 1994, Hess et al. 1994, Merbach et al. 1995) and broad-area (Hess 1993, Hess and Kuhn 1996) semiconductor lasers. Spatio-temporal chaos of optical filaments and its control by time-delayed optical feedback have been studied in terms of dependencies on the injection current and the stripe separation (Münkel et al. 1997). In nonlinear passive optical systems such as CdS crystals, fronts of optical switching from a state of low transmissivity to a state of high transmissivity have been predicted theoretically and verified experimentally to occur (Schmolke et al. 1995). There also exists a variety of electrooptic instabilities, e.g. bistability in quantum-well structures associated with the quantum-confined Stark effect is used in self-electrooptic-effect devices (SEEDs) (Miller 1990, Lentine and Miller 1993, Merbach et al. 1999).

The book is organized as follows. After this introduction to the fundamentals of dynamic systems, in Chapter 2 we review basic concepts of nonlinear charge transport in semiconductors. Particular emphasis is placed upon state-of-the-art low-dimensional semiconductor structures, and the nonlinear feedback between space charges and transport processes inherent in such structures. A hierarchy of modeling approaches is surveyed, ranging from quantum transport to classical drift–diffusion and reaction–diffusion models, and several examples of mechanisms and models for semiconductor instabilities are reviewed.

In Chapter 3 general aspects of pattern formation and oscillatory instabilities in semiconductors, which do not depend upon the particular microscopic transport mechanism are discussed. Important relations between the stability of the spatially uniform operating point and the differential conductance for general elements operated in external circuits are derived, and some specific conclusions for S- and Z-shaped current–voltage characteristics are drawn. The bifurcation of spatially inhomogeneous patterns such as current filaments and field domains is discussed, and the stability of current filaments is analyzed under very general assumptions.

Finally, the propagation of lateral current-density fronts that switch the system between the low-conductivity and the high-conductivity state is treated generally for systems with S- and Z-shaped current–voltage characteristics.

In the following four chapters specific instabilities are considered in more detail. First, in Chapter 4, we focus on the nonlinear dynamics of charge carriers in the regime of low-temperature impurity breakdown. Chaotic current oscillations as well as the nonlinear dynamics of current filaments and longitudinally traveling waves and the nascence of filaments in various contact geometries are treated.

In Chapter 5 the additional action of a magnetic field in a Hall-effect configuration in the regime of impurity breakdown is considered. It leads to more complex temporal and spatio-temporal scenarios and to deformation of the current filaments. Various chaotic instabilities and chaos control are considered.

Chapter 6 deals with vertical transport in a superlattice. In the high-field regime, this system displays complicated multistable current–voltage characteristics and a menagery of instabilities associated with stationary or oscillating field domains. Chaotic instabilities may be induced by applying an ac driving voltage.

Finally, in Chapter 7, spatio-temporal chaos is considered in a class of models of reaction–diffusion type that are representative of a number of layered semiconductor structures such as p–n–p–n thyristors, double-barrier resonant-tunneling structures, and heterostructure hot-electron diodes. Both globally coupled and locally coupled spatio-temporal dynamics are treated. Spatio-temporal spiking scenarios including routes to chaos and methods for the control of spatio-temporal chaos are presented. Transient spatio-temporal chaos and complex mixed Turing–Hopf modes are found and analyzed using sophisticated methods of nonlinear dynamics.

Of course these examples of model systems are not exhaustive. It is hoped, however, that this book will provide the reader with an understanding of the principles of this new field, and give guidance on how to apply these concepts and methods to other semiconductor structures.

Chapter 2

Concepts of nonlinear charge transport in semiconductors

The theory of electric transport in semiconductors describes how charge carriers interact with electric and magnetic fields, and move under their influence. In general, we will consider the regime far from thermodynamic equilibrium, under which linear relations between current and voltage do not hold, and more sophisticated modeling of the microscopic charge-transport processes is required. There are various levels at which semiconductor transport can be modeled, depending upon the specific structures under consideration as well as the operating conditions. In this chapter we will survey a hierarchy of approaches to nonlinear charge transport and discuss the regimes of validity. This hierarchy of models will form the physical basis for the instabilities and spatio-temporal pattern formation processes to be discussed in the subsequent chapters. It will also serve to introduce some systematics and give guidance concerning the confusing variety of nonlinear dynamic models that have been developed and studied in the field of semiconductor instabilities within the recent past.

2.1   Introduction

With the advent of modern semiconductor-growth technologies such as molecular beam epitaxy (MBE) and metal–organic chemical vapor deposition (MOCVD), artificial structures composed of different materials with layer widths of only a few nanometers have been grown, and additional lateral patterning by electron-beam lithography or other lithographic or etching techniques (ion-beam, X-ray, and scanning-probe microscopies) can impose lateral dimensions of quantum confinement in the 10 nm regime. Alternatively, lateral structures can be induced by the Stranski–Krastanov growth mode in strained material systems, which leads to the self-organized formation of islands ("quantum dots") a few nanometers in diameter (Bimberg *et al.* 1999). Thus it has become possible to design and fabricate semiconductor structures whose vertical and lateral dimensions are controlled on an atomic length scale. This has given us the unprecedented capability of being able to tailor devices with extraordinary electric and optical properties. If the geometrical dimensions of the semiconductor structures reach the order of the characteristic physical length scales of transport which will be discussed below, the transport and optical properties are no longer determined by the bare material constants but will heavily depend upon the size and geometry of the device. This has opened up a vast field of research activity that is, on the one hand, of fundamental interest since it pushes the border of physically accessible phenomena in semiconductors to the ultimate quantum limit, but which is also, on the other hand, of utmost importance with respect to applications since the miniaturization and ultra-large-scale integration of electronic components is still in progress. For example, channel lengths of field-effect transistors (FETs) of 100 nm are nowadays standard in mass production megabit chips. While these are based on silicon technology, the class of GaAs–AlGaAs materials is essential for high-speed electronic and optoelectronic nanometer devices such as the high-electron-mobility transistor (HEMT), the modulation-doped field-effect transistor (MODFET), real-space transfer devices, heterostructure hot-electron diodes (HHED), double-barrier resonant-tunneling structures (DBRT), and injection laser diodes. Even single-electron tunneling has been realized in GaAs nanostructures. In all those structures space charges confined on nanometer scales and sharp electric potential gradients lead to strong electric fields and far-from-equilibrium conditions that give rise to a menagery of nonlinear transport phenomena.

There is need for a thorough theoretical understanding of the transport properties of such semiconductor structures. Over the years various approaches and concepts have been developed (Mahan 1990, Beenakker and van Houten 1991, Landsberg 1992, Haug and Koch 1993, Ferry *et al.* 1995, Datta 1995, Haug and Jauho 1996, Ferry and Goodnick 1997, Schöll 1998*b*). In applying these, three issues associated with nonlinear transport in semiconductor structures should be noted.

2.1.1   Device simulation

There is a fruitful mutual interaction of transport theory and device simulation. The limitations of classical drift–diffusion device simulators (Markovich 1986) have been surpassed by todays' nanostructure devices, and new means of device simulation have been investigated. The need for new simulation strategies has stimulated research on novel approaches to transport theory. Conversely, state-of-the-art concepts of charge transport are applied to simulate realistic semiconductor device structures and real electronic materials. For instance, the concept of hydrodynamic simulation using not only the carrier densities but also the mean carrier momenta and energies as dynamic variables has successfully been applied to logic and memory components with complex three-dimensional structures and realistic geometries (Ferry *et al.* 1998). Also, the Monte Carlo method has been employed extensively for realistic simulation of nanostructure devices. Hereby not only the overall carrier densities but also the distribution of carriers over various energy states can be correctly described. A statistical ensemble of carriers with different energies is simulated, taking into account their acceleration in the electric field as well as detailed scattering processes (Jacoboni and Lugli 1989). While the inclusion of complicated realistic boundary conditions is generally difficult within this approach, other methods relying on a discretization of the time as well as the real space and state space in cells ("cellular automata") have been devised as an efficient alternative scheme that can cope with arbitrarily complex geometries and boundary conditions, and pronounced spatial inhomogeneities (Kometer *et al.* 1992).

2.1.2   Quantum transport

With smaller and smaller dimensions, the conventional semiclassical picture of charge transport based on the Boltzmann transport equation must be replaced by fully quantum mechanical concepts. Quantum mechanical effects become important on length and time scales given by the uncertainty relations. On time scales τ shorter than that given by the energy-time uncertainty $\tau \Delta E \geq \hbar$, energies ΔE can not be resolved. If τ is the mean time between collisions, this effect leads to a collisional broadening of the energy levels by ΔE, i.e. the energy is not strictly conserved. If we require, for instance, ΔE to be much smaller than an optical phonon energy of 50 meV, τ should be much larger than 10^{-14} s. Similarly, the Heisenberg uncertainty relation requires $\Delta x \, \Delta p \geq \hbar/2$, where x and p are position and momentum, respectively. In the classical picture electrons are considered as particles with well-defined positions and momenta in between two collisions. This approximation requires that the quantum mechanical wavepacket describing the electron has a momentum uncertainty Δp much less than the average momentum p and a position uncertainty Δx much less than the mean free path. Furthermore, in the classical Boltzmann transport theory collisions are assumed instantaneous in time and pointlike in space. However, if a strong electric field is applied, the electron momentum and energy will change appreciably during the collision, and the transition rates will be modified. This is called the intracollisional field effect. In devices of length 0.1 μm and fields of the order of 10^5 V cm^{-1}, electron energies of the order of 1 eV and average times between collisions of 10^{-14} s can be reached, and a noticeable effect will occur. Thus the conditions of classical transport can easily be violated in semiconductor nanostructures, and quantum transport theories must be applied (Rossi *et al.* 1992).

2.1.3 Nonlinear dynamics

If a bias of typically a few volts is applied across a nanostructure, electric fields up to 10^6 V cm^{-1} can easily arise. Therefore we are dealing with the high-field regime, under which strong nonlinearities and considerable electron heating have to be considered. In the electric field the charge carriers may acquire average kinetic energies many orders of magnitude above the value of $\frac{3}{2}k_B T_L$, k_B being the Boltzmann constant, which characterizes thermal equilibrium with the crystal lattice at temperature T_L. An "electron temperature" T_e may be associated with that kinetic energy, which can reach 10^4 K in real devices (Balkan 1998). Far from thermodynamic equilibrium the carrier dynamics is described by nonlinear equations, and instabilities or nonequilibrium phase transitions may arise (Schöll 1987, Shaw *et al.* 1992, Thomas 1992, Niedernostheide 1995). The viewpoint of a semiconductor as a nonlinear dynamic system has provided a fruitful concept for understanding and predicting complex transport behavior like the self-organized formation of inhomogeneous current-density and field distributions in the form of current filaments or high-field domains, threshold switching, bifurcations of various solution branches leading to multistability and hysteresis, spontaneous current or voltage oscillations, and chaotic dynamics. In fact, the nonlinear spatio-temporal dynamics of carriers has become an active and still growing field of research, in particular in the realm of semiconductor nanostructures.

2.2 Nonequilibrium transport

In order to specify the nonlinear dynamic system which describes the semiconductor far from thermodynamic equilibrium, a physical model of charge transport and nonequilibrium generation–recombination kinetics must be applied. This will in general involve Maxwell's equations coupled with an appropriate set of nonlinear transport equations. In the following, after introducing some specific features of low-dimensional structures, we shall briefly discuss the role of the space-charge dynamics, and then outline various transport regimes which require different modelings.

2.2.1 Low-dimensional semiconductor structures

A typical feature of semiconductor structures fabricated by epitaxial growth and lateral patterning is carrier confinement. It means that the charge carriers can not move freely in all directions of space but are confined by potential barriers that form at interfaces between different materials. If this confinement occurs on a nanometer scale the semiconductor behaves as a system of reduced dimensionality. Depending on whether the confinement occurs in one, two, or even all three spatial directions, the carriers can move only in the remaining two, one, or zero directions, respectively, and hence the electronic system is called two-dimensional, one-dimensional, or zero-dimensional, respectively. Figure 2.1 schematically shows three prototypes of nanostructures with reduced dimensionalities: (a) a *quantum well* formed of a GaAs epitaxial layer sandwiched between two layers of $Al_xGa_{1-x}As$ (two-dimensional), (b) a *quantum wire* with an additional lateral confinement effected, for example, by the V groove shape of the epitaxial layers which leads to enhanced deposition of GaAs in the groove during growth (one-dimensional), and (c) a *quantum dot* grown, for example, by the Stranski-Krastanov growth mode, forming an island on a strained substrate (zero-dimensional).

Figure 2.2 shows the schematic energy diagram of an AlGaAs/GaAs quantum well heterostructure. Because the energy gap of AlGaAs is larger than that of GaAs, there is an energy offset in the conduction band E_c at the interfaces between the two

(a)

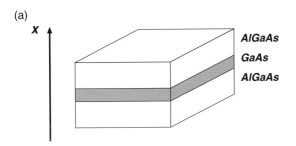

AlGaAs

GaAs

AlGaAs

Figure 2.1. Sketches of low-dimensional semiconductor structures: (a) a quantum well (two-dimensional), (b) a quantum wire (one-dimensional), and (c) a quantum dot (zero-dimensional).

(b)

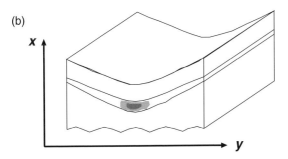

(c)

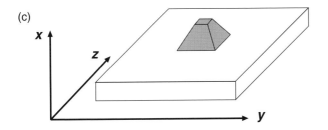

materials, and a quantum well is formed along the confining x-direction. It follows from textbook quantum mechanics (Bastard 1988) that bound states exist in this quantum well, and the lowest energy level is denoted by E_x. However, propagation of the electron is still possible in the remaining two (y- and z-) directions, and therefore the energy E_x represents the bottom of a two-dimensional conduction (sub)band characterized by Bloch vectors k_y and k_z. Depending upon the width and depth of the quantum well, several bound levels may exist, corresponding to several two-dimensional subbands. Similarly, the energy levels in a quantum wire form one-dimensional subbands with Bloch vector k_z.

The electric and optical properties of low-dimensional structures are severely changed with respect to those of the bulk material. In particular, the density of states generally becomes sharper with decreasing dimensionality (Fig. 2.3), and hence the optical and electrooptic properties are improved. For a simple isotropic parabolic band with effective mass m^*, for example, the energy dispersion is $E = \hbar^2 k^2/(2m^*)$. In k-space the values of k_i are quantized with discrete differences $\Delta k_i = 2\pi/L_i$, where L_i is the system size in the i-direction. Hence the number of states per unit volume up to an energy E, i.e. the number of states contained in a sphere in k-space of radius $k(E) = (2m^*E/\hbar^2)^{1/2}$, is given by

$$N(E) = \frac{2}{(2\pi)^3} \frac{4\pi k^3}{3} = \frac{1}{3\pi^2} k^3(E), \tag{2.1}$$

the factor 2 allowing for spin degeneracy. The density of states is obtained as $D(E) = dN(E)/dE$, i.e. for the three-dimensional case

$$D_{3D}(E) = \frac{1}{2\pi^2} \left(\frac{2m^*}{\hbar^2} \right)^{3/2} E^{1/2}. \tag{2.2}$$

For the two-dimensional case with $E = E_\nu + \hbar^2 k^2/(2m^*)$, $k^2 = k_y^2 + k_z^2$, and $\nu = 1, 2, \ldots$ an analogous argument yields for the number of states per unit area in each individual subband

$$N(E) = \frac{2}{(2\pi)^2} \pi k^2 = \frac{1}{2\pi} k^2(E) = \frac{m^*}{\pi \hbar^2} E \tag{2.3}$$

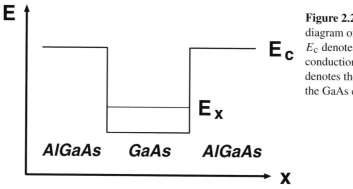

Figure 2.2. The energy diagram of a quantum well. E_c denotes the bulk conduction band, and E_x denotes the ground level in the GaAs quantum well.

and hence

$$D_{2D}(E) = \frac{m^*}{\pi\hbar^2} \sum_\nu \Theta(E - E_\nu), \tag{2.4}$$

where $\Theta(E)$ is the Heaviside function.

For the one-dimensional case with confinement in two directions $E = E_{\nu,\mu} + \hbar^2 k^2/(2m^*)$, and $k^2 = k_z^2$, $\nu, \mu = 1, 2, \ldots$ we obtain for the number of states per unit length in each subband

$$N(E) = \frac{2}{2\pi} 2k = \frac{2}{\pi}\left(\frac{2m^*E}{\hbar^2}\right)^{1/2} \tag{2.5}$$

(a)

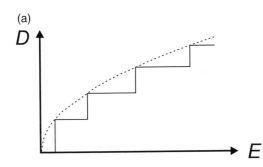

Figure 2.3. The density of states $D(E)$ of low-dimensional semiconductor structures: (a) a quantum well (two-dimensional), the broken line represents the three-dimensional density of states; (b) a quantum wire (one-dimensional); and (c) a quantum dot (zero-dimensional).

(b)

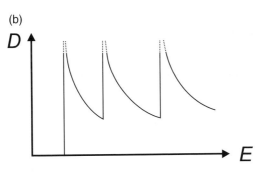

(c)

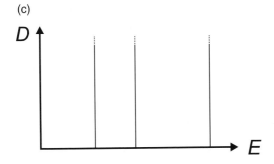

and hence

$$D_{1D}(E) = \frac{1}{\pi}\left(\frac{2m^*}{\hbar^2}\right)^{1/2}\sum_{\nu\mu}(E - E_{\nu\mu})^{-1/2}\Theta(E - E_{\nu\mu}).\qquad(2.6)$$

Finally, for the zero-dimensional case the spectrum is completely discrete, and the density of states becomes a series of δ-peaks:

$$D_{0D}(E) = 2\sum_{\nu\mu\lambda}\delta(E - E_{\nu\mu\lambda}).\qquad(2.7)$$

Another important nanostructure arises if several alternating layers of two materials with different bandgaps are grown as shown in Fig. 2.4 for the GaAs/AlAs system. The energy diagram shows a periodic modulation of the conduction band with a period given by the sum of the widths of the GaAs quantum well and the AlAs barrier. Therefore this artificial structure is called a *superlattice* (Grahn 1995b). For sufficiently thin barriers the quantum wells are so strongly

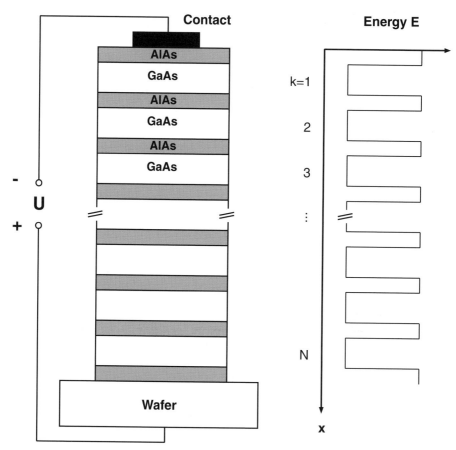

Figure 2.4. A superlattice structure of alternating GaAs and AlAs layers, and the corresponding energy diagram.

coupled that a one-dimensional energy band is formed in the growth direction. However, because the period of the superlattice is much larger than the atomic lattice constant, the resulting Brillouin zones and bandwidths are much smaller than those of atomic lattices, and the band structure obtained is called a miniband. For larger barrier widths the coupling is weaker, and sequential tunneling of electrons between different wells plays a dominant role. Interesting transport phenomena occur in such superlattices if a voltage U is applied, as shown in Fig. 2.4; these will be discussed in detail in Chapter 6.

For example, if a strong bias is applied to a weakly coupled superlattice, the current density j as a function of the electric field $\mathcal{E}$ displays a strongly nonlinear relation, as sketched in Fig. 2.5. With increasing field the current density first rises and then drops again as the overlap between the energy levels (two-dimensional subbands) in adjacent wells decreases, thereby displaying negative differential

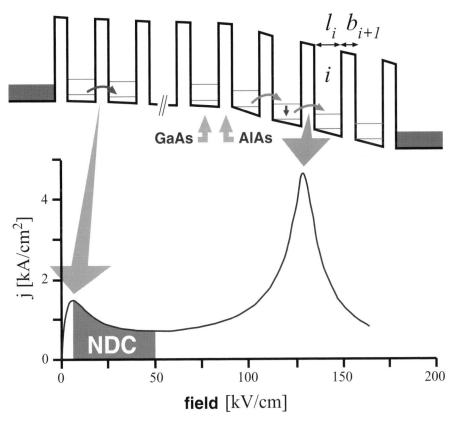

Figure 2.5. A superlattice exhibiting domain formation. The associated current density (j) versus field ($\mathcal{E}$) characteristic exhibits negative differential conductivity (NDC). The low-field domain corresponds to sequential tunneling between equivalent levels of adjacent quantum wells (the low-field peak of the $j(\mathcal{E})$ characteristic), while the high-field domain corresponds to resonant tunneling between different levels of adjacent wells (the high-field peak).

conductivity (NDC). Upon further increase of the field the current density rises again up to a sharp peak that occurs when the ground level in one well is aligned with the second level in the neighboring well, i.e. when the field $\mathcal{E}$ satisfies the relation $e\mathcal{E}d = \Delta E$, where ΔE is the intersubband spacing and d is the superlattice period. Then resonant tunneling between adjacent wells produces a large current. Such an N-shaped $j(\mathcal{E})$ characteristic may give rise to instabilities. When the superlattice is biased into the regime of NDC, the homogeneous field distribution may break up into a low-field domain where the field is near the first peak of the $j(\mathcal{E})$ characteristic and a high-field domain where the field is close to the second, resonant-tunneling peak. This is schematically indicated by the different slopes of the potential drop in the left-hand part and the right-hand part of the superlattice structure depicted in Fig. 2.5. The field distribution must adjust such that the total applied voltage drops between the two contacts. If the space charge available (by doping or by optical generation of carriers) is not sufficient to form a stable domain boundary, self-generated current oscillations may occur instead.

Other features that are associated with transport in strongly coupled superlattices are *Bloch oscillations*. These result from coherent motion of wavepackets through the periodic miniband structure of the superlattice, which leads to terahertz emission of radiation with a frequency given by $\nu = e\mathcal{E}d/h$. This can be understood by observing that a Bloch electron in a potential with period d is accelerated in an applied uniform field $\mathcal{E}$ until it reaches the edge of the Brillouin zone at $k = \pi/d$. After Bragg reflection its quasimomentum is changed to $k = -\pi/d$, and it starts its next cycle through the miniband. The time τ it takes to complete one cycle is given by $\hbar\,\Delta k = e\mathcal{E}\tau$.

A different type of superlattice that has recently attracted much attention is obtained if the electronic system is modulated laterally (i.e. perpendicular to the growth direction) in a periodic array of quantum wires or dots. Such lateral superlattices can be generated by lithographic techniques in combination with etching or addition of metallic gates. The latter structures are designed to change the strength of the modulation of the electrostatic potential in the plane of the two-dimensional electron gas. This leads to one- or two-dimensional lateral superlattices described by one- or two-dimensional minibands. As a special two-dimensional lateral nanostructure, the so-called antidot superlattices have been studied extensively (Schöll 1998*b*). Here the lateral potential modulation is so strong that it is seen by the carriers as strongly repulsive barriers arranged in a two-dimensional periodic array. Therefore the electrons can move everywhere except inside these potential barriers (*antidots*), complementary to the case of normal quantum dots. This system is a realization of the Sinai billiard which, in classical nonlinear dynamics, has been shown to exhibit chaos.

The potential profile of quantum wells and barriers is intimately connected with the charge-transport properties of the nanostructure. Tunneling through barriers provides one transport mechanism, and thermionic emission of hot electrons that have enough kinetic energy to overcome the barrier is another. Both mechanisms depend strongly upon the applied bias. The double-barrier resonant-tunneling diode consists just in two barriers but the principle which produces N-shaped negative differential conductivity of the $j(\mathcal{E})$ characteristic is similar to that of the superlattice shown in Fig. 2.5. The heterostructure hot-electron diode is even simpler in that it requires only one heterojunction. It possesses an S-shaped $j(\mathcal{E})$ characteristic with two stable current states at a given voltage in a certain range, the lower one corresponding to tunneling through and the upper one corresponding to thermionic emission over the barrier.

In conclusion, by controlled layer-by-layer epitaxial growth of heterostructures in combination with lateral patterning, intricate artificial nanostructures with arbitrary shapes of barriers and wells can be designed. Such band-structure engineering can produce novel semiconductor devices with desired transport properties. It is an aim of the current research activity not only to understand the complex and sometimes chaotic spatio-temporal dynamics of charge carriers in such structures, but conversely also to make efficient use of those nonlinear transport properties in specific switching and oscillating electronic devices.

2.2.2 Space-charge dynamics

A full theoretical description of charge transport in semiconductor structures requires a self-consistent solution of the coupled problem of Maxwell's equations for the fields and a suitable set of transport equations for the carriers, as is schematically shown in Fig. 2.6. This is important because the presence of space charges leads to a mutual nonlinear interdependence of these sets of equations. For example, the potential profile including band-bending effects due to space charges can be determined in thermodynamic equilibrium by the simultaneous solution of the Schrödinger equation and Poisson's equation, whereby the electrostatic potential calculated from Poisson's equation for a given carrier-density distribution enters the Schrödinger equation, and, conversely, the free-carrier density is determined by the electron wave function calculated from the Schrödinger equation. A similar procedure must be used in the more interesting case when a bias is applied to the structure,

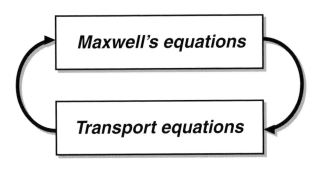

Figure 2.6. The scheme of the self-consistently coupled dynamic equations for the field and carrier distributions.

and an appropriate set of transport equations must be solved simultaneously with Poisson's equation under nonequilibrium conditions. It is important to note that, in time-dependent nonequilibrium situations, the fields do not instantaneously follow the carrier-density distributions, but take some finite dielectric relaxation times due to the build-up or sweep-out of space charges. Therefore the space-charge dynamics introduces some additional dynamic degree of freedom, as we have demonstrated explicitly with the simple example of a homogeneous conductor in Chapter 1, Section 1.4.2.

In general, as discussed above, the potential profile of quantum wells and barriers is reciprocally influenced by the nonequilibrium charge transport through the structure, which may lead to charge accumulation or depletion. These space charges, according to Poisson's equation, induce a feedback between the charge-carrier distribution and the electric-potential distribution governing the transport. This nonlinear coupling is particularly pronounced in case of semiconductor heterostructures and low-dimensional structures, in which abrupt junctions between different materials on an atomic length scale cause conduction-band discontinuities resulting in potential barriers and wells. The local charge accumulation in these potential wells and nonlinear processes of charge transport across the barriers have been found to provide a number of mechanisms for negative differential conductivity, bistability of the current at a given voltage, and nonlinear dynamics (Balkan *et al.* 1993). The effect of space charges can be illustrated most simply by the example of a single heterojunction (Fig. 2.7). If n-doped AlGaAs and intrinsic or p-doped GaAs are joined together, electrons spill over from the higher potential in the AlGaAs into the GaAs, leaving behind positively charged donors. This space charge gives rise to an electrostatic potential that causes the bands to bend near the interface such that a narrow triangular potential well is formed in the GaAs layer at the interface. In this well a thin accumulation layer of electrons constitutes a two-dimensional electron gas. Since the electrons are spatially separated from the host impurities in the AlGaAs barrier, which act as scattering centers, they experience very little scattering, and their mobility in such modulation-doped structures is extremely high,

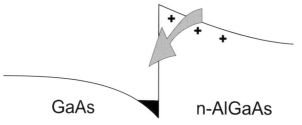

GaAs   n-AlGaAs

Figure 2.7. The conduction-band energy diagram of a modulation-doped n-AlGaAs/GaAs heterojunction. A two-dimensional electron gas is formed in the triangular potential well.

reaching values of up to 10^7 cm^2 V^{-1} s^{-1}. Real-space transfer of electrons back into the AlGaAs barrier with increasing voltage, i.e. by carrier heating, can reduce the mobility and lead to N-shaped negative differential conductivity (Gribnikov *et al.* 1995). If the GaAs is p-doped, the type of majority carriers is even reversed in this so-called inversion layer. Such inversion layers were first demonstrated in Si metal–oxide–semiconductor (MOS) devices, where the oxide takes over the role of the wide-gap AlGaAs, and they are essential to the operation of all field-effect transistors.

2.2.3   Various transport regimes

The transport properties of semiconductor structures depend on a number of characteristic length scales and their relations to the system size. A conductor usually exhibits classical Ohmic behavior only if it is much larger than all of these lengths. The length scales vary widely from one material to another and also depend upon temperature, electric and magnetic fields, impurity concentrations, etc.

- **The mean free path**
  The mean free path l_{m} is the average distance that an electron travels before it experiences elastic scattering that destroys its initial momentum. The dominant mechanism for elastic scattering is impurity scattering. The mean free path is related to the momentum relaxation time τ_m by $l_{\mathrm{m}} = v\tau_m$ with the average carrier velocity v. In high-mobility semiconductors at $T < 4$ K, l_{m} is typically in the 10–100 μm regime.

- **The phase relaxation length**
  The phase-relaxation length l_ϕ is the average distance that an electron travels before it experiences inelastic scattering that destroys its initial coherent state. Typical scattering events, such as electron–phonon and electron–electron collisions, change the energy of the electron and randomize its quantum mechanical phase. Impurity scattering may also contribute to phase relaxation if the impurity has an internal degree of freedom so that it can change its state. For example, magnetic impurities have an internal spin that fluctuates with time. In high-mobility degenerate semiconductors, phase relaxation often occurs on a time scale τ_ϕ that is of the same order as or shorter than the momentum relaxation time τ_m. Then $l_\phi = v_{\mathrm{F}}\tau_\phi$ holds, with the Fermi velocity v_{F}. In low-mobility semiconductors the momentum relaxation time τ_m can be considerably shorter than the phase-relaxation time τ_ϕ, and diffusive motion may occur over a phase-coherent region; then $l_\phi^2 = D\tau_\phi$ with a diffusion constant $D = v_{\mathrm{F}}^2\tau_m/2$.

- **The de Broglie wavelength**
  The de Broglie wavelength $\lambda = 2\pi/k = h/(2m^*E)^{1/2}$ is related to the kinetic energy of an electron. It defines the length scale on which quantum

mechanical effects, i.e. the wave-like nature of the electron, become important. According to Pauli's exclusion principle the electrons in a metal or degenerate semiconductor at $T = 0$ fill up all states up to the Fermi energy E_F corresponding to the Fermi velocity $v_F = (2E_F/m^*)^{1/2}$ or the Fermi wave vector $k_F(E) = (2m^*E_F)^{1/2}/\hbar$. It is determined by the electron density per unit volume $n_{3D} = k_F^3/(3\pi^2)$ or per unit area $n_{2D} = k_F^2/(2\pi)$ in the three- and two-dimensional cases, respectively. For a two-dimensional electron density of 10^{11} cm^{-2} the Fermi wavelength is about 75 nm.

■ **The magnetic length**
In the presence of a magnetic field (inductance B) the electron energy is quantized in Landau levels $E_N = (N + \frac{1}{2})\hbar\omega_c$ where $\omega_c = eB/m^*$ is the cyclotron frequency. The magnetic length $l_B = (\hbar/eB)^{1/2}$ characterizes the extension of the cyclotron orbit. The importance of the magnetic length lies in the fact that it can be tuned over a large range by changing the magnetic field. Thus a magnetic field provides additional means of reducing the effective dimensionality of the system. The magnetic length corresponding to a magnetic field of 1 T is about 25 nm.

■ **The thermal length**
The thermal length $l_T = \hbar v_F/(k_B T)$ is connected with the average excess energy of thermal electrons $k_B T$. An electron traveling at the Fermi velocity can only be located within l_T, due to the energy–time uncertainty relation.

Depending upon the values of those characteristic lengths in comparison with the system size L, several transport regimes can be distinguished.

■ **Classical diffusive transport**
For macroscopic dimensions $L \gg l_m, l_\phi$ the carrier experiences many elastic and inelastic collisions so that the energy and the momentum are relaxed, and the average velocity is given by the electron drift velocity $v = -\mu F$ with the mobility $\mu = e\tau_m/m^*$ following in the simplest case from Drude theory. Diffusion occurs due to gradients in the carrier density, and the diffusion current density is given by $eD \nabla n$, where the diffusion constant D is related to the mobility by the Einstein relation $eD = \mu k_B T$.

■ **Coherent transport**
For system sizes L smaller than the phase relaxation length l_ϕ the quantum mechanical wave function of the charge carriers has a well-defined phase throughout the system. Quantum interference phenomena like Aharonov–Bohm oscillations or universal conductance fluctuations can be observed in transport.

■ **Ballistic transport**

When the system size L becomes smaller than the mean free path l_m, a carrier can cross the device without any scattering. The carrier momentum grows due to the accelerating force of the electric field.

■ **Quantum size effects**

If the system size L in one or several directions is of the order of the de Broglie wavelength λ, size quantization occurs. Propagation in those directions is no longer possible due to quantum confinement, and the density of states is modified as discussed in Section 2.2.1.

2.3   The hierarchy of modeling approaches

In this book various theoretical approaches to charge transport will be used. From a conceptual point of view, these may be grouped into a hierarchy of transport models in which different approximations are used at different levels (Schöll 1998*b*). In this section the levels of that hierarchy will be surveyed, and the principal approximations which lead from one level to the next will be pointed out. Starting from a microscopic quantum transport description we shall thus eventually arrive at macroscopic classical drift–diffusion and reaction–diffusion models.

2.3.1   Quantum transport

At the most fundamental level, the dynamics of electrons should be described by quantum mechanics. The Schrödinger picture or the Heisenberg picture may be used. In the Heisenberg picture, the dynamics of an operator A is given by

$$i\hbar \frac{d}{dt} A = [A, H], \tag{2.8}$$

where H is the Hamiltonian. In the framework of second quantization, which is appropriate to many-body theory, one introduces the creation and annihilation operators of an electron with wave vector k in the jth conduction subband, e.g. of a quantum well or quantum wire, $c_{j,k}^{\dagger}$ and $c_{j,k}$, respectively. Likewise, $d_{j,k}^{\dagger}$ and $d_{j,k}$ are creation and annihilation operators of holes with wave vector k in the jth valence subband.

The Hamiltonian H can be expressed as a sum of single-particle contributions of the form

$$H_0 = \sum_{j,k} E_{j,k} c_{j,k}^{\dagger} c_{j,k}, \tag{2.9}$$

where $E_{j,k}$ is the single-electron energy of an electron with wave vector k in the jth subband and $c_{j,k}^{\dagger} c_{j,k}$ is the corresponding number operator (and analogously for holes), and two-particle contributions describing the Coulomb interaction of the form

$$H_{cc} = \frac{1}{2} \sum_{j_1, j_2, i_2, i_1} \sum_{k_1, k_2, q} V_{j_1 j_2 i_2 i_1}(q) \, c^{\dagger}_{j_1, k_1+q} c^{\dagger}_{j_2, k_2-q} c_{i_2, k_2} c_{i_1, k_1}, \qquad (2.10)$$

where $V_{j_1 j_2 i_2 i_1}(q)$ is the Coulomb matrix element. The Hamiltonian may further include additional coupling terms with external fields, phonons, etc.

In the language of second quantization eq. (2.10) describes elementary processes whereby two electrons from subbands i_1 and i_2 with wave vectors k_1 and k_2 are scattered into subbands j_1 and j_2 with wave vectors $k_1 + q$ and $k_2 - q$, respectively (Fig. 2.8).

In the density-matrix formalism the quantum statistical ensemble of carriers is described by the single-particle density matrices which are given in the k-representation by the ensemble average $\langle \rangle$ of the operators $c^{\dagger}_{j,k} c_{i,k}$, $d^{\dagger}_{j,k} d_{i,k}$, $d^{\dagger}_{j,-k} c_{i,k}$, etc. For simplicity, we shall restrict ourselves to electrons in the following.

All single-particle density matrices can be summarized in the form

$$f_{ji}(k) = \langle c^{\dagger}_{j,k} c_{i,k} \rangle, \qquad i, j = 1, \ldots, n. \qquad (2.11)$$

Note that the two operators are to be taken at equal times. In particular, this comprises

- the *distribution functions* of electrons in the various subbands of the conduction band, given by the intrasubband density matrices:

$$f_{ii}(k) = \langle c^{\dagger}_{i,k} c_{i,k} \rangle; \qquad (2.12)$$

- and the *intersubband polarizations*, given by the intersubband density matrices:

$$f_{ji}(k) = f_{ij}^{*}(k) = \langle c^{\dagger}_{j,k} c_{i,k} \rangle, \qquad j \neq i. \qquad (2.13)$$

The dynamics of these density matrices can be derived from the Heisenberg equation of motion (2.8) for the single-particle operator $A = c^{\dagger}_{j,k} c_{i,k}$ by taking the

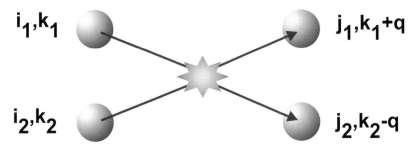

Figure 2.8. A schematic diagram of electron–electron scattering.

ensemble average. The temporal change of the expectation values is composed of two contributions corresponding to the commutator with H_0 and H_{cc}, respectively. The dynamics induced by the interaction with the crystal lattice (H_0)

$$i\hbar \frac{d}{dt} f_{j_1 i_1}(k_1)\Big|_0 = (E_{i_1,k_1} - E_{j_1,k_1}) f_{j_1 i_1}(k_1) \tag{2.14}$$

describes merely a rotation of the polarizations ($j_1 \neq i_1$) in the complex plane with a frequency corresponding to the intersubband splitting energy at the respective k-value.

The Coulomb interaction of the electrons (H_{cc}) leads to a coupling of the single-particle density matrices to two-particle density matrices $\langle c^\dagger_{j'_1,k'_1} c^\dagger_{j'_2,k'_2} c_{j_2,k_2} c_{j_1,k_1} \rangle$. A set of dynamic equations for the temporal evolution of these two-particle density matrices can be set up in the same way using eq. (2.8). However, those equations of motion contain the three-particle density matrices, and so on. In order to obtain a closed set of equations, this infinite hierarchy of many-particle correlations must be truncated, and this is done by replacing the expectation value of a product of field-operator pairs by the product of their expectation values, i.e. by introducing a mean-field (or Hartree–Fock) approximation at some stage. If the two-particle density matrices are retained as independent dynamic variables, the resulting quantum kinetics – with respect to the one-particle densities alone – is not Markovian, i.e. it contains memory effects. Moreover, it includes renormalizations of the single-particle energies and internal fields due to many-body effects, and electronic coherence properties represented by the polarizations and higher-order correlations.

In the preceding survey spatial homogeneity has been assumed, so that the single-particle density matrices do not contain off-diagonal elements with respect to k, because those would not correspond to a momentum-conserving process. Spatially inhomogeneous systems can be described by Wigner functions that are obtained by performing a Fourier transform with respect to the relative momentum. The Wigner function can be conceived as a quantum extension of the semiclassical distribution function $f(r, k, t)$. However, it is not possible to attribute a direct probabilistic interpretation to the Wigner function since it is in general not positive definite.

An alternative approach to quantum transport is via Green's functions. The general concept of a Green's function involves a quantity that describes how a system in a state q' at the initial time t' evolves into a state q at a later time t. With respect to quantum dynamics the Green function $G(q, t, q', t')$ may be defined by the time evolution of a quantum state

$$\Phi(q, t) = i\hbar \int dq' \, G(q, t, q', t') \Phi(q', t'), \tag{2.15}$$

where q stands for the set of system variables. The two solutions G propagating the system forward or backward in time, respectively, are denoted by superscripts r (retarded) and a (advanced), respectively. In the framework of second quantization the single-particle retarded Green functions (with $\hbar = 1$) can be written as

$$G^{\mathrm{r}}_{m,n}(t, t', \mathbf{k}) = -i\Theta(t - t')\langle\{c_{m,\mathbf{k}}(t), c^{\dagger}_{n,\mathbf{k}}(t')\}\rangle, \tag{2.16}$$

where $\{A, B\}$ denotes the anticommutator between the Fermi operators in the Heisenberg picture. Note that the two times are usually not equal, in contrast to the density-matrix theory. Whereas in thermodynamic equilibrium the occupation of states is given by the Fermi distribution, in the theory of nonequilibrium Green functions the lesser functions

$$G^{<}_{m,n}(t, t', \mathbf{k}) = i\langle c^{\dagger}_{n,\mathbf{k}}(t')c_{m,\mathbf{k}}(t)\rangle \tag{2.17}$$

have to be considered as well. When they are evaluated at $t = t'$, they equal the density matrix and determine the electron densities as well as the polarizations. While the density-matrix formalism is particularly useful for systems with long phase-relaxation times since the coherent temporal evolution of the energetically resolved electron distribution can be directly obtained from a coupled set of differential equations, the Green function approach can efficiently deal with strong scattering processes. The advantage of the Green function approach is that nonperturbational techniques allow a partial summation of some interactions up to infinite order. This technique results in integro-differential equations that display also quantum correlations and memory effects but are generally difficult to solve.

2.3.2   The semiclassical Boltzmann equation

At a semiclassical level the carrier kinetics under the influence of electric and magnetic fields, $\mathcal{E}$ and B, respectively, is described by the Boltzmann transport equation for the semiclassical carrier distribution function $f(\mathbf{r}, \mathbf{k}, t)$, where $\mathbf{r}$ is the spatial coordinate and $\mathbf{k}$ is the wave vector:

$$\frac{\partial f}{\partial t} + \mathbf{v}_{\mathrm{g}} \cdot \nabla_{\mathbf{r}} f + \frac{q}{\hbar} \left(\mathcal{E} + \mathbf{v}_{\mathrm{g}} \times B\right) \nabla_{\mathbf{k}} f = \left(\frac{\partial f}{\partial t}\right)_{\mathrm{coll}}. \tag{2.18}$$

Here $\mathbf{v}_{\mathrm{g}} = \hbar^{-1} \nabla_{\mathbf{k}} E(k)$ is the group velocity. The collision integral $(\partial f/\partial t)_{\mathrm{coll}}$ includes in principle all scattering processes such as phonon, impurity, and electron–electron scattering, generation, and recombination processes. For intraband single-electron processes,

$$\left(\frac{\partial f}{\partial t}\right)_{\mathrm{coll}} = \int (W(\mathbf{k}', \mathbf{k}) f(\mathbf{r}, \mathbf{k}', t) - W(\mathbf{k}, \mathbf{k}') f(\mathbf{r}, \mathbf{k}, t)) z \, d^3 k' \tag{2.19}$$

holds, where $W(\mathbf{k}, \mathbf{k}')$ is the probability per unit time for a transition from state $\mathbf{k}$ to $\mathbf{k}'$, and z is the density of states in k-space. The carrier charge is $q = \pm e$ for holes and electrons, respectively.

The Boltzmann equation relies on the assumptions that

(i) the distribution function varies little over the de Broglie wavelength,

(ii) the carrier density is sufficiently low that only binary collisions occur,

(iii) the time between successive collisions is much longer than the duration of a collision, and

(iv) the density gradients over the range of the interparticle potential are small.

The set of quantum kinetic equations of the density-matrix theory can be reduced to the semiclassical Boltzmann transport equation for the distribution function $f(\mathbf{k}, t)$ under the following approximations. The general procedure consists in the adiabatic elimination of variables involving quantum mechanical correlations by means of a Markov approximation under the assumption that initially the system was uncorrelated. In particular, interband and intersubband polarizations and two-particle density matrices can be eliminated by formally integrating the respective dynamic equations. This yields non-Markovian convolution integrals in time, whose kernels contain oscillating terms of the form $\exp(-i\Omega t)$, where $\hbar\Omega = E_{\text{in}} - E_{\text{out}}$ represents the total energy balance of the interaction process. Assuming that the distribution functions and field amplitudes occurring also in the integral are sufficiently slowly varying functions of time, the Markov limit can be performed by taking these functions out of the integral, and approximating the remaining oscillating time integral by the energy-conserving delta function $\delta(\Omega)$ in the infinite-time limit. Thus the semiclassical rates of the interaction processes are regained in the familiar form of Fermi's golden rule in first-order quantum mechanical perturbation theory:

$$W(\mathbf{k}', \mathbf{k}) = \frac{2\pi}{\hbar}|M|^2\delta(E_{\text{in}} - E_{\text{out}}), \tag{2.20}$$

where M is the matrix element of the interaction process. Applying this procedure to electron–photon, electron–phonon, or electron–electron interactions yields the respective collision integrals of the Boltzmann equation.

For example, for the electron–electron density correlations $s_{ijlm}(k_1, k_2, q; t)$ corresponding to the electron–electron scattering process in Fig. 2.8, the equations of motion derived from (2.8) are of the form (Prengel and Schöll 1999)

$$i\hbar\frac{d}{dt}s_{ijlm}(k_1, k_2, q; t) = \hbar\Omega_{ijlm}(k_1, k_2, q; t)\,s_{ijlm}(k_1, k_2, q; t)$$
$$+ y_{ijlm}(k_1, k_2, q; t), \tag{2.21}$$

where

$$\hbar\Omega_{ijlm}(k_1, k_2, q) = E_m(k_1 + q) + E_l(k_2 - q) - E_j(k_2) - E_i(k_1), \tag{2.22}$$

$$y_{ijlm}(k_1, k_2, q; t) = V^*_{mlji}(k_1, k_2, q)\mathcal{F}^{mlji}_{ijlm}(k_1, k_2, q; t), \tag{2.23}$$

with the product of distribution functions

$$\mathcal{F}_{ijlm}^{mlji}(k_1, k_2, q) \equiv f_{ii}(k_1) f_{jj}(k_2)\big[1 - f_{mm}(k_1 + q)\big]\big[1 - f_{ll}(k_2 - q)\big] \qquad (2.24)$$
$$- f_{mm}(k_1 + q) f_{ll}(k_2 - q)\big[1 - f_{ii}(k_1)\big]\big[1 - f_{jj}(k_2)\big].$$

$$(2.25)$$

The inhomogeneous differential equation (2.21) can be formally integrated. If the distribution functions are slowly varying in time and if we assume that the Coulomb interaction is turned on adiabatically at $t_0 = -\infty$, a Markov and adiabatic approximation can be applied, and we obtain

$$s(t) = -i\pi \left(\frac{1}{i\pi} \frac{\mathcal{P}}{\hbar\Omega} + \delta(\hbar\Omega) \right) y(t) \approx -i\pi \delta(\hbar\Omega) y(t), \qquad (2.26)$$

where the principal value part $\mathcal{P}$ has been neglected in the last step, and all subscripts and arguments except time have been dropped for simplicity.

Therefore by (2.8)

$$\frac{d}{dt} f_{ii}(k_1) = \frac{2}{\hbar} \sum_{jlm} \sum_{k_2, q} \mathrm{Im}\big(V_{ijlm}(q) s_{ijlm}(k_1, k_2, q)\big)$$
$$= -\sum_{jlm} \sum_{k_2, q} W_{ijlm}(k_1, k_2, q) \mathcal{F}_{ijlm}^{mlji}(k_1, k_2, q), \qquad (2.27)$$

where

$$W_{ijlm}(k_1, k_2, q) = \frac{2\pi}{\hbar} V_{ijlm} V_{mlji}^* \delta\big[E_m(k_1 + q) + E_l(k_2 - q) - E_j(k_2) - E_i(k_1)\big]. \qquad (2.28)$$

This is similar to a Boltzmann scattering term for the process shown in Fig. 2.8, where the δ-function ensures conservation of energy.

The Boltzmann equation describes charge transport on a semiclassical kinetic level. The most powerful method to solve it is the Monte Carlo approach (Jacoboni and Lugli 1989). It consists in a simulation in (r, k) space of the individual carrier motion, subject to scattering effects selected stochastically. The distribution function is obtained either by taking a time average over the trajectory of a single carrier, or by sampling over an ensemble of equivalent carriers (ensemble Monte Carlo simulation).

2.3.3   Hydrodynamic balance equations

At the hydrodynamic level of description the detailed kinetics in k-space is averaged over, and only slow, macroscopic quantities such as the carrier density

$$n(r, t) = \int f(r, k, t) z\, d^3k, \qquad (2.29)$$

the mean momentum per carrier

$$\boldsymbol{p}(\boldsymbol{r}, t) = \langle \hbar \boldsymbol{k} \rangle, \tag{2.30}$$

and the mean energy per carrier

$$w_n(\boldsymbol{r}, t) = \langle E(\boldsymbol{k}) \rangle \tag{2.31}$$

are considered as dynamic variables. Here the brackets denote the semiclassical ensemble average

$$\langle A \rangle = n^{-1} \int A(\boldsymbol{k}) f(\boldsymbol{r}, \boldsymbol{k}, t) z \, d^3 k. \tag{2.32}$$

The resulting hydrodynamic balance equations have the form of continuity equations for the carrier density, the mean momentum density, and the mean energy density. For simplicity we confine our attention to a nondegenerate, isotropic parabolic band structure $E(\boldsymbol{k}) = \hbar^2 k^2 / (2m^*)$.

The mean group velocity $\boldsymbol{v}(\boldsymbol{r}, t) = \langle \boldsymbol{v}_g \rangle$ is then related to the carrier temperature

$$T_e = \frac{m^*}{3k_B} \langle (\boldsymbol{v}_g - \langle \boldsymbol{v}_g \rangle)^2 \rangle \tag{2.33}$$

and the mean momentum and energy by

$$\boldsymbol{p} = m^* \boldsymbol{v}, \qquad w_n = \frac{m^*}{2} v^2 + \frac{3}{2} k_B T_e. \tag{2.34}$$

Hydrodynamic balance equations for n, $\boldsymbol{p}$, and w_n (*moments* of the Boltzmann equation) are obtained by multiplying (2.18) by appropriate powers of $\boldsymbol{k}$ and integrating over the first Brillouin zone. Since each balance equation is coupled to the next higher moment, we obtain an infinite hierarchy of moment equations, which can be truncated by the following approximations.

■ The electron temperature is a scalar,

$$\langle (v_{gi} - \langle v_{gi} \rangle)(v_{gj} - \langle v_{gj} \rangle) \rangle = \frac{3}{m^*} k_B T_e \delta_{ij}, \tag{2.35}$$

i.e. the momentum-flux tensor reduces to the scalar electron pressure $n k_B T_e$.

■ The heat flux $\boldsymbol{j}_Q = \frac{1}{2} n m^* \langle (\boldsymbol{v}_g - \langle \boldsymbol{v}_g \rangle)^2 (\boldsymbol{v}_g - \langle \boldsymbol{v}_g \rangle) \rangle$ is approximated by Fourier's law

$$\boldsymbol{j}_Q = -\kappa \, \boldsymbol{\nabla}_r T_e \tag{2.36}$$

with thermal conductivity κ.

- The collision integrals are expressed in terms of generation–recombination rates, momentum-relaxation rates, and the energy-relaxation rate, respectively, via

$$\int \left(\frac{\partial f}{\partial t}\right)_{\text{coll}} z\, d^3k = \phi(n, w_n), \tag{2.37}$$

$$\int \hbar k \left(\frac{\partial f}{\partial t}\right)_{\text{coll}} z\, d^3k = -n\frac{\boldsymbol{p}}{\tau_m}, \tag{2.38}$$

$$\int E(\boldsymbol{k}) \left(\frac{\partial f}{\partial t}\right)_{\text{coll}} z\, d^3k = -n\frac{w_n - w_{n0}}{\tau_e(w_n)}, \tag{2.39}$$

with the mean-energy-dependent energy-relaxation time τ_e, momentum-relaxation time τ_m, and $w_{n0} = \frac{3}{2}k_B T_L$.

The following closed set of hydrodynamic equations is obtained:

$$\frac{\partial n}{\partial t} + \boldsymbol{\nabla}_r(n\boldsymbol{v}) = \phi(n, w_n), \tag{2.40}$$

$$\frac{\partial \boldsymbol{p}}{\partial t} + (\boldsymbol{v}\,\boldsymbol{\nabla}_r)\boldsymbol{p} + \frac{1}{n}\boldsymbol{\nabla}_r(nk_B T_e) - q(\boldsymbol{\mathcal{E}} + \boldsymbol{v}\times\boldsymbol{B}) = -\frac{\boldsymbol{p}}{\tau_m}, \tag{2.41}$$

$$\frac{\partial w_n}{\partial t} + (\boldsymbol{v}\,\boldsymbol{\nabla}_r)w_n + \frac{1}{n}\boldsymbol{\nabla}_r(nk_B T_e\boldsymbol{v}) - \frac{\kappa}{n}\Delta T_e - q\boldsymbol{v}\boldsymbol{\mathcal{E}} = -\frac{w_n - w_{n0}}{\tau_e(w_n)}. \tag{2.42}$$

2.3.4   Classical drift–diffusion theory

If momentum and energy relaxation occur faster than all other processes, $\boldsymbol{p}$ and w_n can be eliminated adiabatically, and the carrier densities remain as the only dynamic variables on this slow time scale. Transport may then be described within classical drift–diffusion theory. Specifically, from the momentum-balance equation (2.41) $\boldsymbol{p}$ can be eliminated adiabatically by setting $d\boldsymbol{p}/dt \equiv \partial\boldsymbol{p}/\partial t + (\boldsymbol{v}\boldsymbol{\nabla}_r)\boldsymbol{p} = 0$. Thus (2.41) leads to a current density as a function of field

$$\boldsymbol{j} = -en\boldsymbol{v} = en\mu_B\boldsymbol{F}_{\text{tot}} + en\mu\mu_B(\boldsymbol{B}\times\boldsymbol{F}_{\text{tot}}) + en\mu^2\mu_B\boldsymbol{B}(\boldsymbol{B}\boldsymbol{F}_{\text{tot}}) \tag{2.43}$$

with the mobility $\mu = (e/m^*)\tau_m$, $\mu_B = \mu/[1 + (\mu B)^2]$, and

$$\boldsymbol{F}_{\text{tot}} = \boldsymbol{\mathcal{E}} + \frac{1}{en}\boldsymbol{\nabla}_r(nk_B T_e), \tag{2.44}$$

where we have restricted ourselves to electrons, and a constant momentum relaxation time. More sophisticated treatments distinguish between drift and Hall mobility (Hüpper *et al.* 1992).

Neglecting the spatial variations of w_n, and hence of T_e, and setting $B = 0$, the usual drift–diffusion expression for the current density

$$\boldsymbol{j} = en\mu\boldsymbol{\mathcal{E}} + eD\,\boldsymbol{\nabla}_r n \tag{2.45}$$

is recovered, if the Einstein relation $\mu k_B T_e = eD$ is used.

Furthermore, if energy relaxation occurs fast, temporal and spatial derivatives in (2.42) can be neglected, yielding a local energy–field relation:

$$w_n = w_{n0} + e\tau_e\mu\mathcal{E}^2. \tag{2.46}$$

2.3.5 Reaction–diffusion models

By further reductions it is often possible to derive even simpler reaction–diffusion equations for the dynamic evolution of the transport variables. For a wide class of semiconductor systems the internal state can be characterized by a single variable $a(x, y, t)$ that corresponds to the internal degree of freedom relevant for the nonlinearity of the charge transport. It may be an interface charge density (as in the heterostructure hot-electron diode, see Section 2.4.5), the electron concentration in a quantum well (as in double-barrier resonant-tunneling structures, see Section 2.4.6), or a potential drop within the semiconductor structure (such as the emitter voltage in p–n–p–i–n structures, see Section 2.4.7, or the gate–base potential in a gate-driven thyristor, see Section 2.4.8). The spatial variables x and y denote the two coordinates perpendicular to the current flow, while the z-coordinate, i.e. the spatial dependency in the direction of the current flow, has been eliminated from the transport equations. The reduced dynamic description in terms of the parameter a can be derived from a full three-dimensional transport model, the Poisson equation, and continuity equations via adiabatic elimination of fast variables, as has been done for various semiconductor systems (Volkov and Kogan 1969, Osipov and Kholodnov 1973, Wacker and Schöll 1994a, Niedernostheide et al. 1994a, Gorbatyuk and Rodin 1992a, 1992b, Glavin et al. 1997, Mel'nikov and Podlivaev 1998, Meixner et al. 2000a). This equation takes the form of a reaction–diffusion equation (1.38):

$$\tau_a \frac{\partial a(x, y, t)}{\partial t} = f(a, u) + l^2\left(\frac{\partial^2 a}{\partial x^2} + \frac{\partial^2 a}{\partial y^2}\right). \tag{2.47}$$

Here τ_a and l are the relaxation time and diffusion length of the variable a, respectively. Generally, the combination of lateral diffusion and drift (due to the lateral electric field induced by an inhomogeneous charge distribution) effectively results in a modified diffusion term with an a-dependent diffusion coefficient $D(a) \equiv l^2(a)/\tau_a$ (see Section 2.4 below). The function $f(a, u)$ depends upon the internal variable a and the voltage u dropping across the semiconductor. It mimics a nonlinear reaction term, as is familiar from chemical reaction systems. Physically, it describes the local transport or generation–recombination kinetics of the charge carriers.

A further generic reduction in terms of so-called *amplitude equations* is often possible. Near a supercritical Hopf bifurcation, for example, a reaction–diffusion system may be expanded in terms of the complex amplitude A of the oscillations, yielding the complex Ginzburg–Landau equation (Cross and Hohenberg 1993, Mikhailov 1994, Walgraef 1997)

$$\frac{\partial A}{\partial t} = \mu A + (1 + i\alpha)\,\Delta A - (1 + i\beta)|A|^2 A \tag{2.48}$$

with real parameters μ, α, and β.

The hierarchy of transport equations is summarized in Fig. 2.9.

2.4   Models for current instabilities

In this section we shall survey physical mechanisms that give rise to negative differential conductivity (NDC) and current instabilities. We shall derive a number of models based on various levels of the hierarchy presented in Section 2.3, yielding N-shaped, S-shaped, and Z-shaped current–voltage characteristics. Those mechanisms are effective in a number of semiconductor devices that are used as microwave oscillators (the Gunn diode, resonant-tunneling diode, real-space transfer device, and impact-ionization avalanche transit time (IMPATT) diode) or electronic

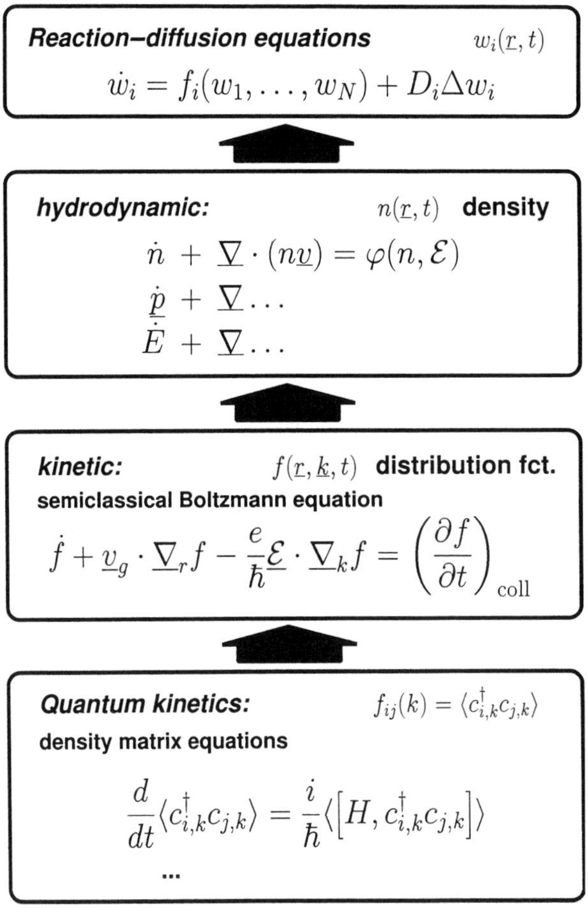

Reaction–diffusion equations $w_i(\underline{r}, t)$

$$\dot{w}_i = f_i(w_1, \ldots, w_N) + D_i \Delta w_i$$

Figure 2.9. The hierarchy of transport equations.

hydrodynamic: $n(\underline{r}, t)$ **density**

$$\dot{n} + \underline{\nabla} \cdot (n\underline{v}) = \varphi(n, \mathcal{E})$$
$$\dot{p} + \underline{\nabla} \ldots$$
$$\dot{E} + \underline{\nabla} \ldots$$

kinetic: $f(\underline{r}, \underline{k}, t)$ **distribution fct.**
semiclassical Boltzmann equation

$$\dot{f} + \underline{v}_g \cdot \underline{\nabla}_r f - \frac{e}{\hbar}\underline{\mathcal{E}} \cdot \underline{\nabla}_k f = \left(\frac{\partial f}{\partial t}\right)_{\text{coll}}$$

Quantum kinetics: $f_{ij}(k) = \langle c_{i,k}^\dagger c_{j,k}\rangle$
density matrix equations

$$\frac{d}{dt}\langle c_{i,k}^\dagger c_{j,k}\rangle = \frac{i}{\hbar}\langle [H, c_{i,k}^\dagger c_{j,k}]\rangle$$

$\cdots$

switches (the thyristor, p–i–n diode, and heterostructure hot-electron diode) (Shaw *et al.* 1992). Furthermore, there is a high potential of expected applications of such active spatially extended media for information processing and pattern recognition (Mikhailov 1989, Haken *et al.* 1994). The mechanism can be dominated either by bulk properties as in classical NDC systems, or by heterojunctions and potential barriers and wells as in state-of-the-art low-dimensional semiconductor structures.

The classical *bulk-dominated* NDC includes three major classes of electronic instabilities. The nonmonotonic shape of $j(\mathcal{E})$, which leads to $dj/d\mathcal{E} < 0$ in some range of local electric fields $\mathcal{E}$, can be due to a nonlinearity

- of the mobility (*drift instability*),

- of the carrier density (*generation–recombination instability*), or

- of the electron temperature (*electron-overheating instability*).

An important class of instabilities arises in *heterostructures* grown of alternating layers of different materials, as discussed in Section 2.2.1. In such layered and low-dimensional structures NDC is often due to *tunneling* across or *thermionic emission* over potential barriers (Balkan *et al.* 1993); those processes exhibit a strong nonlinear dependency upon the distribution of the voltage U across the structure.

2.4.1 Drift instability

The best known drift instability is the *Gunn effect* (Gunn 1964, Shaw *et al.* 1979). It is used in Gunn diodes to generate and amplify microwaves at frequencies typically beyond 1 GHz (These devices are called "diodes", since they are two-terminal devices, but no p–n junction is involved!). The mechanism is based upon intervalley transfer of electrons in k-space from a state of high mobility to a state of low mobility under a strong electric field ($\mathcal{E} > 3 \text{ kV cm}^{-1}$) in direct semiconductors like GaAs and other III–V compounds (Ridley and Watkins 1961). At low electric fields the electrons are essentially in the minimum of the central Γ-valley, which corresponds to a low effective mass m^* and hence a high mobility. As the field $\mathcal{E}$ is increased, the electrons are heated up, and gain enough energy to be transferred to the satellite valley with a higher minimum energy, but larger effective mass, and hence lower mobility. As more and more electrons are transferred, the average mobility μ decreases strongly so that the current density $j = en\mu(\mathcal{E})\mathcal{E}$ decreases with increasing field, as a result of negative differential mobility $d(\mu\mathcal{E})/d\mathcal{E} < 0$ (NDM). When most electrons are in the upper valley, j increases again. Thus an NNDC $j(\mathcal{E})$ characteristic is produced. Depending upon the cathode boundary condition, this may lead to the formation of moving domains that show up as transit-time oscillations of the current.

Another mechanism for negative differential mobility is due to the anisotropy of equivalent valleys (Asche *et al.* 1982, Shaw *et al.* 1992), for example in indirect semiconductors like n-Ge and n-Si. Electrons in these valleys have the same density of states but different effective masses in a fixed direction of the electric field. While the Gunn instability occurs for a current parallel to the applied field, this instability involves off-diagonal elements of the differential conductivity tensor. Consider in the simplest case two equivalent valleys 1 and 2 with anisotropic axes pointing in different directions and assume that the conductivity effective mass in valley 1 is larger than that in valley 2. The rate at which electrons absorb energy from the electric field, i.e. the rate at which they are heated, is inversely proportional to the effective mass, so electrons in valley 2 are heated more than are those in valley 1. Thus there will be a net transfer of electrons by intervalley scattering from the hotter valley to the cooler valley. The decrease in the population of the higher-mobility valley 2 then leads to negative differential conductivity (NNDC).

2.4.2   Generation–recombination instability

The large class of generation–recombination (GR) instabilities is distinguished by a nonlinear dependency of the steady-state carrier concentration n upon the field $\mathcal{E}$, which yields a nonmonotonic current density versus field relation $j = en(\mathcal{E})\mu\mathcal{E}$ of either NNDC or SNDC type. This dependency is due to a redistribution of electrons between the conduction band and bound states (and possibly the valence band, in case of ambipolar currents) with increasing field. The microscopic transition probabilities of the carriers between different states, and hence the GR coefficients, generally depend upon the electric field. A particularly strong dependency is expected for the rate constants of the following GR processes: *field-enhanced trapping* and *Poole–Frenkel emission* from impurity centers, which often leads to NNDC, and *impact ionization*, which is the key process for SNDC.

Field-enhanced trapping occurs in gold-doped n-Ge (Bonch-Bruevich *et al.* 1975). The gold atoms form deep impurity levels, corresponding to singly or doubly charged negative ions. Trapping of electrons in these levels requires the penetration of a Coulomb potential barrier. Therefore the trapping coefficient increases with the field $\mathcal{E}$ while the (thermal) emission of electrons is practically field-independent, as long as $\mathcal{E}$ is well below the threshold for field ionization. Thus the free-carrier density decreases with rising field: $dn/d\mathcal{E} < 0$, and the differential conductivity $dj/d\mathcal{E} = e\mu(n + \mathcal{E}\, dn/d\mathcal{E})$ may become negative. In larger fields the ionization co-efficient increases and causes the carrier density to rise again with the field, leading to a positive-differential-conductivity branch. Thus an N-shaped $j(\mathcal{E})$ characteristic is produced. Self-generated low-frequency oscillations (LFO) in semi-insulating GaAs at room temperature (Northrup *et al.* 1964, Sacks and Milnes 1970, Knap *et al.* 1988, Maracas *et al.* 1989, Samuilov 1995) have been attributed to traveling field domains induced by field-enhanced trapping, and these have indeed been observed experimentally by electrooptic techniques (Piazza *et al.* 1997). Chaotic bifurcation sequences have been simulated on the basis of domain motion in a model for a recombination instability (Oshio 1998). Field-enhanced emission from deep trap levels due to the lowering of the ionization energy by an electric field, i.e. the *Poole–Frenkel effect* (Reggiani and Mitin 1989), has also been proposed as a possible mechanism for NNDC (Schöll 1989), accounting in particular for very low frequencies in the range 50–100 Hz.

Impact ionization of carriers from impurity levels (shallow donors, acceptors, or deep traps) or across the bandgap (avalanching) is another process that may lead to NDC. If a free carrier has gained enough kinetic energy in the electric field, it may transfer this energy in a collision to a bound carrier, which is then released to the conduction band (or the valence band, in case of a bound hole); thereby an additional free carrier is generated, which may, in turn, impact ionize other carriers. Such a positive-feedback (*autocatalysis*) leads to a rapid increase of the free-carrier density. The impact-ionization coefficient strongly increases with $\mathcal{E}$ above a threshold that is necessary to heat up the carriers, so that they have enough energy for ionization. Models of this type are relevant for a variety of materials and in various temperature ranges (Schöll 1987), for instance in low-temperature impurity breakdown (Schöll 1982a) and in threshold switching (Landsberg *et al.* 1978, Adler *et al.* 1980). These models can explain SNDC of the $j(\mathcal{E})$ characteristic in the regime of low-temperature impurity breakdown, and self-generated current oscillations including chaotic behavior. In crossed electric and magnetic fields impurity-impact ionization can trigger a dynamic Hall instability (Hüpper and Schöll 1991). These matters will be discussed in Chapters 4 and 5.

Two important devices are also based upon GR-induced bulk negative differential conductivity, but the coupling with junction effects is essential in these cases: p–i–n diodes and IMPATT diodes (Shaw *et al.* 1992). The *p–i–n diode* involves double injection of electrons and holes, and field-enhanced trapping. It consists of an intrinsic (undoped) layer adjacent to a p-doped and an n-doped region on either side. If the n-layer is connected to a negative and the p-layer is connected to a positive voltage, electrons and holes are injected from the n-side and the p-side, respectively, into the central intrinsic (i) layer. We assume that the i-region contains deep acceptor-like recombination centers with a large cross-section (or capture coefficient) for hole capture and a much smaller cross-section for electron capture, and that these are completely occupied by electrons in thermal equilibrium. At low injection currents almost all of the injected holes will be captured by the recombination centers near the injecting p–i junction, while the injected electrons freely traverse most of the i-layer. Thus there is a "recombination barrier" to the passage of holes. The resulting electron current is limited by the space charge that the injected electrons build up. At high injection currents, for which the injected electron and hole concentrations exceed that of the recombination centers, all centers have trapped a hole, and the excess holes as well as the electrons traverse the i-layer. They are approximately equal in concentration, if the injection level is sufficiently high. The current is then carried by a quasineutral semiconductor plasma. Thus, at a given value of the applied voltage, there are two stable steady states: A low-current state, in which the recombination centers (except for a narrow region near the p–i junction) are occupied by the electrons and the current is a single-carrier space-charge-limited (SCL) current, and a high-current state, in which the recombination centers are filled with holes, and the current is carried by an injected plasma. In between, there is an NDC state in which both the single-carrier SCL current and the semiconductor plasma extend some way into the i-region from both sides (from the i–n and the p–i junction, respectively), and they are connected by regions of more complex carrier and field distributions.

2.4.3   Electron-overheating instability

Negative differential conductivity can also arise from changes in the nature of the dissipation of energy and momentum of the carriers as they are heated up by the electric field (Volkov and Kogan 1969, Bonch-Bruevich *et al.* 1975). In such an electron-overheating instability the energy and momentum relaxation times (τ_e and τ_m, respectively) depend in a nonlinear way upon the average energy per carrier w_n, which can be related to an effective electron temperature T_e by $w_n = \frac{3}{2}k_B T_e$. The mobility (in the simplest approximation $\mu = e\tau_m/m^*$) is thus a nonlinear function of T_e, which is specified by the particular scattering mechanism, e.g. by acoustic phonons, optical phonons, or impurities. The electron temperature T_e as a function of $\mathcal{E}$ follows in the steady state from a balance of the electric power density $j\mathcal{E} = en\mu(T_e)\mathcal{E}^2$ with the power density $n(w_n - w_{n0})/\tau_e(T_e)$ dissipated by the electrons to the lattice. From this the differential conductivity $dj/d\mathcal{E}$ with $j = en\mu(T_e(\mathcal{E}))\mathcal{E}$ can be calculated; both SNDC and NNDC are possible, but Z- and loop-shaped current–field characteristics may also occur (Schöll *et al.* 1987). Regular (Kerner and Osipov 1980) and spatially chaotic (Dubitskij *et al.* 1986) stationary current filaments have been studied on the basis of this mechanism. Self-generated oscillations are also found if the optical-phonon-induced nonlinearity of $\tau_e(T_e)$ and $\tau_m(T_e)$ is coupled with impurity-impact ionization (Hüpper *et al.* 1989). SNDC in the regime of impurity-impact ionization can be modeled by an increase in the mobility due to electron heating, accompanied by the feedback amplification of impact ionization of the shallow donors (Kozhevnikov *et al.* 1995). Analogous examples of overheating instabilities in *low-dimensional structures* have also been given (Ridley 1991, Balkan *et al.* 1993).

There are many other bulk NDC mechanisms, such as the acousto-electric effect (Bonch-Bruevich *et al.* 1975), plasma instabilities involving a magnetic field, e.g. helicon waves (Požela 1981), and electrothermal instabilities (Shaw *et al.* 1992), but these will not be treated here.

2.4.4   Real-space transfer instability

A variety of instabilities can arise due to the specific transport properties of heterostructures. First, we shall discuss a mechanism for NDC that is the real-space analog of intervalley transfer in the Gunn effect.

The theory of NDC due to the electron transfer in real space in a modulation-doped semiconductor heterostructure under transport parallel to the interface was developed independently by Gribnikov (1973) and Hess *et al.* (1979). The effect was demonstrated experimentally by Keever *et al.* (1981) for modulation-doped GaAs/Al$_x$Ga$_{1-x}$As heterostructure layers. New transistor concepts were also based upon real-space transfer (Kastalsky *et al.* 1989): The charge-injection-effect transistor (CHINT) and the negative-resistance field-effect transistor (NER-FET). The high-electron-mobility transistor (HEMT) also uses modulation-doped GaAs/Al$_x$Ga$_{1-x}$As layers as the conducting channel. For reviews of these real-space transfer instabilities and device concepts see Hess (1988), Balkan *et al.* (1993), and Gribnikov *et al.* (1995).

Figure 2.7 shows the energy-band diagram of a modulation-doped GaAs/AlGaAs heterostructure. The AlGaAs layer is heavily n-doped, while the GaAs is undoped. There is a band-offset ΔE_c between the conduction bands in the two layers. In thermodynamic equilibrium the electrons fall into the lower GaAs well, where they experience strongly reduced impurity scattering if they are separated from their parent donors by more than 20 nm. Thus, for a layer thickness of typically 40 nm, the mobility μ_1 in the GaAs layer will be high ($\geq$5000 cm^2 V^{-1} s^{-1} at room temperature) and the mobility μ_2 of the electrons in the AlGaAs will be low ($\approx$500 cm^2 V^{-1} s^{-1}). Application of a low field *parallel* to the layer interface will result in current due primarily to electrons in the GaAs, since both the carrier density and the mobility are much larger there than they are in the AlGaAs. A high electric field, however, will result in heating of the high-mobility electrons to energies far above their thermal-equilibrium values. The low-mobility carriers in the AlGaAs will not be heated significantly since the power input per electron is equal to $e\mu_2\mathcal{E}^2$, which is small. As a result, the electrons in the GaAs acquire a high kinetic energy and are thermionically emitted into the AlGaAs, where their mobility is much lower due to strongly enhanced impurity scattering. Thus, during the transfer from the high-mobility GaAs layer to the low-mobility AlGaAs layer, the sample exhibits NNDC, as in the Gunn effect. The physical mechanism differs, however, in that the electrons are transferred in real space, rather than in momentum space. The switching speed is higher owing to the fast thermionic emission time ($\approx$10^{-11} s), and the static and dynamic characteristics can easily be controlled over a wide range by varying the material parameters such as the layer widths L_1 and L_2, the doping concentration N_D, and the fraction of aluminum x (which determines the band offset ΔE_c). Thus, a much larger peak-to-valley current ratio than that in the Gunn effect can be achieved by choosing small ratios μ_2/μ_1 and L_1/L_2.

A quantitative understanding of the effect can be obtained from analytic two-electron-temperature models (Shichijo *et al.* 1980) or by Monte Carlo simulations (Sakamoto *et al.* 1989). Here we shall describe a simple dynamic approach (Schöll and Aoki 1991, Döttling and Schöll 1992) based on the coupled nonlinear dynamics of the real-space electron transfer and of the space charge in the AlGaAs layer which controls the interface potential barrier Φ_B (Fig. 2.10). Real-space transfer of the electrons in the GaAs layer leads to an increase of the carrier density in the AlGaAs, which diminishes the positive space charge controlling the band bending. Subsequently, the potential barrier Φ_B decreases with some delay due to the finite dielectric relaxation time. This leads to an increase in the backward thermionic emission current, which decreases the carrier density in the AlGaAs. Hence the space charge and Φ_B are increased. This, in turn, decreases the backward thermionic emission current, which completes the cycle. This mechanism explains, in addition to the static NNDC characteristics, self-generated oscillations of a real-space transfer oscillator.

It can be modeled by the following set of differential equations for the carrier densities n_1 and n_2 in the two layers, the dielectric relaxation of the applied field $\mathcal{E}_\parallel$, and the potential barrier Φ_B. The spatially averaged carrier density in the GaAs layer $n_1 \equiv \int_{-L_1}^{0} n(x, t)\,dx/L_1$ as a function of time is governed by the equation of continuity:

$$\frac{dn_1}{dt} = \frac{1}{eL_1}(J_{1\to2} - J_{2\to1}), \tag{2.49}$$

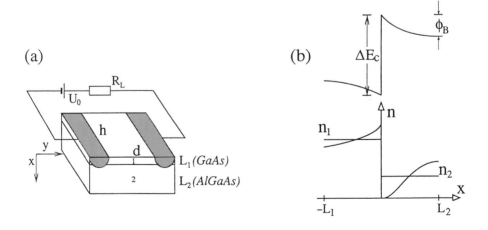

Figure 2.10. (a) A schematic sample and circuit configuration of a modulation-doped GaAs/Al$_x$Ga$_{1-x}$As heterostructure with heterolayer widths L_1 and L_2, and lateral dimensions h and d. (b) The energy-band diagram (top) and carrier density (bottom) versus the vertical coordinate x of the heterolayer (schematic).

where

$$J_{1\to2} = -en_1 \left(\frac{k_B T_1}{2\pi m_1^*}\right)^{1/2} \exp\left(-\frac{\Delta E_c}{k_B T_1}\right),$$ (2.50)

$$J_{2\to1} = -en_2 \left(\frac{k_B T_2}{2\pi m_2^*}\right)^{1/2} \exp\left(-\frac{\Phi}{k_B T_2}\right)$$ (2.51)

are the thermionic emission current densities given by Bethe's theory (Hess 1988), and m_i^* and T_i are the effective masses and the electron temperatures, in the two layers, denoted by subscripts $i = 1$ and 2, respectively. The thermionic-emission theory corresponds to the following physical picture. Electrons in the GaAs with energy less than ΔE_c can not propagate into the adjacent AlGaAs layer; all electrons with higher energies are emitted across the barrier without collisions. This is correct only within a certain region which is of the order of the mean free path of electrons. If the GaAs/AlGaAs layers are wider, the thermionic emission current represents the current only close to the interface, and diffusive currents play a major role. Conservation of the total number of carriers provided by the donor density N_D in the AlGaAs requires

$$n_1 L_1 + n_2 L_2 = N_D L_2$$ (2.52)

with $n_2 \equiv \int_0^{L_2} n(x,t)\,dx/L_2$.

The energy transfer between the heterolayers is described by the energy-balance equations containing Joule's heat, convective, diffusive, and electron-pressure-induced heat flows, and energy loss due to polar scattering of optical phonons (Aoki *et al.* 1989). Since the energy relaxation occurs on a fast time scale, the mean energies, and hence the carrier temperatures, can be eliminated adiabatically from the energy-balance equations. The mean energies $w_{ni} = \frac{3}{2}k_B T_i$ as a function of the applied electric field $\mathcal{E}_\parallel$ are then roughly approximated as

$$w_{n1} \approx w_{n0} + e\tau_e \mu_1 \mathcal{E}_\parallel^2, \qquad w_{n2} \approx w_{n0},$$ (2.53)

with the energy relaxation time τ_e.

The dielectric relaxation of the parallel electric field is given by the current equation

$$\epsilon \frac{d\mathcal{E}_\parallel}{dt} = -\sigma_L(\mathcal{E}_\parallel - \mathcal{E}_0) - \frac{en_1\mu_1 L_1 + en_2\mu_2 L_2}{L_1 + L_2}\mathcal{E}_\parallel,$$ (2.54)

where ϵ is the permittivity, $\sigma_L = [h(L_1 + L_2)R_L/d]^{-1}$ is connected to the load resistance R_L, and $U_0 = \mathcal{E}_0 d$ is the applied bias voltage.

The interface potential barrier $\Phi_B \equiv -e\int_0^{L_2} \mathcal{E}_\perp(x,t)\,dx$ is governed by the space-charge dynamics in the AlGaAs layer and the resulting internal electric field

$\mathcal{E}_\perp$. The dynamics of Φ_B can be derived by spatial averaging of the transverse current equation, using Poisson's equation (Schöll and Aoki 1991):

$$\frac{d\Phi_B}{dt} = \frac{e}{\epsilon}\left(-\mu_2 N_D \Phi_B + \mu_2 \frac{e^2}{2\epsilon}L_1^2 n_1^2 - eL_1 L_2 \dot{n}_1\right), \tag{2.55}$$

where we have neglected the diffusive contributions, which is appropriate if L_2 is less than or comparable to the mean free path of the electrons.

The static $j(\mathcal{E})$ characteristic is given by

$$j = \frac{1}{L_1 + L_2}(en_1\mu_1 L_1 + en_2\mu_2 L_2)\mathcal{E}, \tag{2.56}$$

where n_1 and n_2 are determined by the time-independent simultaneous solution of (2.49)–(2.55). The result gives NNDC. Both spatially homogeneous current oscillations (Aoki *et al.* 1989, Schöll and Aoki 1991, Döttling and Schöll 1992) and periodically or chaotically moving high-field domain (Döttling and Schöll 1993, 1994) have been predicted from the dynamic model which can be extended to include inhomogeneous propagating domains in the longitudinal *y*-direction. In the NNDC regime current oscillations of 2–200 MHz have been observed experimentally under ac-driven (Coleman *et al.* 1982) and dc-driven (Balkan and Ridley 1989, Vickers *et al.* 1989) conditions. Theoretical analyses have established that the oscillations are relatively robust even against weak disorder introduced by fluctuations of the well and the barrier widths, the doping concentration, and the alloy fraction (Rudzick and Schöll 1995).

Another mechanism for current oscillations under parallel transport in heterostructures is caused by the injection of electrons from the contact into the barrier (Hendriks *et al.* 1991). Parallel transport and real-space transfer have also been shown to give rise to hot-electron light emitting and lasing in semiconductor heterostructures (HELLISH) (Straw *et al.* 1995, Balkan *et al.* 1996, Naundorf *et al.* 1998).

2.4.5   The heterostructure hot-electron diode

Another class of instabilities occurs under *vertical* electric transport in layered semiconductor structures. The simplest example is the heterostructure hot-electron diode (HHED), which is a two-terminal semiconductor device consisting of a GaAs and an adjacent $Al_xGa_{1-x}As$ layer with layer widths of typically 2000 Å (Fig. 2.11). Both layers are undoped or slightly n-doped, so that large electric fields can be supported. If a bias is applied across the heterojunction, the HHED exhibits S-shaped negative differential conductance (SNDC), based on bistability between tunneling and thermionic emission across the energy barrier, as shown by Hess *et al.* (1986). A similar device consisting of a multilayer structure was introduced by Belyantsev *et al.* (1986); the first ideas of such a structure had been published by Gribnikov and Mel'nikov (1966). Some early attempts to calculate the static current–voltage characteristic of the single-heterostructure device (Higman *et al.* 1987, Arnold *et al.* 1988) were made, but only Tolstikhin (1986) reproduced the S-shape found experimentally. Voltage oscillations in semiconductor heterostructures consisting of GaAs/AlGaAs layers have been found experimentally by several groups (Belyantsev *et al.* 1986, Kolodzey *et al.* 1988, Krotkus *et al.* 1998), and the SNDC has been shown to persist up to room temperature (Stasch *et al.* 1996). The oscillations have been explained as circuit-induced or, especially for multi-heterostructure devices, caused by a combination of dielectric relaxation and energy relaxation of the electrons (Belyantsev *et al.* 1988). Picosecond switching between the low- and high-conductivity states has been observed (Krotkus *et al.* 1999).

Here we review a simple dynamic model for the HHED (Wacker and Schöll 1991, 1994*a*, 1994*b*) that extends the ideas of Higman *et al.* (1987) by considering the dynamic degrees of freedom of the electric fields in both layers. Additionally we include tunneling from the two-dimensional electron gas between the two layers (which is found to be the dominant mechanism of transport across the barrier on the lower branch of the current–voltage characteristics) and transport in higher bands of GaAs as suggested from Monte Carlo (MC) simulations (Arnold *et al.* 1989). More sophisticated MC simulations confirming the S-shaped current–voltage character-istic and the appearance of voltage oscillations under current-controlled conditions with an external capacitance in parallel, as predicted by the simple model (Wacker and Schöll 1991), were first reported by Belyantsev *et al.* (1993) and for a system with a thick barrier, by Reklaitis *et al.* (1997).

As can be seen in Fig. 2.11, the two different stable current states in the HHED differ by virtue of the distribution of the total voltage U across the two layers. This

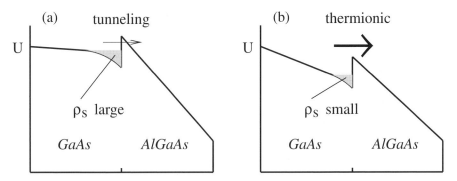

Figure 2.11. Schematic potential profiles of the heterostructure hot-electron diode (HHED). Bistability occurs in the HHED for fixed sample voltage U. The low (a) and the high (b) current states are distinguished by different distributions of the voltage U, corresponding to (a) tunneling and (b) thermionic emission, and are characterized by a large and a small interface charge density ρ_s between the layers, respectively.

can be distinguished by the interface charge density $\rho_s = \epsilon_2\mathcal{E}_2 - \epsilon_1\mathcal{E}_1$ at the interface, where ϵ_i and $\mathcal{E}_i$ are the permittivities and the electric fields near the junction in the respective layers. The subscripts 1 and 2 denote the GaAs and the AlGaAs layers of widths L_1 and L_2, respectively. Dynamic equations for $\rho_s(x, y, t)$ and $U(t)$ will be derived in the following (Wacker and Schöll 1992). To allow for filamentation, ρ_s is assumed to vary spatially in the (x, y)-plane parallel to the heterojunction, while U is spatially independent for ideal contacts.

The dynamics of $\rho_s(x, y, t)$ is given by the continuity equation

$$\frac{\partial\rho_s(x, y, t)}{\partial t} = j_1 - j_2 - \left(\frac{\partial j_x^{\|}}{\partial x} + \frac{\partial j_y^{\|}}{\partial y}\right), \tag{2.57}$$

where j_1 and j_2 denote the current densities in the z-direction in the GaAs and the AlGaAs layer, respectively, and

$$j^{\|} = \mu_{\|}|\rho_s|\mathcal{E}_{\|} \tag{2.58}$$

is the interface current density (in units of A cm^{-1}) in the two-dimensional electron gas defined by ρ_s, $\mu_{\|}$ is its mobility, and $\mathcal{E}_{\|}$ is the electric field parallel to the interface. From Maxwell's equations in the quasistationary case, $\oint dr \cdot \mathcal{E} = 0$, it follows for the closed path sketched in Fig. 2.12 that $\mathcal{E}_1 L_1 + \mathcal{E}_{\|}\, \delta x - (\mathcal{E}_1 + \delta\mathcal{E}_1)L_1 = 0$ and hence $\mathcal{E}_{\|} = L_1 \nabla\mathcal{E}_1(x, y)$. Because $U = U_1(\mathcal{E}_1) + U_2(\mathcal{E}_1, \mathcal{E}_2)$ and $\rho_s = \epsilon_2\mathcal{E}_2 - \epsilon_1\mathcal{E}_1$, the electric fields $\mathcal{E}_1$ and $\mathcal{E}_2$ can be written as functions of ρ_s and U, and thus $j^{\|}$ can be expressed in terms of ρ_s as (Wacker and Schöll 1994a)

$$j^{\|} = -\frac{\mu_{\|}f(U, \rho_s)|\rho_s|}{\frac{4}{3}\epsilon_1/L_1 + \epsilon_2/L_2}\nabla\rho_s. \tag{2.59}$$

with

$$f(U, \rho_s) = 1 + \frac{\epsilon_2}{L_2}\frac{\partial}{\partial\rho_s}(U - \tfrac{3}{4}L_1\mathcal{E}_1 - L_2\mathcal{E}_2).$$

In a simpler approximation (Wacker and Schöll 1992) $U = L_1\mathcal{E}_1 + L_2\mathcal{E}_2$, whence

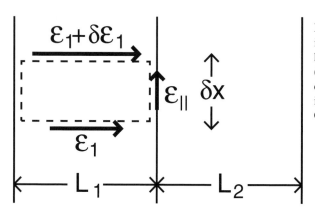

Figure 2.12. Electric fields inside the heterostructure hot-electron diode (schematic). The interface charge is distributed inhomogeneously in the GaAs/AlGaAs interface.

$$j^{\parallel} = -\frac{\mu_{\parallel}|\rho_s|}{\epsilon_1/L_1 + \epsilon_2/L_2} \nabla \rho_s. \tag{2.60}$$

and

$$\frac{\partial}{\partial t}\rho_s(x, y, t) = j_1 - j_2 + \frac{\mu_{\parallel}}{\epsilon_1/L_1 + \epsilon_2/L_2} \nabla \cdot (|\rho_s| \nabla \rho_s). \tag{2.61}$$

To derive the dynamic equation for U we consider the case that the device is operated under current-controlled conditions with a capacitance C_{ext} in parallel. Then the voltage U across the sample is not fixed but must be treated as an additional variable. Its dynamics can be determined by the argument given in Section 1.5.2. On defining the intrinsic capacitance C by $C^{-1} = \int_0^L dz \, (A\epsilon(z))^{-1}$, we obtain eq. (1.63):

$$\frac{dU}{dt} = \frac{1}{C + C_{ext}}\left(I_0 - \frac{C}{A}\int_A dx\, dy \int_0^L dz\, \frac{j_z}{\epsilon(z)}\right), \tag{2.62}$$

where I_0 is the total current and z is the position of the intersection of the surface with the sample of cross-section A. Note that the dielectric constant is only z-dependent due to the geometry of the structure.

Assuming that there is no charge accumulation within the two layers except at the interface located at $z = L_1$, the integrals $\int_A dx\, dy\, j_z$ do not depend on z within the intervals $[0, L_1)$ and $(L_1, L]$. Since $\epsilon(z)$ is constant within each of these intervals, we find

$$\frac{dU}{dt} = \frac{1}{C + C_{ext}}\left[I_0 - \int_A dx\, dy\left(\frac{C}{C_1}j_1 + \frac{C}{C_2}j_2\right)\right], \tag{2.63}$$

where $C_1^{-1} = \int_0^{L_1} dz\, (A\epsilon(z))^{-1}$ and $C_2^{-1} = \int_{L_1}^L dz\, (A\epsilon(z))^{-1}$ are the inverse intrinsic capacitances of the two layers with $C = 1/(C_1^{-1} + C_2^{-1})$ denoting the total internal capacitance.

Because the dynamics of $\rho_s(t)$ and $U(t)$ are slow compared with those of other intrinsic processes like, for instance, electron heating, we may assume that the current densities j_1 and j_2 can be written as explicit local functions of $U(t)$ and $\rho_s(x, y, t)$. Therefore, eqs. (2.61) and (2.63) constitute a closed dynamic system.

Various transport models can be used for the current densities j_1 and j_2. In early work (Wacker and Schöll 1992) a displaced heated Maxwellian calculated from bulk properties was used for the electron distribution in the GaAs layer. This approach has the shortcoming of neglecting the influence of ballistic effects, which are important if voltages of about 0.3 V are applied across distances of 2000 Å. The appearance of ballistic peaks in the distribution function was verified in MC simulations (Belyantsev et al. 1993) for the HHED. Therefore Belyantsev et al. (1993) proposed pure ballistic transport in the GaAs layer, as a first analytic approximation. This approach can be refined by including electron transfer from the Γ-valley (effective mass m_{Γ}) into the L-valley (band offset ΔE_L)

and energy loss due to polar optical scattering (phonon energy $\hbar\omega_{\mathrm{o}}$) (Wacker and Schöll 1994a).

First we determine the current density j_1 in the GaAs layer. Let the z-axis be perpendicular to the heterojunction. At the contact ($z = 0$) of the GaAs layer we set the electric potential $\varphi(0) = 0$. Neglecting scattering, the ballistic energy of the electrons in the GaAs layer is given by $E_{\mathrm{bal}}(z) = e\varphi(z)$. If $E_{\mathrm{bal}}(z) < \Delta E_{\mathrm{L}}$, then polar optical scattering is the main scattering mechanism. It yields an average of about two scattering events in the whole GaAs layer. Since small angles of scattering are preferred, the velocity v of the electrons is nearly unchanged if $E_{\mathrm{bal}} \gg \hbar\omega_{\mathrm{o}}$, which is justified for high electric fields. Therefore we can assume that $m_{\Gamma}v(z)^2/2 = e\varphi(z)$ holds as long as $e\varphi(z) < \Delta E_{\mathrm{L}}$. If $e\varphi(z_{\mathrm{L}}) = \Delta E_{\mathrm{L}}$ at some point z_{L} in the first layer, two effects have to be taken into account for $z > z_{\mathrm{L}}$. First, the effective-mass approximation breaks down at energies of about ΔE_{L}. We set $v(z) = \sqrt{2\,\Delta E_{\mathrm{L}}/m_{\Gamma}} = \text{constant}$ for the electrons in the Γ-valley at $z > z_{\mathrm{L}}$, which is about the maximum velocity in the Γ-valley. Second, scattering into the L-valley becomes possible. Assuming a constant intervalley scattering rate, the current in the Γ-valley is $j_{\Gamma 1}(z) \approx j_1[1 - (z - z_{\mathrm{L}})/L_{\mathrm{S}}]$ with the scattering length $L_{\mathrm{S}} \approx 1000$ Å, as indicated by MC simulations (Tang and Hess 1982). The continuity equation yields for the current in the L-valley $j_{\mathrm{L}1}(z) = j_1 - j_{\Gamma 1}(z)$. Owing to the larger effective mass and the more effective deformation-potential scattering the electron velocity v_{L} is much smaller in the L-valley. We use $v_{\mathrm{L}} = 10^7$ cm s^{-1}. By combining these equations with Poisson's equation it is possible to derive explicit expressions for $U_1(\mathcal{E}_1)$ and $j_1(\mathcal{E}_1)$, where $U_1 = -\varphi(L_1)$ and $\mathcal{E}_1 = -\varphi'(L_1)$.

The current density j_2 in the AlGaAs layer is given by the electrons crossing the energy barrier if we neglect any currents in the opposite direction, which is justified since $|\mathcal{E}_2| \gg |\mathcal{E}_1|$. A fraction of the injected electrons from the Γ-valley is able to cross the barrier either thermionically or by tunneling: $j_{\Gamma 2} = j_{\Gamma 1}(L_1) \int dE\, f_{\Gamma}(E) D_{\Gamma}(E)$, where $D_{\Gamma}(E)$ is the WKB tunneling coefficient for the barrier at the given electron energy E. The distribution function $f_{\Gamma}(E)$ (normalized with respect to unity) exhibits a pronounced peak at the ballistic energy $E = -eU_1$. It is broadened due to the energy loss in the first layer by the emission of optical phonons. The transmission current $j_{\mathrm{L}2}$ from the L-valley is calculated analogously. We assume that the reflected electrons remain in the GaAs layer and reach the two-dimensional electron gas between the layers. These quasibound states have a finite lifetime due to tunneling through the barrier. The tunneling current j_{T} is calculated as by Wacker and Schöll (1992). The total current in the AlGaAs layer is now given by $j_2(\mathcal{E}_1, \mathcal{E}_2) = j_{\Gamma 2} + j_{\mathrm{L}2} + j_{\mathrm{T}}$. Assuming an average velocity $v_{\mathrm{S}} = 10^7$ cm s^{-1} in the AlGaAs layer, we find a charge density $\rho_2 = j_2/v_{\mathrm{S}}$ in this layer, which influences the voltage drop $U_2 = L_2\mathcal{E}_2 + \rho_2 L_2^2/(2\epsilon_2)$ in the second layer (Belyantsev et al. 1993). This effect is important since it reduces the voltage interval where bistability between the two current states occurs.

Although different transport models give slightly different quantitative results (Wacker and Schöll 1992, 1994a), the general properties of the globally coupled reaction–diffusion system (2.61) and (2.63) are not affected. Therefore, one can derive a generic simplified, more widely applicable model that is independent of the particular transport mechanism effective in j_1 and j_2 (Wacker and Schöll 1994b). In a first order approximation we set $j_1 = \sigma \mathcal{E}_1$ with a constant conductivity σ, and assume constant fields $\mathcal{E}_1$ and $\mathcal{E}_2$ in the two layers. It follows from $\rho_s = \epsilon_2 \mathcal{E}_2 - \epsilon_1 \mathcal{E}_1$ and $U = L_1 \mathcal{E}_1 + L_2 \mathcal{E}_2$ that $\mathcal{E}_1$ is given by $\mathcal{E}_1 = (\epsilon_2 U - L_2 \rho_s)/(\epsilon_1 L_2 + \epsilon_2 L_1)$. A fraction γ of j_1 is crossing the barrier. We use $\gamma = \mathcal{E}_1^2/(\mathcal{E}_1^2 + \mathcal{E}_0^2)$ describing the onset of thermionic emission when $\mathcal{E}_1 \approx \mathcal{E}_0$, where the order of $\mathcal{E}_0$ is given by the barrier height divided by L_1. Additionally, there is a tunneling current from the interface charge states, described by a constant tunneling coefficient $\mathcal{T}_0$. Together this yields $j_2 = \gamma j_1 + \mathcal{T}_0 \rho_s$.

In order to obtain a simple description we approximate $\nabla \cdot (|\rho_s| \nabla \rho_s) \approx |\overline{\rho_s}| \Delta \rho_s$ in eq. (2.61), where $\overline{\rho_s}$ is the spatial and temporal average of $\rho_s(x, y, t)$, and set $(C/C_1) j_1 + (C/C_2) j_2 \approx j_1$ in eq. (2.63), which holds exactly for stationary states due to $j_1 = j_2$.

Introducing dimensionless variables $u = U/U_s$, $a = \rho_s/V_s$, and $j_0 = I_0/I_s$ with the scalings $U_s = L_1 \mathcal{E}_0 C_1/C$, $V_s = \epsilon_1 \mathcal{E}_0 C_2/C$, and $I_s = A\sigma \mathcal{E}_0$, and normalizing time by using $\tau_s = \epsilon_1 C_2/(\sigma C)$ and space by using $l_s = \sqrt{\mu L_1 |\overline{\rho_s}|/\sigma}$, we obtain the equations

$$\frac{\partial a}{\partial t} = f(a, u) + \Delta a, \tag{2.64}$$

$$\frac{du}{dt} = \alpha \left(j_0 - \langle j(a, u) \rangle \right), \tag{2.65}$$

where

$$f(a, u) = \frac{u - a}{(u - a)^2 + 1} - \mathcal{T} a \tag{2.66}$$

is the nonlinear transport function,

$$j(a, u) = u - a \tag{2.67}$$

is the normalized current density, $\langle j \rangle$ denotes the spatial average

$$\langle j \rangle = \int dx \, dy \, j(x, y)/A, \tag{2.68}$$

$\alpha = C_2/(C + C_{\text{ext}})$ defines the time scale of dielectric relaxation, and $\mathcal{T} = \mathcal{T}_0 \tau_s$ is the normalized tunneling coefficient. In case of the HHED we obtain typical scales of the order of $\tau_s \simeq 1$ ps, $l_s \simeq 3$ μm, $U_s \simeq 0.8$ V, and $I_S/A \sim 50$ kA cm^{-2} for $L_1 = L_2 = 2000$ Å. The intrinsic capacitances are then $C_1/A \approx 5.8 \times 10^{-4}$ F m^{-2}, $C_2/A \approx 5.2 \times 10^{-4}$ F m^{-2}, and $C/A \approx 3 \times 10^{-4}$ F m^{-2}.

SNDC and homogeneous relaxation oscillations (Wacker and Schöll 1991), stationary current filaments, and periodic or chaotic spatio-temporal spiking are found in distinct regimes of the parameter space, as will be discussed in Chapter 7 (Wacker and Schöll 1994a, 1994b, Bose et al. 1994, 2000). Since similar spiking behavior has been observed experimentally in various other bistable devices, e.g. p–i–n- and p–n–p–i–n structures (Symanczyk et al. 1991a, Niedernostheide et al. 1992a, 1996b), and in the impurity-breakdown regimes of n-GaAs (Brandl and Prettl 1991) and p-Ge (Peinke et al. 1992), the generic model might indeed be more generally applicable to semiconductor structures. It will be used as a prototype of a globally coupled bistable system with an S-shaped current–voltage characteristic in the general discussion of stability of filaments and front dynamics in Sections 3.4 and 3.5.

2.4.6   Double-barrier resonant-tunneling diode

Resonant tunneling is an important mechanism for instabilities in low-dimensional semiconductor structures. The effect was described by Kazarinov and Suris (1971, 1972) and Tsu and Esaki (1973). The gross features of the current–voltage characteristics can be understood from the arguments given in Section 2.2.1 for resonant sequential tunneling in superlattices (Fig. 2.5). The current between two adjacent quantum wells is maximum if there is maximum overlap between the occupied states in one well and the available unoccupied states in the other, i.e. if the energies are in resonance. For low fields equivalent levels in adjacent wells are approximately in resonance. With increasing field the available states in the collecting well are lowered with respect to the emitting well, and hence the current density drops as the overlap between the energy levels decreases, thereby displaying NDC. Upon further increase of the field the current density rises again up to a sharp peak when the ground energy level in one quantum well is aligned with the second level in the neighboring well. Thus resonant tunneling produces NNDC in coupled quantum wells and superlattices (Grahn 1995b, Wacker 1998), see Chapter 6. The same mechanism remains in effect for resonant tunneling through a double-barrier structure with an embedded quantum well in between. If this structure is sandwiched between a highly doped emitter and a collector region, resonant tunneling from the emitter to the quantum-well level is analogous to tunneling between two adjacent wells. If the two-dimensional quantum well is replaced by an array of quantum dots, resonant tunneling may occur in just the same way. Such quantum-dot structures, which may be manufactured by self-organized growth in strained material systems (Bimberg et al. 1999), have recently become of great interest, and resonant tunneling in such structures, e.g. InAs dots embedded in an AlGaAs double barrier sandwiched between GaAs emitter and collector layers, has indeed been observed (Itskevich et al. 1996, Narihiro et al. 1997, Suzuki et al. 1997, Hapke-Wurst et al. 1999).

In the following we shall present in some detail the derivation of a reaction–diffusion model of type (2.47) for the double-barrier resonant-tunneling diode (DBRT). The basic mechanism which gives rise to negative differential conductance and to an N-shaped current–voltage relation (NNDC) is resonant tunneling between an emitter and a collector via a discrete level in a quantum well embedded between two barriers (Fig. 2.13). However, the situation becomes more complicated if the nonlinear feedback between space charges and transport processes is taken into account. An intrinsic bistability superimposed on the NNDC characteristic occurs due to the dynamic charge accumulation within the potential well. The built-up charge leads to an electrostatic feedback mechanism that increases the energy of the quasibound state, supporting resonant-tunneling conditions for higher applied voltages. This may result in bistability and hysteresis such that a high-current and a low-current state coexist for the same applied voltage u and the current–voltage characteristic becomes Z-shaped rather than the conventional N-shaped characteristic associated in general with resonant tunneling (Capasso *et al.* 1986, Goldman *et al.* 1987, Joosten *et al.* 1991, Martin *et al.* 1994). It has recently been pointed out (Wacker and Schöll 1995, Glavin *et al.* 1997, Mel'nikov and Podlivaev 1998, Feiginov and Volkov 1998) that such bistability provides a basis for lateral pattern formation in the DBRT. Here we derive a nonlinear reaction–diffusion model for

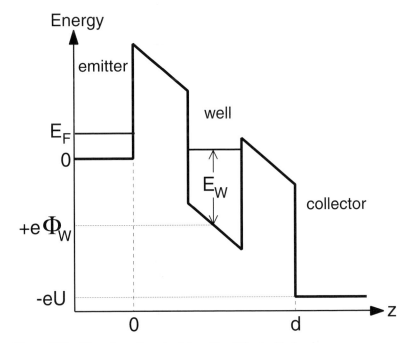

Figure 2.13. The schematic potential profile of the double-barrier resonant-tunneling structure (DBRT). E_F and E_w denote the Fermi level in the emitter, and the energy level in the quantum well, respectively. U is the voltage applied across the structure.

the DBRT in the bistable regime (Meixner *et al.* 1999, 2000a). It will be used in Section 3.5 as an example of a system with a Z-shaped current–voltage characteristic in simulations of globally coupled front dynamics.

We consider a symmetric resonant-tunneling structure and assume incoherent sequential tunneling for the vertical (along the z-axis) transport (Fig. 2.13). We characterize the internal state of the device in terms of the built-up electron concentration $n(x, y, t)$ (per unit area) in the well. The continuity equation for $n(x, y, t)$ has the form

$$\frac{\partial n}{\partial t} + \frac{1}{e}\nabla j_\parallel = \frac{1}{e}\big(J_{ew}(n, U) - J_{wc}(n)\big), \qquad \nabla \equiv e_x \frac{\partial}{\partial x} + e_y \frac{\partial}{\partial y}, \quad (2.69)$$

where $e < 0$ is the electron charge, U is the applied voltage, $J_{ew}(x, y)$ and $J_{wc}(x, y)$ are the local densities of the emitter–well and the well–collector currents per unit area, respectively, and $j_\parallel$ is the current density (per unit length) in the well parallel to the interface.

The emitter–well current density J_{ew} can be evaluated to lowest order in the coupling in the spirit of Mahan (1990). Assuming conservation of the transverse momentum k during the tunneling process and neglecting the broadening of the states in the emitter, we obtain

$$J_{ew} = \frac{2e}{L_x L_y}\sum_k \sum_q \frac{2\pi}{\hbar}|H_q^2|\Theta(E_F - E_k - E_q)$$

$$\times \frac{1}{2\pi}A_w(E_q + E_k - E_w - e\phi_w, k)(1 - f_w), \qquad (2.70)$$

where q and k are the vertical and in-plane wave vectors of the electrons in the emitter, respectively. $E_q = \hbar^2 q^2/(2m)$ and $E_k = \hbar^2 k^2/(2m)$ are the corresponding kinetic energies (here m is the effective electron mass) and H_q is the matrix element of the emitter–well transition. $\Theta(E)$ is the step function describing the occupation of the emitter up to the Fermi energy E_F for zero temperature. $A_w(E, k)$ is the spectral function of the bound state in the well (binding energy E_w), which has still the transverse degree of freedom k in the (x, y)-plane. ϕ_w is the electric potential at the bottom of the well (the potential of the emitter is taken to be zero). Finally, f_w is the filling factor for the states in the well, which we estimate by a homogeneous distribution of the electron density up to the Fermi energy of the emitter

$$f_w(n) = \frac{n}{\rho_0(E_F - E_w - e\phi_w)}, \qquad (2.71)$$

where $\rho_0 \equiv m/(\pi\hbar^2)$ is the two-dimensional density of states (per unit area). In the following we assume that $A_w(E, k) = \Gamma/[(E - E_k)^2 + \Gamma^2/4]$, where $\Gamma \equiv \Gamma_L + \Gamma_R + \Gamma_{scatt} = $ constant is the total broadening of the quasibound well state with terms resulting from escape via the emitter–well barrier and the well–collector barrier and scattering in the well, respectively. Then we obtain in the continuum limit ($\sum_k \sum_q \to L_x L_y L_z/(2\pi)^3 \int d^2k \int dq$)

$$J_{ew} = \frac{e}{2\pi\hbar}\rho_0 \int_0^{E_F} dE_q\,(E_F - E_q)\frac{2m|H_q^2|L_z}{\hbar^2 q}\frac{\Gamma}{(E_q - E_w - e\phi_w)^2 + \Gamma^2/4}(1 - f_w). \tag{2.72}$$

In the following $\Gamma_L \equiv 2m|H_q^2|L_z/(\hbar^2 q)$ is taken to be constant for simplicity.

For $n = 0$ the energy of the bottom of the well in a symmetric structure is given by $\phi_w = -U/2$. Owing to the built-up charge ϕ_w depends also on the electron concentration $n(x, y)$ in the well. Assuming that transverse variations of $n(x, y)$ are smooth in the sense that their characteristic wavelength is much larger than the effective thickness of the structure d, we can represent the corresponding correction locally as $\Delta\phi_w(n) = en/C_{int}$, where C_{int} is the effective capacitance per area of the well. This yields (Glavin *et al.* 1997, Mel'nikov and Podlivaev 1998)

$$\phi_w = -\frac{U}{2} + \frac{en}{C_{int}}, \qquad C_{int} = \frac{4\epsilon}{d}, \tag{2.73}$$

where $\epsilon = \epsilon_0 \epsilon_r$ denotes the permittivity.

With these ingredients the evaluation of eq. (2.72) gives the final formula

$$J_{ew}(n, U) = \tag{2.74}$$
$$\frac{e}{\hbar}\Gamma_L\rho_0\left[\Delta(n, U)\frac{\arctan(2\Delta/\Gamma) - \arctan(2\Omega/\Gamma)}{\pi} - \frac{\Gamma}{4\pi}\ln\left(\frac{\Delta^2 + (\Gamma/2)^2}{\Omega^2 + (\Gamma/2)^2}\right)\right]$$
$$\times\left(1 - \frac{n}{\rho_0\Delta}\right),$$

$$\Delta(n, U) \equiv E_F - E_w + \frac{eU}{2} - \frac{e^2 n}{C_{int}}, \qquad \Omega(n, u) \equiv \frac{eU}{2} - \frac{e^2 n}{C_{int}} - E_w,$$

where Δ and Ω denote the energy of the quasibound state with respect to the Fermi level and the bottom of the conductance band in the emitter, respectively. Since the bottom of the conductance band of the collector is much lower than the quasibound state (Fig. 2.13) if a negative bias U is applied, the well–collector current can be taken to be proportional to n:

$$J_{wc} = \frac{e}{\hbar}\Gamma_R n, \tag{2.75}$$

where $\Gamma_R/\hbar$ is the escape rate via the well–collector barrier.

The in-plane current density in the well is described in the drift–diffusion approximation:

$$j_\parallel = |e|n\mu\mathcal{E}_\parallel - eD_0\,\nabla n, \tag{2.76}$$

where μ and D_0 are the mobility and the diffusion coefficient in the well, respectively. $\mathcal{E}_\parallel = -\nabla\phi_w$ is the in-plane electric field in the well. Taking into account

(2.73), we conclude that the in-plane coupling in the DBRT is effectively due to a concentration-dependent diffusion term

$$j_{\parallel} = -e D(n) \, \nabla n, \qquad D(n) \equiv D_0 + \frac{|e| \mu n}{C_{\text{int}}}. \tag{2.77}$$

By substituting (2.77) into (2.69) we arrive at the following reaction–diffusion equation which describes the internal dynamics in the DBRT:

$$\frac{\partial n}{\partial t} = \frac{1}{e} \left(J_{\text{ew}}(n, U) - J_{\text{wc}}(n) \right) + \nabla (D(n) \, \nabla n). \tag{2.78}$$

A more general form of the effective diffusion coefficient $D(n)$ has been derived by Cheianov *et al.* (2000). The current densities J_{ew} and J_{wc} are given by (2.74) and (2.75), respectively.

The dynamic equation for U has been derived in the general form in Section 1.5.2 from Kirchhoff's and Ampère's laws with proper accounting for the displacement currents within the semiconductor element:

$$RC \frac{dU}{dt} = U_0 - U - R \frac{\tilde{C}}{L_x L_y} \int_0^{L_x} \int_0^{L_y} \int_0^d dx \, dy \, dz \, \frac{j_z(z)}{\epsilon(z)}, \tag{2.79}$$

$$\frac{1}{\tilde{C}} = \int_0^d \frac{dz}{L_x L_y \epsilon(z)},$$

where $j_z(z)$ is the vertical component of the current density, $\epsilon(z)$ is the permittivity of the device, and $\tilde{C}$ is an effective intrinsic sample capacitance. For the symmetric DBRT considered here, we have $j_z(z) = J_{\text{ew}}$ for $0 < z < d/2$ and $j_z(z) = J_{\text{wc}}$ for $d/2 < z < d$, and therefore we find

$$RC \frac{dU}{dt} = U_0 - U - R \int_0^{L_x} \int_0^{L_y} dx \, dy \, \frac{J_{\text{ew}}(n, U) + J_{\text{wc}}(n)}{2}. \tag{2.80}$$

In the following we restrict ourselves to one transverse spatial coordinate x. We transform to dimensionless variables according to $a = n/(\rho_0 E_F)$, $u = eU/\Gamma$, $\tilde{J} = \hbar J/(e\rho_0 E_F \Gamma)$, $\tilde{t} = t/\tau_a$, and $\tilde{x} = x/\sqrt{D\tau_a}$, choosing $\tau_a = \hbar/\Gamma$ (and subsequently omit the tilde). Typical units of time, voltage, and current density are on the order of $\tau_a = 3$ ps, $\Gamma/e = 2$ mV and $(e\rho E_F \Gamma)/\hbar \approx 70$ kA cm^{-2}, respectively. The length scale $l = \sqrt{(D\tau_a)}$ scales with the square root of the effective transverse diffusion constant D, which we assume to be constant for simplicity. From (2.78) and (2.80) we obtain the dimensionless equations

$$\frac{\partial a}{\partial t} = f(a, u) + \frac{\partial^2 a}{\partial x^2}, \tag{2.81}$$

$$\varepsilon \frac{du}{dt} = U_0 - u - r \langle j(a, u) \rangle, \tag{2.82}$$

$$\varepsilon \equiv \frac{\tau_u}{\tau_a}, \qquad \tau_u \equiv RC, \qquad \langle j \rangle \equiv \frac{1}{L_x} \int_0^{L_x} j \, dx. \tag{2.83}$$

with

$$f(a, u) \equiv \tilde{J}_{\text{ew}}(a, u) - \tilde{J}_{\text{wc}}(a) \tag{2.84}$$

$$j(a, u) \equiv \frac{\tilde{J}_{\text{ew}}(a, u) + \tilde{J}_{\text{wc}}(a)}{2} \tag{2.85}$$

$$r \equiv R L_x L_y e^2 \rho_0 E_{\text{F}} / \hbar. \tag{2.86}$$

These equations are of the general form of a globally coupled reaction–diffusion system of activator–inhibitor type (1.38) and (1.58). They furnish an example of a Z-shaped current–voltage characteristic, which will be used in Section 3.5 to discuss the dynamics of lateral switching fronts.

2.4.7   Layered p–n–p–n structures

Multilayered structures based on homojunctions with alternating p- and n-doping represent another class of systems of activator–inhibitor type that exhibit SNDC and complex spatio-temporal patterns. Starting from the basic p^+–n^-–p–n^+ structure (Varlamov and Osipov 1970), where + and − denote high and low doping, respectively, various modifications such as silicon p^+–n^+–p–n^-–n^+ devices (Niedernostheide *et al.* 1992*b*) and n^+–p^+–n–p^-–p^+ devices (Gorbatyuk and Niedernostheide 1996) have been studied.

The basic p^+–n^-–p–n^+ structure can be conceived as a double-injection device composed of a p^+–n^-–p transistor and an n^-–p–n^+ transistor (Sze 1998). The bias voltage is always applied such that, in the low-conductivity state, the middle p–n junction is reverse biased, and the two neighboring junctions are forward biased. Thus the hole current injected from the p^+ emitter into the p layer flows as base current into the n^-–p–n^+ transistor and enhances the injection of electrons from the n^+ emitter. These electrons, upon reaching the n^- layer which is the base of the p^+–n^-–p transistor, in turn enhance injection of holes from the p^+ emitter. This autocatalytic process may lead to a situation in which the middle p–n junction also becomes forward biased, and the device switches to a high-conductivity state, thus exhibiting SNDC. The basic p^+–n^-–p–n^+ structure does not possess a mechanism of internal inhibition. However, it can be changed by modifying the doping profiles.

An essential modification consists in a five-layer p^+–n^+–p–n–$n^-$$n^+$ structure (also referred to as a p–n–p–i–n structure) with an additional, very wide and low-doped (typically 10^{13} cm^{-3}) n^- layer, resulting in less injection of electrons into the p base (Fig. 2.14(a)). These structures can be considered as two-layer systems with the particular feature that an autocatalytic multiplication of charge carriers may occur in one of the layers, whereas the second layer provides an inhibiting mechanism.

Note that two-layer models have also been proposed for Si p–i–n diodes (Radehaus *et al.* 1987, Kerner *et al.* 1987, Symanczyk *et al.* 1991*b*, Minarsky and Rodin 1997), Si and SiC p–n diodes (Gafiichuk *et al.* 1990, Kerner *et al.* 1987, Vashchenko *et al.* 1991), heterostructure hot-electron diodes, cf. Section 2.4.5 (Wacker and Schöll 1994*a*, 1994*b*), n^+–n$^-$–n$^+$ GaAs thin films in the regime of band–band impact ionization (Vashchenko *et al.* 1997), n^+–p^+–n–p^-–p^+ structures (Gorbatyuk and Niedernostheide 1996, 1999), and thyristors, see Section 2.4.8 (Gorbatyuk and Rodin 1990, 1995).

In the following we describe the two-layer model of the p^+–n^+–p–n$^-$–n$^+$ structure. The relatively high doping of the bases results in high fields at the n^+–p junction and band–band impact ionization as an autocatalytic process. In such silicon p^+–n^+–p–n$^-$–n$^+$ devices, several generic oscillation modes of current filaments can arise due to the competition between the activating and the inhibiting mechanism. Besides periodically oscillating and traveling filaments (Niedernostheide *et al.* 1992*a*, 1992*b*, 1994*a*, 1994*b*, 1997, Wierschem *et al.* 1995) – both representing fundamental generic oscillation modes of filaments – irregular (Niedernostheide *et al.* 1993, 1996*b*) and, under a driving ac bias, quasiperiodic and chaotic (Niedernostheide 1995, Niedernostheide *et al.* 1996*a*, Niedernostheide and Kleinkes 1999) filament motion has been observed in experiments and also found in numerical solutions of the activator–inhibitor equations derived from the two-layer model (Niedernostheide 1995).

When the p^+–n^+–p–n$^-$–n$^+$ device is connected to a dc bias via a load resistor (Fig. 2.14(a)) such that the p^+ layer is positively biased with respect to the n$^-$ layer, the n^+–p junction is reverse-biased leading to a low-conductivity branch in

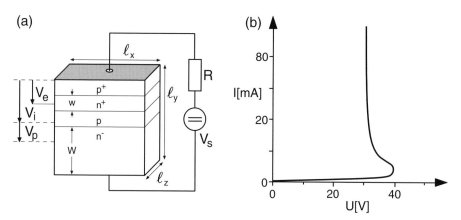

Figure 2.14. (a) A schematic p^+–n^+p–n$^-$–n$^+$ layered structure, operated in a load circuit with a resistor of resistance R and a bias voltage V_s. Typical dimensions of the sample are $l_x = 1$–5 mm, $l_y = 0.4$–0.8 mm, $l_z = 0.4$–1 mm, $w = 1$–10 μm, and $W = 800$ μm (after Niedernostheide *et al.* (1994*a*)). (b) A typical measured current–voltage characteristic $I(U)$ (after Schöll *et al.* (1998*a*)).

the current–voltage characteristic $I(U)$ (Fig. 2.14(b)) at low sample currents. For a sufficiently large voltage drop across the sample, avalanche multiplication of charge carriers in the space-charge region of that junction takes place by interband impact ionization. The electron–hole generated pairs are separated in the high-field zone of the junction so that the electrons and holes move toward the p^+–n^+ junction and the p–n^- junction, respectively. Those carriers reaching the outer junctions induce an additional carrier injection there, leading to a further increase in number of carriers and causing the appearance of a branch with negative differential resistance in the $I(U)$ characteristic.

For a simple description of the device one may consider the positive-feedback loop caused mainly by the injection of holes from the p^+–n^+ junction. Then, it is useful to imagine the device being composed of two parts, a p^+–n^+–p transistor and a p–n^- diode (Niedernostheide *et al.* 1992*b*). In order to gain insight into the relevant physical processes, let us consider the two device parts in more detail. Suppose that the voltage drop V_c across the n^+–p junction is sufficiently large that charge-carrier multiplication takes place. Then, a small fluctuation of the voltage V_e of the p^+–n^+ emitter junction has the following consequences. Owing to the essentially exponential dependency of the emitter current density on the emitter voltage, a fluctuation of V_e causes an additional injection of holes into the n^+ layer. Those holes which reach the high-field zone of the n^+–p junction generate electron–hole pairs that are separated due to the high electric field. The electrons move toward the p^+–n^+ junction and induce an additional injection of holes and, consequently, an increase of V_e. Because of this activating property of the transistor part, the emitter voltage V_e may be called an activator.

The coupling of the transistor part to the p–n^- diode leads to a counteraction to the autocatalytic increase of V_e: The current fluctuation caused by the fluctuation of V_e leads to an increase in injection of holes into the n^- layer and therefore to an augmentation both of the mean hole concentration p in the n^- layer and of the voltage drop $V_p \propto \ln p$ across the p–n^- diode. Keeping the device voltage V constant means that the voltage drop $V_i = V - V_p$ across the transistor decreases if V_p increases. This inhibiting process in the p–n^- diode limits the autocatalytic process and, consequently, p can be viewed as the inhibiting variable.

The following set of reaction–diffusion equations for the activator V_e and the inhibitor p can be derived by treating the p^+–n^+–p transistor as an avalanche transistor and the p–n^- diode as a Shockley diode (Niedernostheide *et al.* 1992*b*, 1994*a*):

$$\frac{\partial V_e}{\partial t} = D_e \frac{\partial^2}{\partial x^2} V_e - q(V_e, p), \tag{2.87}$$

$$\frac{\partial p}{\partial t} = D_p \frac{\partial^2}{\partial x^2} p - Q(V_e, p). \tag{2.88}$$

The diffusion coefficient $D_e = w\sigma_b/C_e$ is determined by the capacitance per unit area C_e of the p^+–n^+ junction and the width w and the conductivity σ_b of the n^+ base. D_p is the diffusion coefficient of holes in the n^- layer, and x denotes the spatial coordinate perpendicular to the main current flow. The nonlinear functions

$$q(V_e, p) = \frac{1}{C_e}\left(j_e(V_e) - j_c(V_e, p)\right),\tag{2.89}$$

$$Q(V_e, p) = \frac{j_c(V_e, p)}{eW} - \frac{p - p_{n0}}{\tau},\tag{2.90}$$

where p_{n0} is the equilibrium value of the mean hole density, τ is the hole lifetime, and W is the width of the n^- layer, arise from the various current-transport processes:

The current density $j_e^{tot}(x, t)$ of the p^+–n^+ emitter is the sum of the displacement current density $C_e\, \partial V_e/\partial t$ and the current density arising from diffusion of injected holes into the n^+ base and the recombination current density

$$j_e^{tot}(x, t) = C_e\frac{\partial V_e}{\partial t} + j_e(V_e),\tag{2.91}$$

$$j_e(V_e) = j_s\left[\exp\left(\frac{eV_e}{k_B T}\right) - 1\right] + j_r\left[\exp\left(\frac{eV_e}{2k_B T}\right) - 1\right]\tag{2.92}$$

where j_s and j_r, denote the diffusion and recombination saturation current densities, respectively, of the p^+–n^+ junction.

The current density $j_c(V_e, p)$ of the n^+–p collector is the sum of the collector saturation current density j_{sc} and the transfer current density of holes injected from the p^+–n^+ emitter, both being multiplied by the multiplication factor

$$M = \left[1 - \left(\frac{V_i - V_e}{V_b}\right)^3\right]^{-1}\tag{2.93}$$

due to the avalanche effect, and the leakage current density $V_c(x, t)/\rho_L$, where ρ_L and V_b denote the leakage resistance and the breakdown voltage of the n^+–p junction, respectively, and β is the base transport factor:

$$j_c(V_e, p) = Mj_{sc} + \beta M j_s\left[\exp\left(\frac{eV_e}{k_B T}\right) - 1\right] + \frac{V_i - V_e}{\rho_L}.\tag{2.94}$$

To take account of the current spreading in the n^+ base, the emitter and collector current densities are connected by

$$C_e\frac{\partial V_e}{\partial t} + j_e = j_c + w\sigma_b\frac{\partial^2}{\partial x^2}V_e,\tag{2.95}$$

which yields eq. (2.87).

The p–n^- junction is considered as an ideal Shockley diode. The continuity equation for the hole density, averaged over the current flow direction, is given by eq.

(2.88), and the mean hole density p and the voltage drop across the p–n$^-$ junction $(V - V_i)$ are connected by

$$p = p_{n0} \exp\left(\frac{e(V - V_i)}{k_B T}\right).$$

(2.96)

The SNDC characteristic and a variety of spatio-temporal patterns in one (Niedernostheide *et al.* 1992*b*, 1994*a*, Niedernostheide and Kleinkes 1999) and two (Niedernostheide *et al.* 1997) spatial dimensions have been obtained by numerically solving the set of reaction–diffusion equations (2.87) and (2.88).

2.4.8 Thyristor structures

As discussed above, current instabilities and complex spatio-temporal dynamics are widespread in multilayered semiconductor structures. There is a high potential of expected applications of such active spatially extended media for information processing and pattern recognition (Mikhailov 1989, Haken *et al.* 1994); this implies spatially inhomogeneous dynamic modes of operation and therefore the prospects essentially depend on the invention of efficient methods for controlling nonlinear spatio-temporal patterns. This can be achieved by connecting a control electrode (gate) to the p–n–p–n structures as shown in Fig. 2.15. Such bistable gate-turn-off (GTO) thyristor structures not only exhibit a great variety of spatio-temporal patterns, e.g. stationary filaments (Varlamov and Osipov 1970, Gorbatyuk and Rodin 1992*a*, 1992*b*, 1997), front propagation (D'yakonov and Levinstein 1978, Meixner *et al.* 1998*a*), Turing patterns (Gorbatyuk *et al.* 1989, Gorbatyuk and Rodin 1990, Gorbatyuk and Rodin 1995, Gorbatyuk and Niedernostheide 1999, Meixner *et al.* 1997*b*), but also possess unique features with respect to controllability. These features are introduced by a spatially distributed microelectronic gate that allows one to influence the internal state quasi-uniformly over the whole cross-section of the device (Gorbatyuk and Rodin 1992*a*, 1992*b*, 1997). In this case, in addition to the global coupling through the main circuit, which occurs in all spatially distributed semiconductor systems (Bass *et al.* 1970, Schimansky-Geier *et al.* 1991) and has been introduced in Section 1.5.2, a global coupling through the gate circuit arises (Gorbatyuk and Rodin 1992*a*, 1992*b*, 1997, Meixner *et al.* 1998*a*, 1998*b*). The external control circuit imposes a global constraint on the internal dynamics and provides means for control over spatio-temporal patterns. Active external circuits with negative resistance and capacitance have recently been implemented in experimental studies of multistable structures (Martin *et al.* 1994). This makes the control of semiconductor systems via a global constraint even more flexible since it allows one to arrange global constraints of both activating and inhibiting types.

A bistable p–n–p–n structure can be switched from the low-conductivity to the high-conductivity state via the propagation of a lateral current-density front. Such a switching front triggers double injection from the cathode and the anode, increasing the concentration of excess carriers by up to ten orders of magnitude. This results in a dramatic increase both of conductivity and of emission of light, providing a basis for numerous electric and optical applications. Originally these applications were seen mainly in the field of power electronics, but nowadays p–n–p–n structures are attracting attention as prominent examples of controllable solid-state-based active media that can serve as hardware for electric (Ruwisch *et al.* 1996) and optical (Radehaus and Willebrand 1995) pattern-recognition systems. An implementation of the principles of autowave holography for information processing (Balkarei and Elinson 1991) demands controllable distributed media that exhibit both front propagation and solitary patterns (Balkarei *et al.* 1987). Multilayered thyristor structures are promising candidates for single-crystal realizations of such media.

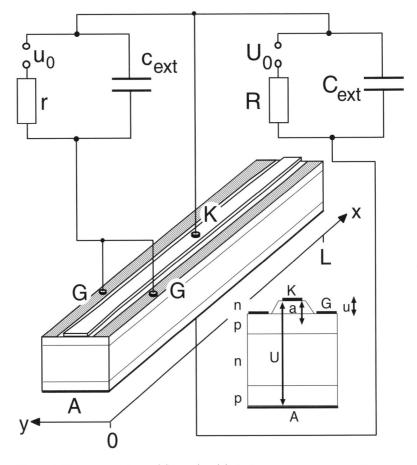

Figure 2.15. A gate-driven p^{++}–n–p^{+}–n^{++} thyristor structure operated in an external main circuit and a gate circuit (schematic). The potential drops between the cathode K and the p^{+} layer, the gate G, and the anode A are denoted by a, u, and U, respectively (see the inset showing a cross-sectional view of the structure).

In lasing structures, control of switching fronts implies control of the active area of the semiconductor laser, which could be of great importance for optical systems.

In Fig. 2.15 we consider a p–n–p–n structure whose design is similar to that of the single element of a modern gate-turn-off thyristor. It consists of a main circuit (K-A) connected to a bias voltage U_0 via a load resistor R and a parallel capacitance C_{ext}, and a control circuit (G) including an applied voltage u_0, resistance r, and external capacitance c_{ext}. For appropriate parameters the structure can be switched between the high-current and the low-current state by a trigger front moving along the x-direction. The dynamics of the switching process can be controlled by means of the spatially extended gate G. The gate-driven p–n–p–n structure can be described by the reaction–diffusion equation

$$\tau_a \frac{\partial a(x,t)}{\partial t} = l^2 \frac{\partial^2 a(x,t)}{\partial x^2} + f(a,u,U) \tag{2.97}$$

with a nonlinear local kinetic function (Gorbatyuk and Rodin 1992a)

$$f(a,u,U) \equiv -\alpha a + \exp a - \beta \exp(-U + 2a) + \gamma U + \kappa u. \tag{2.98}$$

Here $a(x,t)$, $u(t)$, and $U(t)$ are the p-base potential, gate potential and cathode–anode voltage, respectively, as indicated in the inset of Fig. 2.15 (a, u, and U are measured in units of $k_B T/e$ and are therefore dimensionless). The characteristic length l and the coefficients α, β, γ, and κ of the local kinetic function are determined by the structural parameters; τ_a is the characteristic relaxation time of a. We are interested in the x-dependency of the internal state $a(x,t)$ of the semiconductor structure, assuming it to be homogeneous along the y-direction. Neumann boundary conditions are imposed on $a(x,t)$.

The nonlinear function $f(a,u,U)$ can be derived from charge conservation and transport equations by similar arguments to those in Section 2.4.7 (Gorbatyuk and Rodin 1992a, 1992b). The potentials u and U play the roles of parameters with respect to the bistable medium. Physically, bistability of a p–n–p–n structure is associated with a change of the central (collector) p^+–n junction bias, which is negative in the off state and becomes positive as the structure is switched on. The function $f(a,u,U)$ combines linear terms, corresponding to the leakage current of the p–n junctions which makes a major contribution to charge transport in the off state, and highly nonlinear exponential terms, corresponding to injection currents of the collector and emitter p–n junctions, respectively. The latter dominate in the on state, but the small prefactor β makes them negligible in the off state. That leads to the sharp rise of $f(a,u,U)$ as a function of a (Fig. 2.16). Since the first term in (2.98) takes into account Ohmic conductances both of the gate contact and of the emitter n^{++}–p^+ junction, whereas the last term contains the latter contribution only, one generally has $\alpha > \kappa$, which is important for the following analysis.

The global constraints corresponding to the main circuit and the control circuit, respectively, are given by

$$\tau_U \frac{dU}{dt} = U_0 - U - R \int_0^L J(a, U) \, dx, \tag{2.99}$$

$$\tau_U \equiv RC, \qquad C \equiv C_{\text{int}} + C_{\text{ext}}. \tag{2.100}$$

$$\tau_u \frac{du}{dt} = u_0 - u - r \int_0^L j(a, u) \, dx, \tag{2.101}$$

$$\tau_u \equiv rc, \qquad c \equiv c_{\text{int}} + c_{\text{ext}}. \tag{2.102}$$

Here L is the system length along the x-direction. The characteristic relaxation times τ_U and τ_u are determined by the sums of the differential capacitances of the external circuits C_{ext} and c_{ext} and the differential cathode–anode and cathode–gate capacitances C_{int} and c_{int}, respectively. For the current density per unit length between the cathode and anode $J(a, U)$, and that between the gate and the cathode $j(a, u)$ we assume (Meixner *et al.* 1998b)

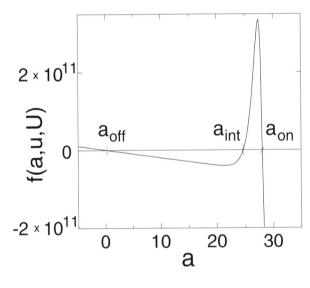

Figure 2.16. The (dimensionless) local kinetic function $f(a, u, U)$ of a gate-driven thyristor structure. The off, intermediate, and on states are denoted by a_{off}, a_{int}, and a_{on}, respectively. The values of u and U are chosen as $u = 0$ and $U = 30.75$, satisfying the equal-areas rule $\int_{\text{off}}^{\text{on}} f(a, u, U) \, da = 0$. The numerical parameters $\alpha = 2 \times 10^9$, $\beta = 14.5$, $\gamma = 10^7$, and $\kappa = 10^9$ correspond to a realistic p–n–p–n structure (Varlamov and Osipov 1970). All potentials a, u, and U are in units of $k_B T / e$ (after Meixner *et al.* (1998a)).

$$J(a, U) = \sigma_U U + J_S(\exp a - 1), \tag{2.103}$$

$$j(a, u) = \sigma_u(u - a) \tag{2.104}$$

with linear conductivities σ_U and σ_u, and saturation current density J_S.

The $a(u, U)$ dependency resulting from the null-cline $f(a, u, U) = 0$ is bistable in a certain range of the parameters u and U. The $a(u)$ dependency for $U = $ constant (Fig. 2.17(a)), and the $a(U)$ dependency for $u = $ constant are both S-shaped. The steady-state current–voltage characteristic $j(u) \equiv j(a(u), u)$ of the gate circuit which results from the dependencies $a(u)$ and $j(a, u)$ is Z-shaped (Fig. 2.17(b)). The current–voltage characteristic $J(U) \equiv J(a(U), U)$ of the main circuit which results from the dependencies $a(U)$ and $J(a, U)$ is S-shaped. Note that, in contrast to the DBRT (cf. Section 2.4.6), which also has a Z-shaped current–voltage characteristic, gate-driven p–n–p–n structures are three-terminal devices, and the Z-shaped cathode–gate characteristic is used to control the S characteristic of the main circuit. Depending upon the control circuit, a rich menagery of pattern-formation processes, namely accelerated, decelerated, and oscillatory lateral current-density fronts, is found (Meixner *et al.* 1998a, 1998b).

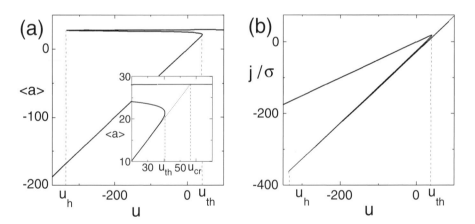

Figure 2.17. (a) The null-isocline $a(u)$ given by the local kinetic function $f(a, u, U) = 0$ and (b) the Z-shaped local current-density versus voltage characteristic $j(u)$ for $U = 30.75$ (a, u, U and j/σ are in units of $k_B T/e$). The lower branch (off) of the $a(u)$ relation corresponds to the upper branch of $j(u)$ and vice versa. The holding voltage u_h and the threshold voltage u_{th} denote the left- and right-hand boundaries of the bistability regime. The inset in (a) shows the turning point at u_{th} on an enlarged scale (after Meixner *et al.* (1998a)).

Chapter 3

Pattern formation and oscillatory instabilities in semiconductors

Self-organized pattern formation is closely connected with negative differential conductivity in semiconductors. In this chapter we develop a general framework for the analysis of the formation and stability of patterns such as current filaments, field domains, and fronts. Special emphasis is placed upon the interaction with an external circuit and the resulting global coupling which strongly affects the stability.

3.1 Introduction

In semiconductors, spatially homogeneous states of negative differential conductivity are in general unstable against spatio-temporal fluctuations, which may give rise to self-organized pattern formation. If the $j(\mathcal{E})$ characteristic is N-shaped (NNDC), such as in the Gunn effect, inhomogeneous electric-field profiles in the form of a *high-field domain* may arise. If the $j(\mathcal{E})$ characteristic is S-shaped, such as in threshold switching, the current flow may become inhomogeneous over its cross-section and form a *current filament*. These spatial structures may be static or time-dependent. In the latter case, current oscillations can arise due to domains moving in the direction of the current flow, or filaments "breathing" transversally to the current flow (Schöll 1987).

At this point a word of warning is indicated. First, negative differential conductivity does not always imply instability of the steady state, and positive differential conductivity does not always imply stability. For example, SNDC states can be stabilized by a heavily loaded circuit (and experimentally observed!), and, on the other hand, the Hopf bifurcation of a limit-cycle oscillation can occur on a $j(\mathcal{E})$ characteristic with positive differential conductivity. Second, there is no one-to-one correspondence between SNDC and filaments, or between NNDC and domains. Finally, it is important to distinguish between the local $j(\mathcal{E})$ characteristic and the global $I(U)$ relation, which may exhibit negative differential conductance even if the $j(\mathcal{E})$ relation does not exhibit NDC, and vice versa (Schöll 1987).

Moving field domains due to the Gunn effect (Shaw *et al.* 1979, 1992) or a recombination instability (Bonch-Bruevich *et al.* 1975) were extensively investigated experimentally and theoretically in the 1960s and 1970s. More recently, attention has focussed on field domains in superlattices (Wacker 1998). Current filaments have attracted interest in connection with low-temperature impurity breakdown in n-GaAs and p-Ge (Schöll 1982*a*, Aoki and Yamamoto 1983, Brandl *et al.* 1989, Schöll 1987, Peinke *et al.* 1992) and with p–i–n diodes (Jäger *et al.* 1986, Purwins *et al.* 1987, Symanczyk *et al.* 1991*b*) and p–n–p–i–n structures (Niedernostheide *et al.* 1992*b*, 1993, 1996*a*), and thyristors (Gorbatyuk and Rodin 1990, 1992*a*, 1997) at room temperature. They have been observed directly by a number of spatially resolved sophisticated experimental techniques ranging from invasive methods such as scanning electron microscopy (Mayer *et al.* 1988, Wierschem *et al.* 1995), scanning laser microscopy (Brandl *et al.* 1989, Spangler *et al.* 1994, Kukuk *et al.* 1996), and potential-probe measurements (Baumann *et al.* 1987, Niedernostheide *et al.* 1992*a*) to more elaborate noninvasive techniques such as measurement of quenched photoluminescence (Eberle *et al.* 1996).

Domains and filaments can be theoretically described as special nonuniform solutions of the basic semiconductor transport equations, subject to appropriate boundary conditions. A linear stability analysis around the spatially homogeneous steady state (fixed point) for small space- and time-dependent fluctuations of the electromagnetic field and the relevant transport variables, e.g. carrier concentrations or charge densities, yields conditions for the onset of domain-type or filamentary instabilities. This linear mode analysis can be put into the form of a linear response $\delta j = \sigma(\lambda)\,\delta\mathcal{E}$ between the field fluctuation $\delta\mathcal{E}$ and the current-density fluctuation δj, where the dynamic conductivity tensor $\sigma(\lambda)$ contains all eigenmodes λ (Thomas 1992). If the classical drift–diffusion approach (Selberherr 1984, Markovich 1986, Sze 1998) is used as a transport model, a careful analysis gives a variety of unstable modes due to coupled drift, diffusion, and generation–recombination processes that can lead to the bifurcation of stationary current filaments or traveling field domains (Schöll 1987). Important modifications occur through global coupling to an external load circuit. Some general aspects of this linear mode analysis will be discussed in this chapter.

The fully developed self-organized spatial patterns must be computed from the full *nonlinear* transport equations. Although a full numerical solution of these equations provides the most detailed information about the spatial profiles, it does not give immediate physical insight. Therefore there is a need for approximations. In particular, it is desirable to derive simple analytic relations that provide some relevant information about the domains and filaments, such as the peak field and the propagation velocity of the domains, or the filament radius as a function of the material parameters and the applied bias, without explicitly solving the differential equations for the profiles. This is the purpose of *equal-areas rules*, which may be visualized geometrically in analogy with the Maxwell rule in thermodynamics that describes the coexistence of a vapor and a liquid phase with a planar boundary (Schöll and Landsberg 1988). Mathematically, equal-areas rules are definite first integrals of the spatial profiles. A simple equal-areas rule for traveling Gunn domains associated with NNDC was first derived by Butcher (1965). Conditions for plane or cylindrical current filaments associated with SNDC were developed in the form of equal-areas rules both for unipolar and for ambipolar generation–recombination mechanisms (Schöll 1982*b*, 1986*a*). Extensions to time-dependent filaments (Schöll and Drasdo 1990, Kunz and Schöll 1992) and to domains in superlattices (Schwarz and Schöll 1996, Wacker *et al.* 1997) have also been given. We shall see later in this chapter that, in a more general context, filaments, domains, and fronts can often be described by reduced dynamic equations whose fixed points are given by the common equal-areas rules.

By way of example, let us discuss the prototype of a bistable system, which consists in a single reaction–diffusion equation for the variable a

$$\frac{\partial a}{\partial t} = D \, \Delta a + f(a), \tag{3.1}$$

with a cubic nonlinearity (Fig. 3.1)

$$f(a) = -q(a - a_1)(a - \bar{a})(a - a_3), \qquad q > 0, \qquad a_1 < \bar{a} < a_3 \tag{3.2}$$

and diffusion constant D. This system has the stable fixed points a_1 and a_3, and the unstable fixed point $a_2 = \bar{a}$. It can be derived from a chemical-reaction scheme

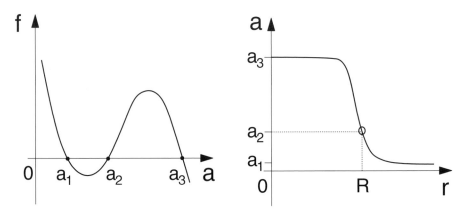

Figure 3.1. A bistable nonlinear kinetic function $f(a)$ and the corresponding spatial front profile $a(r)$ (schematic).

for the species of concentration a which is subject to autocatalytic reproduction (cf. impact ionization in semiconductors, eqs. (1.40)–(1.43), Section 1.4.2):

$$2a + b \rightleftharpoons 3a, \tag{3.3}$$

$$a \rightleftharpoons c. \tag{3.4}$$

The profound analogy of this bistable nonequilibrium system with phase transitions in equilibrium (e.g. the Van der Waals gas) was pointed out by Schlögl (1972). The condition for spatial coexistence of the two stable phases a_1 and a_3 can be derived on a one-dimensional infinite spatial domain by multiplying eq. (3.1) by $\partial a / \partial x$ and integrating over $\int_{-\infty}^{\infty} dx$ subject to the boundary conditions

$$a(-\infty) = a_3, \qquad a(\infty) = a_1. \tag{3.5}$$

In the steady state this yields

$$\int_{a_1}^{a_3} f(a) \, da = \frac{D}{2} \left[\left(\frac{\partial a}{\partial x} \right)^2 \right]_{-\infty}^{\infty} \tag{3.6}$$

and hence, because

$$\frac{\partial a}{\partial x} (\pm \infty) = 0,$$

$$\int_{a_1}^{a_3} f(a) \, da = 0. \tag{3.7}$$

This condition may be visualized as an equal-areas rule that fixes $\bar{a}$ to the value $\bar{a} = \frac{1}{2}(a_3 - a_1)$ since it requires that the two areas above and below the a-axis in Fig. 3.1 must be equal. Thus it is analogous to Maxwell's rule in thermodynamics that determines the pressure $P = P_{co}$ under which the liquid (volume V_1) and the vapor (volume V_3) can coexist at temperature T such that $\int_{V_1}^{V_3} (P(V, T) - P_{co}) \, dV = 0$; this is a typical feature of a first-order equilibrium phase transition. Hence the nonlinear function $f(a)$ corresponds to a nonmonotonic isotherm $P(V, T)$ in the (P, V) diagram.

If the equal-areas rule is not satisfied, a front between the two locally stable phases a_1 and a_3 propagates with constant velocity v either into phase a_1 ($v > 0$) or into phase a_3 ($v < 0$), inducing a nonequilibrium phase transition to the respective absolutely stable phase. The front velocity v can be derived by the same argument as the equal-areas rule above, assuming a self-similar spatial profile $a(x, t) = a_0(x - vt)$ satisfying the boundary conditions (3.5). In the co-moving frame $\xi = x - vt$ (3.1) becomes

$$D \frac{d^2 a_0}{d\xi^2} + v \frac{d a_0}{d\xi} + f(a_0) = 0. \tag{3.8}$$

Multiplying eq. (3.8) by $d a_0 / d\xi$ and integrating over $\int_{-\infty}^{\infty} d\xi$ yields

$$v = \frac{\int_{a_1}^{a_3} f(a)\, da}{\int_{-\infty}^{\infty} \left(\frac{\partial a}{\partial x}\right)^2 dx}.$$ (3.9)

The direction of motion of the front is thus determined by the sign of the integral $\int_{a_1}^{a_3} f(a)\, da$, since the denominator is always positive.

It should be noted that these results (3.7) and (3.9) hold generally for an arbitrary bistable nonlinear kinetic function f. For the cubic nonlinearity (3.2), additionally, an exact front solution of (3.1) subject to the boundary conditions (3.5) can be found (Montroll and Shuler 1958):

$$a_0(x - vt) = a_1 + \frac{1}{2}(a_3 - a_1)\left[1 - \tanh\left(\frac{2(x - vt)}{W}\right)\right]$$ (3.10)

with front velocity

$$v = \sqrt{\frac{Dq}{2}}(a_1 + a_3 - 2\bar{a})$$ (3.11)

and interface thickness

$$W = 4\sqrt{\frac{2D}{q}}\frac{1}{a_3 - a_1}.$$ (3.12)

The stability of this solution was proven by Magyari (1982) and Schlögl et al. (1983).

The above argument for phase coexistence with a flat interface can be extended to spherically symmetric "droplets" of phase a_3 embedded in phase a_1 in d spatial dimensions. Such droplets can coexist with phase a_1 if their radius R is given by the critical radius

$$R_c^0 = \frac{(d-1)D}{v} = (d-1)D\frac{\int_0^{\infty}\left(\frac{\partial a}{\partial r}\right)^2 dr}{\int_{a_1}^{a_3} f(a)\, da}.$$ (3.13)

This follows from the reaction–diffusion equation (3.1) in d dimensions under spherical symmetry with radial coordinate r:

$$\frac{\partial a}{\partial t} = D\frac{\partial^2 a}{\partial r^2} + \frac{d-1}{r}D\frac{\partial a}{\partial r} + f(a).$$ (3.14)

We assume that the droplet solution $a(r, t) = a_0(r - R(t))$, where $R(t)$ denotes the position of the interface dividing the two phases, satisfies the boundary conditions

$$a(0, t) \approx a_3, \quad a(\infty, t) = a_1, \quad \frac{\partial a}{\partial r}(0, t) = 0.$$ (3.15)

Transforming again to the co-moving frame $\xi = r - R(t)$, eq. (3.14) becomes

$$D \frac{d^2 a_0}{d\xi^2} + \left(\dot{R} + \frac{d-1}{r} D \right) \frac{da_0}{d\xi} + f(a_0) = 0. \tag{3.16}$$

Multiplying eq. (3.16) by $da_0/d\xi$ and integrating over $\int_{-R}^{\infty} d\xi$ yields

$$\dot{R} = (d-1)D \left(\frac{1}{R_c^0} - \frac{1}{R} \right) \tag{3.17}$$

with (3.13), where we have used the approximation that $da_0/d\xi$ assumes essential values only around the interface at $r \approx R$ (Fig. 3.1). This dynamic equation describes the temporal evolution of the droplet radius $R(t)$; it contains the critical droplet radius R_c^0 as an unstable fixed point: Larger droplets grow ($\dot{R} > 0$), while smaller ones decay ($\dot{R} < 0$).

The droplet growth can be stabilized by imposing a global constraint

$$-k\langle a \rangle \quad \text{with} \quad k > 0 \tag{3.18}$$

as an additive term on the right-hand side of eq. (3.14) (Schimansky-Geier *et al.* 1991). Hereby eq. (3.17) is amended by replacing R_c^0 by a time-dependent critical radius

$$R_c(t) = (d-1)D \frac{\displaystyle\int_0^{\infty} \left(\frac{\partial a}{\partial x} \right)^2 dr}{\displaystyle\int_{a_1}^{a_3} f(a)\, da - k(a_3 - a_1)a_3 V(t)/V_0}. \tag{3.19}$$

Here $V(t)/V_0 \sim R(t)^d$ denotes the fraction of the spatial domain V_0 which is occupied by the phase a_3, i.e. the droplet, and $a_1 \approx 0$ has been used. The denominator in (3.19), which is analogous to the supersaturation of the vapor in thermodynamics, will decrease for a growing droplet. Accordingly, the critical radius increases as well. This is the typical behavior of Ostwald ripening in a van-der-Waals gas. Here it is the effect of the inhibitory global interaction in a nonequilibrium system. It can be shown that now eq. (3.17), for sufficiently large initial supersaturation $X_0 = \int_{a_1}^{a_3} f(a)\, da$, has a second fixed point $R_0 > R_c^0$, which is stable while R_c^0 is still unstable. If now a supercritical domain is formed ($R > R_c^0$), it will grow only until it reaches the stable size R_0. If more than one droplet is present, the global interaction induces a competition (winner-takes-all dynamics) as a result of which only the initially largest droplet survives.

This simple model has served to illustrate concepts and features of pattern formation in bistable nonequilibrium systems that are found in many of the more sophisticated semiconductor models that will be studied in detail in that which follows. Phase coexistence, moving fronts and droplets occur as plane current-density layers, electric-field domains or cylindrical current filaments ($d = 2$), and their spatio-temporal dynamics, including their global interactions with the load circuit, can be understood on the basis outlined above.

3.2   **Stability and differential conductance**

In this section we shall establish an important general relation between the stability of
the steady state and the shape of the current–voltage characteristic and the load line
(Schöll 1989, Wacker and Schöll 1995). The essential features of the current–voltage
characteristic are expressed by the differential conductance dI/dU, or the differential
conductivity $\sigma_{\text{diff}} = dj/d\mathcal{E}$ in case of uniform steady states. The relation that we
are going to derive is remarkable since it does not depend upon the specific electric
transport mechanism. It is the purpose of this section to demonstrate that the apparently very different individual behaviors of nonlinear and bistable semiconductors
can be understood within a unified framework in terms of a few, very simple principles
without recourse to the underlying microscopic physics. The intention is to provide the
applied scientist who finds complicated multivalued current–voltage characteristics in
some semiconductor device with natural criteria to identify the stability of each of the
various branches.

Bistability between two conducting states at a given sample voltage U can be
manifested in various types of current–voltage characteristics, as shown in Fig. 3.2.
Here the full lines give branches that are usually stable, while those shown by the

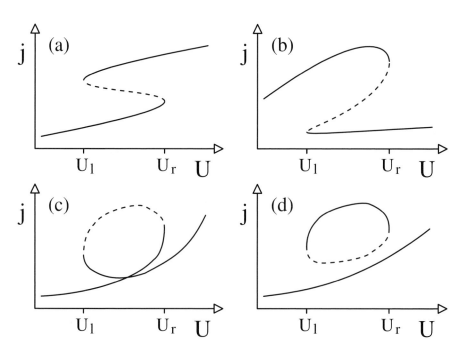

Figure 3.2. Various different types of multistable current–density voltage
characteristics, which can be S-shaped (a), Z-shaped (b), loop-shaped (c), or
disconnected (d). The full lines are usually stable under voltage-controlled
conditions, while the dotted lines can be stabilized only under special conditions.

dashed lines are unstable under voltage-controlled conditions. The most prominent type is the S-type characteristic shown in Fig. 3.2(a), which occurs, for example, in the regime of impact-ionization breakdown (Schöll 1987), p–n–p–i–n diodes (Niedernostheide *et al.* 1992*b*), the heterostructure hot-electron diode (HHED) (Hess *et al.* 1986, Belyantsev *et al.* 1986, Wacker and Schöll 1991), and quantum-dot structures (Wu *et al.* 1992). The Z-shaped case (Fig. 3.2(b)) is widely discussed for the double-barrier resonant-tunneling diode (DBRT) (Goldman *et al.* 1987). It also appears in the post-breakdown regime of p-germanium (Schöll *et al.* 1987), and in gate-driven thyristor structures (Gorbatyuk and Rodin 1992*a*). In the first two of these Z-systems the stable branches of the loop-shaped case (Fig. 3.2(c)) have also been observed, under various conditions (Schöll *et al.* 1987, Leadbeater *et al.* 1989). An unconnected characteristic, as shown in Fig. 3.2(d), occurs in the case of magnetoconductivity in n-Si (Asche *et al.* 1984) and has been predicted for real-space-transfer transistors (Luryi and Pinto 1991). The current–voltage characteristics of doped superlattices even exhibit multistability among a large number of different conducting states under the same operating conditions, as has been shown experimentally and theoretically (Kastrup *et al.* 1994). These examples clearly highlight the physical relevance of various types of bistability for various devices.

All of these devices exhibit at least two stable states with different conductivities for a given voltage in the range $U_l < U < U_r$. Thus, there must exist an internal degree of freedom that determines the state in which the device is operating. This internal degree of freedom has to be identified with an internal physical quantity, such as the charge density in a quantum well, the electron temperature, the occupation of traps or impurities, the density of free carriers, or combinations of several physical quantities. Of course the nature of this physical quantity depends strongly on the semiconductor element considered.

In the following we assume that this internal degree of freedom can be represented by a single physical quantity, which we will denote by a. For instance, a is the electron temperature, if heating effects essentially determine the bistability. This means that all further degrees of freedom are not essential for the bistability and can be eliminated adiabatically for given a and U. This is called the slaving principle in terms of nonlinear dynamics (Haken 1987). This elimination is not possible for the quantity a, since otherwise different stationary values would not be possible for a given voltage U. The current density in the sample is then given by some equation

$$j = j(a, U), \tag{3.20}$$

which has to be derived from a transport model for the individual device considered, as discussed in Chapter 2. Since a is an independent dynamic variable, it is governed by some dynamic equation

$$\frac{da(t)}{dt} = f(a, U) \tag{3.21}$$

resulting from the internal mechanism leading to the instability. Throughout this section we shall neglect spatial dependencies; those will be studied in Sections 3.3–3.5.

In case of various heterostructure devices a can be identified as the charge density in a quantum well. In this case f is simply given by the difference of the incoming and outgoing currents. For example, such a microscopic model has been derived for the DBRT (Joosten *et al.* 1991), the HHED (Wacker and Schöll 1992), and a quantum-dot structure (Wacker 1994), see Section 2.4. Here it is essential to include Poisson's equation to take care of the self-consistency with the electric potential. Similar treatments with different quantities a and different types of $f(a, U)$ have been performed for other systems such as p–n–p–i–n diodes (Niedernostheide *et al.* 1992*b*) and low-temperature impact-ionization breakdown (Schöll 1987). (In the last case a is associated with the free-carrier density and $f(a, U)$ is related to the generation–recombination dynamics, see Chapter 4.)

Rather than studying single specific models, here we want to give some insight into the generic features associated with bistability within a unified point of view. Thus, at this point we do not specify the quantity a but discuss the influence of the dynamics of $a(t)$ given by eq. (3.21) upon the global behavior of the device in general terms. There exist, of course, special situations that can not be described in terms of a single internal variable and that have to be treated separately. Examples of generation–recombination models that involve more than one dynamic variable are provided in Chapter 4. Nevertheless, the approach taken here is considered to be applicable to most nonlinear semiconductor devices.

3.2.1   Stability at fixed voltage

For a given voltage U the stationary states a^* of the internal quantity a are given by the relation $f(a^*, U) = 0$. In the case of multistability we have several stationary states $a_1^* < a_2^* < a_3^* < \cdots$ in the range $U_l < U < U_r$. For simplicity we restrict ourselves to three stationary states in the following, as shown in Fig. 3.3. Cases in which multistability between more than two stable states occurs can be treated analogously.

The curve $f(a, U) = 0$ separates the regions with $f(a, U) > 0$ where $a(t)$ increases with time from regions with $f(a, U) < 0$ where $a(t)$ decreases. Since $|a|$ should not grow unlimitedly with time for large $|a|$, the function $f(a, U)$ must be positive or negative for small or large a, respectively, as depicted in Fig. 3.3. Regarding small perturbations, the stationary state a^* is stable if $f_a(a^*, U) < 0$. Here we use the notation

$$f_a(a^*, U) \equiv \left. \frac{\partial f(a, U)}{\partial a} \right|_{a=a^*}. \tag{3.22}$$

Because $f(a, U)$ changes sign from plus to minus at $a = a_1^*$ for fixed U with increasing a, we find $f_a(a_1^*, U) < 0$. Therefore a_1^* is a stable state. Then $f(a, U)$

must change sign from minus to plus at the next zero a_2^*. Thus, $f_a(a_2^*, U) > 0$ and a_2^* is an unstable fixed point under voltage-controlled conditions. Similarly, a_3^* will be another stable fixed point. Therefore, we conclude that there must be at least three stationary states if two *stable* states are observed under voltage-controlled conditions (i.e. if bistability occurs).

Now let us consider the behavior of the solutions a_i^* as functions of U. For $U < U_l$ we have only one stable solution a^*. As can be seen from Fig. 3.2, this solution must develop continuously into one of the stable states in the bistable range if U is increased. Let this stable solution be a_1^*. (If this were larger than the second stable solution, one could transform $a \to -a$, so that $a_1^* < a_3^*$ is again fullfilled.) Regarding the situation in Figs. 3.2(a)–(c), we find that only the solution a_3^* persists for $U > U_r$ and we obtain a multivalued relation $a^*(U)$ that has the shape presented in Fig. 3.3(a). For the situation shown in Fig. 3.2(d) the shape of $a^*(U)$ is plotted in Fig. 3.3(b).

If we now use the relation $j = j(a^*(U), U)$ from eq. (3.20), we obtain the original current-density voltage characteristic from Fig. 3.2 with the additional unstable branch (the dotted line). If we have the situation shown in Fig. 3.3(a) then the characteristic from Fig. 3.2(a) or Fig. 3.2(b) is obtained if j is monotonically increasing or decreasing in a, respectively. If j is not monotonic in a the characteristic from Fig. 3.2(c) may occur. Of course, more complex characteristics are possible, too. Owing to the stability regarding fluctuations in a, the branches of the characteristic are

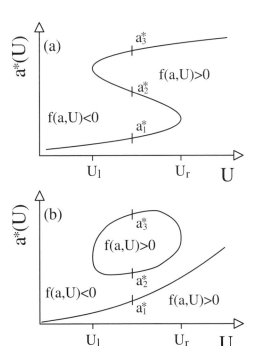

Figure 3.3. Typical shapes of the stationary solutions $a^*(U)$ determined from the equation $f(a^*, U) = 0$ for the connected (a) and disconnected (b) cases.

stable (full line) or unstable (dotted line) for voltage-controlled conditions, as depicted in Fig. 3.2.

3.2.2   Operation via a load resistor

In the following we consider the important case that the voltage U across the sample is not fixed, but the device is operated in a circuit as shown in Fig. 3.4. The capacitance C parallel to the sample is given by the sum of the internal device capacitance, the externally applied capacitance, and parasitic wire capacitances. Note that the resistance R is usually thought to be an external resistance, but it may also be the linear resistance of a part of the sample that is not associated with the bistable behavior (e.g. a linear contact resistance).

Now the temporal behavior of $U(t)$ is determined by the circuit, cf. eq. (1.56)

$$\frac{dU(t)}{dt} = \frac{1}{C}\left(\frac{U_0 - U}{R} - I(a, U)\right),\tag{3.23}$$

where $I(a, U)$ is the total current through the device. If the current density j is constant in the direction of transport (which defines the z-axis), it is simply given by the integral over the cross-section A of the current flow:

$$I = \int_A dx\, dy\, j(a(x, y), U).\tag{3.24}$$

If there is charge accumulation in the sample (such as in the DBRT or the HHED) I is given by a more complicated integral expression as shown in Section 1.5.2, eq. (1.64). In this section we assume for simplicity that j is also homogeneous over the sample's cross-section, so that $I = Aj$ holds. Nevertheless, the general case can be treated analogously.

The stationary points (a^*, U^*) are given by the conditions $f(a^*, U^*) = 0$ and $(U_0 - U^*)/R = Aj(a^*, U^*)$. This depicts the operating point which is the intersection of the current–voltage characteristic $j = j(a^*(U), U)$ with the load line

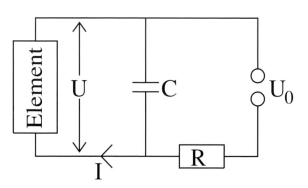

Figure 3.4. The bistable element embedded in a circuit with a load resistor of resistance R and a capacitor of capacitance C.

$j = (U_0 - U)/(RA)$. To determine its stability we consider the temporal behavior of small fluctuations:

$$\delta a(t) \equiv a(t) - a^* = \delta a_0\, e^{\lambda t}, \tag{3.25}$$

$$\delta U(t) \equiv U(t) - U^* = \delta U_0\, e^{\lambda t}, \tag{3.26}$$

where λ is determined by the eigenvalue equation

$$\lambda \begin{pmatrix} \delta a \\ \delta U \end{pmatrix} = A \begin{pmatrix} \delta a \\ \delta U \end{pmatrix} \tag{3.27}$$

with the Jacobian matrix

$$A = \begin{pmatrix} f_a & f_U \\ -j_a \frac{A}{C} & -\frac{1}{RC} - j_U \frac{A}{C} \end{pmatrix} \Bigg|_{(a^*,U^*)}, \tag{3.28}$$

where the subscripts a and U denote partial derivatives. We obtain the eigenvalues

$$\lambda_{1,2} = \frac{\operatorname{tr} A}{2} \pm \sqrt{\frac{(\operatorname{tr} A)^2}{4} - \det A} \tag{3.29}$$

where $\operatorname{tr} A$ is the trace and $\det A$ is the determinant of A. The operating point is stable if the real parts of both eigenvalues are negative. This is equivalent to $\det A > 0$ and $\operatorname{tr} A < 0$. This means

$$\text{stable fixed point} \Leftrightarrow (\det A > 0) \wedge (\operatorname{tr} A < 0). \tag{3.30}$$

The trace of the Jacobian matrix is given by

$$\operatorname{tr} A = f_a - \frac{A}{C} j_U - \frac{1}{RC}. \tag{3.31}$$

A straightforward calculation of the determinant (using $da^*(U)/dU = -f_U/f_a$) yields

$$\det A = -f_a \left(A\, \frac{dj(a^*(U), U)}{dU} + \frac{1}{R} \right) \frac{1}{C}, \tag{3.32}$$

which can be written as

$$\det A = \lambda_1 \lambda_2 = -\frac{f_a}{\tau_U^{\text{diff}}}, \tag{3.33}$$

where f_a determines the stability of the internal dynamic variable for fixed voltage, and

$$\tau_U^{\text{diff}} = \left[\left(\frac{dI}{dU} + \frac{1}{R} \right) \frac{1}{C} \right]^{-1} \tag{3.34}$$

is the differential relaxation time of the circuit.

This important general relation (3.33) connects the differential conductance with the stability of the fixed point, which is determined by the sign of the real parts of the eigenvalues λ_i of the Jacobian matrix $\mathcal{A}$.

Thus, a criterion for the stability of the operating point can be deduced from the difference between the slope dI/dU of the current–voltage characteristic $I(a^*(U), U)$ and the slope $-1/R$ of the load line $I = U_0 - U/R$. This has to be combined with the sign of f_a, which is negative or positive on branches that are stable and unstable under voltage-controlled conditions, respectively. The various possibilities for positive R and C are shown in Fig. 3.5.

A further examination of eq. (3.33) shows that a change of the sign of $\det \mathcal{A}$ (leading to a *saddle-node bifurcation* in terms of nonlinear dynamics (Haken 1987), see Section 1.2) occurs if the slopes of the load line and the characteristic are identical:

$$\det \mathcal{A} = 0 \quad \Leftrightarrow \quad \frac{dI}{dU} = -\frac{1}{R} \quad \text{or} \quad f_a = 0. \tag{3.35}$$

These saddle-node bifurcations mark the onset of the range of bistability of the steady states as a function of the control parameter U_0 or R. For general load lines, with increasing U_0, for instance, the condition $dI/dU = -1/R$ marks the point where the load line becomes tangent to the current–voltage characteristic. In the current-controlled case ($R \to \infty$, i.e. a horizontal load line) the saddle-node bifurcations occur for vanishing differential conductance $dI/dU = 0$, i.e. at the maximum and minimum of an N-shaped current–voltage characteristic. In the limit of the voltage-controlled case ($R \to 0$) the differential conductance would tend to infinity; in fact it turns out that the saddle-node bifurcation is given by vanishing $f_a = 0$ at the turning points of an S-shaped or Z-shaped current–voltage characteristic.

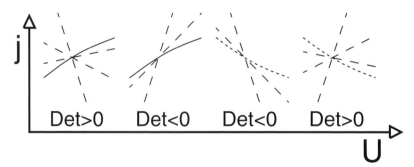

Figure 3.5. Signs of the determinant of the Jacobi matrix for various slopes of the load line (dashed lines). The $j(U)$ characteristic is depicted with a full line for branches that are stable ($f_a < 0$) and with a dotted line for branches that are unstable ($f_a > 0$) under voltage-controlled conditions. The operating point can be stable only for $\det \mathcal{A} > 0$. Here $C > 0$ is assumed. For $C < 0$ the sign of $\det \mathcal{A}$ is changed.

3.2.3 Conditions of oscillatory current instabilities

Next, we consider Hopf bifurcations of limit cycles. They mark one possible onset of oscillatory instabilities. (There are others, associated, for example, with global bifurcations, see Section 1.2.) Note that, in this subsection, we do not restrict ourselves to bistable current–voltage characteristics but consider the general case which includes N-shaped and monotonic characteristics as well. A Hopf bifurcation occurs if a pair of complex-conjugate eigenvalues crosses the imaginary axis.

This is equivalent to $\det A > 0$ and $\operatorname{tr} A = 0$:

$$\det A = -f_a \left(\frac{dI}{dU} + \frac{1}{R} \right) \frac{1}{C} > 0, \tag{3.36}$$

$$\operatorname{tr} A = f_a - \frac{A}{C} j_U - \frac{1}{RC} = 0. \tag{3.37}$$

Equations (3.37) can be satisfied for $C > 0$ by either

$$\frac{dI}{dU} + \frac{1}{R} < 0 \quad \text{and} \quad f_a = \frac{A}{C} j_U + \frac{1}{RC} > 0 \tag{3.38}$$

or

$$\frac{dI}{dU} + \frac{1}{R} > 0 \quad \text{and} \quad f_a = \frac{A}{C} j_U + \frac{1}{RC} < 0. \tag{3.39}$$

Physically, the differential conductance

$$\frac{dj(a^*(U), U)}{dU} = j_a \frac{da^*}{dU} + j_U$$

is composed of a contribution from the internal variable a, and a contribution $j_U = \partial j(a, U)/\partial U$ from the direct voltage or field dependency of the current. If, for example, the local current density is given by a drift current density $j(n, U) = env(\mathcal{E})$, where the carrier density n is the internal variable a and v is the drift velocity, and a homogeneous electric field $\mathcal{E} = U/L$ is assumed, then $j_U \sim dv/d\mathcal{E}$ is proportional to the differential mobility. In the current-controlled case $(R \to \infty)$ a Hopf bifurcation can then occur *either* for negative differential conductivity $(dj/d\mathcal{E} < 0)$, positive differential mobility $(j_U > 0)$, and an instability of the internal variable $(f_a > 0)$, cf. eq. (3.38), *or* for positive differential conductivity $(dj/d\mathcal{E} > 0)$, negative differential mobility $(j_U < 0)$, and stability of the internal variable $(f_a < 0)$, cf. eq. (3.39). Examples of these two cases are provided by impurity-impact-ionization breakdown (see Chapter 4) and by field-enhanced emission (the Poole–Frenkel effect) in semi-insulating GaAs (Schöll 1989), respectively.

If we assume that $j_U > 0$ holds (i.e. the negative differential conductance

$$\frac{dj(a^*(U), U)}{dU} = j_a \frac{da^*}{dU} + j_U$$

is caused by the influence of the quantity a, $f_a > 0$), we find for $R, C > 0$ the following result:

$$\operatorname{tr} \mathcal{A} > 0 \quad \text{if} \quad f_a > 0 \quad \text{and} \quad C > C_{\text{crit}} \equiv \left(j_U A + \frac{1}{R} \right) \frac{1}{f_a}. \tag{3.40}$$

Therefore the trace becomes positive upon increasing C on the middle (NDC) branch of a bistable $j(U)$ characteristic. For $\det \mathcal{A} > 0$ we find a pair of complex-conjugate eigenvalues $\lambda_{1/2}$ with a positive real part above the critical value C_{crit}. This depicts an oscillatory instability (an unstable focus) of the operating point via a Hopf bifurcation, which leads to self-generated limit-cycle oscillations if no other stable stationary point is reached in course of the temporal evolution.

The above considerations can easily be generalized (Schöll 1989) to the case in which there is not only one internal variable a but a set of N dynamic variables $q \equiv (q_1, \ldots, q_N)$ satisfying the nonlinear transport equations

$$\frac{dq}{dt} = F(q, U), \tag{3.41}$$

where $F \equiv (F_1, \ldots, F_N)$ is a set of nonlinear functions, cf. eq. (1.5) in Section 1.2. Let us note that the dynamic circuit equation (3.23) is equivalent to the dielectric relaxation equation

$$\epsilon \frac{d\mathcal{E}}{dt} = J - \sigma(q, \mathcal{E})\mathcal{E} \tag{3.42}$$

in the special case of a uniform electric field $\mathcal{E} = U/L$. This follows by setting $C = \epsilon A/L$ equal to the intrinsic geometrical capacitance, $(U_0 - U)/(RA) = J$, and $j(q, U) = \sigma(q, U)\mathcal{E}$ with conductivity σ. Then the differential relaxation time of the circuit becomes Maxwell's differential dielectric relaxation time

$$\tau_U^{\text{diff}} = \frac{\epsilon}{\sigma_{\text{diff}} + L/RA} \tag{3.43}$$

with differential conductivity $\sigma_{\text{diff}} = dj/d\mathcal{E}$ (Schöll 1989).

The linear stability analysis of the fixed points of the system (3.41) and (3.23) proceeds as in the case $N = 1$ and yields the eigenvalue equation

$$\lambda \begin{pmatrix} \delta q \\ \delta U \end{pmatrix} = \mathcal{A} \begin{pmatrix} \delta q \\ \delta U \end{pmatrix}, \tag{3.44}$$

where the internal part of the Jacobian matrix is given by the $N \times N$ block matrix

$$\mathcal{A}_{ij} = \tilde{A}_{ij} \equiv \left. \frac{\partial F_i}{\partial q_j} \right|_* \quad \text{for} \quad i, j \le N \tag{3.45}$$

and the $(N + 1)$th column and row of $\mathcal{A}$ are defined as before. Using an expansion of $\det \mathcal{A}$ in terms of $\det \tilde{A}$, one obtains the important relation

$$\det \mathcal{A} = \prod_{i=1}^{N+1} \lambda_i = -\frac{\det \tilde{A}}{\tau_U^{\text{diff}}} \tag{3.46}$$

which generalizes eq. (3.33). Again, $\det \mathcal{A}$ factorizes into a part that is determined by the transport equations (3.41) and the inverse differential relaxation time of the circuit.

The eigenvalues $\lambda_1, \ldots, \lambda_{N+1}$ of the Jacobian matrix $\mathcal{A}$ are given by the roots of the characteristic polynomial

$$\sum_{i=1}^{N+1} g_i \lambda^i = 0 \tag{3.47}$$

with

$$g_0 = (-1)^{N+1} \det \mathcal{A}, \quad g_N = -\operatorname{tr} \mathcal{A}, \quad g_{N+1} = 1. \tag{3.48}$$

For $N = 2$ the conditions for a Hopf bifurcation can be easily evaluated:

$$g_1 g_2 = g_0, \quad g_0 > 0, \quad g_1 > 0, \quad g_2 > 0. \tag{3.49}$$

One finds from eqs. (3.49)

$$g_0 = \det \tilde{A} \left(\frac{dI}{dU} + \frac{1}{R} \right) \frac{1}{C}, \tag{3.50}$$

$$g_2 = -\operatorname{tr} \tilde{A} + \frac{A}{C} j_U - \frac{1}{RC}. \tag{3.51}$$

An inspection of (3.51) shows that again *negative* differential conductivity is *not* necessary for the Hopf bifurcation conditions (3.49). The condition $g_0 > 0$ may be satisfied by

$$\frac{dI}{dU} + \frac{1}{R} < 0 \quad \text{and} \quad \det \tilde{A} < 0 \tag{3.52}$$

(a saddle-point instability in the subspace of internal variables $\boldsymbol{q}$) or

$$\frac{dI}{dU} + \frac{1}{R} > 0 \quad \text{and} \quad \det \tilde{A} > 0 \tag{3.53}$$

(a node or focus in $\boldsymbol{q}$-space). Physical examples of three-variable models that exhibit Hopf bifurcations on current–voltage characteristics with negative (Schöll 1986b) or positive (Hüpper et al. 1989) differential conductivity have been provided.

The existence of limit-cycle oscillations is a general precondition for the occurrence of further bifurcations that lead to more complicated, quasiperiodic or chaotic oscillations. Of course, this can be established only by numerical integration of the full nonlinear dynamic equations. An important conclusion of this subsection is that *oscillatory instabilities* may occur not only in regimes of negative differential conductivity but also *in regimes of positive differential conductivity*, i.e. on monotonic current–voltage characteristics.

3.2.4 Application to S- and Z-type current–voltage characteristics

The general considerations following from eqs. (3.31) and (3.32) can easily be applied to various devices. As an example we demonstrate the consequences for elements with S-shaped and Z-shaped current–voltage characteristics.

First, let us consider a device exhibiting an S-shaped characteristic as sketched in Fig. 3.2(a), such that the differential conductance $dj(a^*(U), U)/dU$ is positive on the upper and the lower branch and negative on the middle branch. Then $\det A$ is always positive on the upper and lower branches, while its sign on the middle branch depends on the slope of the load line. If the resistance is large enough that the (negative) slope of the load line is larger than the (negative) differential conductance (which corresponds effectively to current-controlled conditions), the determinant is positive and the middle branch can be stable. The second stability condition $\text{tr}\,A < 0$ is satisfied unless the device is operated in the middle branch and $C > C_{\text{crit}}$ holds. Current filamentation may occur if the device is operated on the middle branch and its spatial width is larger than L_{crit}, as will be discussed in Section 3.3.2 below.

In conclusion, the middle branch is stable for current-controlled conditions, a small total capacitance, and a small sample width. If the capacitance is increased self-generated oscillations are likely to occur, if the device does not jump to another fixed point. (The latter case seems to occur in experiments (Wu *et al.* 1992) with a quantum-dot structure (Wacker 1994).) For a large sample width and sufficiently small capacitance, stable current filaments should form. They constitute a new branch in the $I(U)$ characteristic, which is usually not directly connected with the upper and lower spatially homogeneous branches.

For devices with Z-shaped characteristics as shown in Fig. 3.2(b) the differential conductance $dj(a^*(U), U)/dU$ becomes negative on the upper and the lower branch of the characteristic near U_r and U_l, respectively. In this regime the determinant becomes negative for a large load resistance (a flat load line). Thus, these regimes become unstable for effectively current-controlled conditions.

If we assume that $dj(a^*(U), U)/dU > 0$ on the middle branch (where $f_a > 0$) eq. (3.32) indicates that the determinant is always negative there for $R, C > 0$. Thus, for a conventional load resistance and capacitance the middle branch of a Z-shaped characteristic can not be stabilized. On the other hand, eq. (3.32) shows that the determinant can be positive if either $C > 0$ and $-dj/dU > 1/(RA)$, or $C < 0$ and $-dj/dU < 1/(RA)$ holds. The condition $\text{tr}\,A < 0$ yields

$$\frac{1}{RC} > -\frac{Aj_u}{C} + f_a = -\frac{A}{C}\left(\frac{dj}{dU} - j_a\frac{da^*}{dU}\right) + f_a. \tag{3.54}$$

On the middle branch we have $f_a > 0$ and usually $da^*/dU < 0$ like in Fig. 3.3(a). Furthermore, $j_a < 0$ holds since $j(a, U)$ is monotonically decreasing in a in the Z case as mentioned in Section 3.2.1. Thus, we find

$$-\frac{A}{C}\left(\frac{dj}{dU} - j_a\frac{da^*}{dU}\right) + f_a > -\frac{A}{C}\frac{dj}{dU} \tag{3.55}$$

and the condition (3.54) is incompatible with the condition for a positive determinant (3.32) in case of $C > 0$. Therefore we obtain the inequalities

$$C < 0, \tag{3.56}$$

$$-A\frac{dj}{dU} < \frac{1}{R} < -A\frac{dj}{dU} + Aj_a\frac{da^*}{dU} + Cf_a \qquad (3.57)$$

as necessary and sufficient conditions for the stability of the middle branch. Considering that C is the sum of an internal sample capacitance $C_s > 0$ and an external capacitance C_{ext} of the circuit gives the condition

$$-C_s > C_{\text{ext}} > -A\frac{j_a}{f_a}\frac{da^*}{dU} - C_s. \qquad (3.58)$$

Indeed, such an external circuit with both $C_{\text{ext}} < 0$ and $R < 0$ has been realized experimentally (Martin *et al.* 1994), and it was possible to stabilize the middle branch of the Z-shaped characteristic of the DBRT.

Regarding spatially inhomogeneous fluctuations, the middle branch should become unstable for large sample cross-sections. As discussed above, in this case either an additional filamentary branch or some oscillatory behavior should appear in the device. In conclusion, the stabilization of the middle branch of a Z-shaped characteristic is possible only for negative capacitance C_{ext}, negative effective load resistance R, and a sufficiently small sample cross-sections.

In the preceding subsections we were able to derive the conditions for stability for various cases of bistability. Although the allowed range for the slope of the load line has been obtained directly from the geometrical considerations sketched in Fig. 3.5, we have made only qualitative statements about the restrictions on the sample cross-section and external capacitance. In order to obtain quantitative conclusions for these quantities, one should specify the functions $j(a, U)$ and $f(a, U)$. This has been done in Section 2.4, for instance, for the double-barrier resonant-tunneling structure (DBRT) and the heterostructure hot-electron diode (HHED). There we explicitly obtained the current density

$$j(a, U) = \frac{L_1 j_1 + L_2 j_2}{L_1 + L_2}$$

and the nonlinear kinetic function $f(a, U) = j_1 - j_2$ with appropriate current densities j_1 and j_2 in the emitting and collecting layers of widths L_1 and L_2. The difference in that the HHED exhibits an S-shaped and the DBRTD an Z-shaped characteristic is reflected only in the different forms of the functions $j_1(a, U)$ and $j_2(a, U)$.

Let us conclude this section with some final remarks. The stability of S-shaped elements has often been discussed regarding circuits with inductive and capacitive elements (Shaw and Gastman 1971). These approaches assume that there is a specific relation $j(U)$ between the current and the voltage in the element, which can be multivalued, but nevertheless does exist. Thus, they neglect any internal temporal degree of freedom. The oscillatory behavior is caused by the $\mathcal{L}$–C resonator and is amplified by the negative differential conductivity of the device. Similarly, current filamentation is treated by assuming that the voltage along the sample is varied by inductive effects within the device (see Chapter 7.7 in Shaw *et al.* (1992)). The use of the relation $j(U)$ implies that the internal degree of freedom is always at its stationary value $a^*(U)$. This has two severe implications. First, the operating point is required to be stable with respect to the internal dynamics. Second, the internal dynamics must relax to this stable point on a time scale that is much faster than the relevant time scales considered in the problem. This demonstrates that such an approach using $j(U)$ can be valid only if the stability with respect to the internal dynamics has already been proved. As shown above, the properties associated with this internal degree of freedom exhibit the well-known typical features like self-sustained oscillations and current filamentation. This indicates that the combination of external circuit conditions with the internal dynamics is the essential reason for the instabilities found experimentally.

In case of the DBRT the dynamics is often discussed in terms of quantum inductances (Brown *et al.* 1989) whose origin seems not to be clearly understood. This approach has been used by Martin *et al.* (1994) to derive conditions for stabilizing the middle branch of the DBRTD that are very similar to our conditions. Noteborn *et al.* (1993) have shown that the results of the phenomenological quantum-conductance approach correspond to a microscopic dynamics that has just the structure presented in this section. This indicates that the quantum inductance is nothing other than a kind of equivalent circuit for the internal dynamics in a certain range of operation. Indeed, every inductance $\mathcal{L}$ adds an additional temporal degree of freedom to the circuit through the differential equation $dI/dt = U/\mathcal{L}$.

In the major part of this section we have considered the simplest case of a single relevant dynamic variable and a circuit containing only capacitive elements. If there is more than one relevant internal degree of freedom the situation becomes more complicated and chaotic behavior may occur in the homogeneous system, too (Schöll 1989). Another important feature is the inclusion of noise. This can lead to a suppression of the metastable states near the bifurcations so that even the full lines of the characteristics shown in Fig. 3.2 will not be observed in full length.

In conclusion, we have shown that a simple approach in the framework of a single internal degree of freedom can reproduce many features seen experimentally. This shows that the theoretical modeling of such devices must include a dynamic treatment of at least one internal variable.

3.3   Current filaments and field domains

The stability analysis in the preceding section has been restricted to the case of small *time-dependent* fluctuations in the steady state. As a result we have obtained conditions for bifurcations of fixed points and limit-cycle oscillations. Now we will consider the response of the system to small *space- and time*-dependent fluctuations of the homogeneous steady state. Under certain conditions, in particular when negative differential conductivity occurs, these fluctuations can grow. Following the general linear stability analysis in Chapter 1, Section 1.2.3, we may expect the bifurcation of spatial patterns, e.g. Turing structures and standing or traveling waves. In case of charge transport, the spatial patterns may be modulated either in the direction of current flow or perpendicular to that direction. This implies an important distinction between domain-like and filamentary instabilities (Fig. 3.6). In general, a variety of coupled electromagnetic and transport modes exists, depending upon the various orientations of the electric-field fluctuation as well as the wave vector relative to the uniform electric field. A detailed analysis of these linear modes and possible bifurcations has been performed for the case of generation–recombination instabilities (Schöll 1987), and some of the results will be summarized in Chapter 4 for the model of low-temperature impurity-impact ionization. Here we shall give only simple general arguments for the bifurcation of domains and current filaments in case of NNDC and SNDC, respectively. Although the analogy of field domains and current filaments, based upon the duality of NNDC and SNDC characteristics, respectively, was emphasized in the early investigations (Ridley 1963, Volkov and Kogan 1969, Bonch-Bruevich *et al.* 1975), there remain essential differences since the formation of field domains can already be understood effectively within a one-dimensional treatment, whereas an analysis of current filamentation requires the consideration of at least one transverse direction in addition to the longitudinal coordinate.

3.3.1   Domain instability

Some of the earliest studies of such NNDC and SNDC instabilities in semiconductors were published by Böer *et al.* for CdS (Böer *et al.* 1961a, 1961b) and Si (Böer and Williges 1961). They discovered moving high-field domains in single crystals of CdS (Böer 1959), even before the observation of Gunn-domain oscillations in GaAs had been reported (Gunn 1963, 1964), and developed an electrooptic method to measure these field inhomogeneities (Böer *et al.* 1959) based on the first experimental evidence of the Franz–Keldysh effect. Well-known mechanisms for field-domain formation

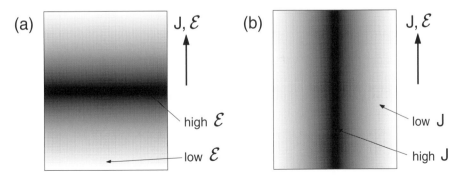

Figure 3.6. Sketches of (a) a high-field domain, and (b) a high-current filament, where j is the current density and $\mathcal{E}$ is the electric field.

in bulk semiconductors are provided by field-assisted trapping in CdS (Böer 1959) and semi-insulating GaAs (Northrup *et al.* 1964, Samuilov 1995, Piazza *et al.* 1997), intervalley transfer (Ridley and Watkins 1961, Gunn 1963, Kroemer 1964), and impurity-impact-ionization breakdown (Kahn *et al.* 1991, 1992*b*, 1992*a*, Bonilla *et al.* 1994*a*, 1997*b*, Bergmann *et al.* 1996) in bulk semiconductors. More recently, field domains in low-dimensional semiconductor structures have also been studied. Bragg scattering in superlattices has been proposed as a mechanism for traveling domains in superlattices (Büttiker and Thomas 1977, 1978). Periodically or chaotically moving domains induced by real-space transfer of electrons in modulation-doped semiconductor heterostructures have been predicted theoretically (Döttling and Schöll 1993, 1994, Borisov *et al.* 1999). Vertical high-field transport in GaAs/AlAs superlattices has been shown experimentally (Esaki and Chang 1974, Choi *et al.* 1987, Helgesen and Finstad 1990, Grahn *et al.* 1991, Zhang *et al.* 1994, Merlin *et al.* 1995, Kwok *et al.* 1995) to be associated with NNDC and field-domain formation induced by resonant tunneling between adjacent quantum wells. While the existence of static domain states and of multistable current–voltage characteristics consisting in many branches has been explained for structurally perfect samples by phenomenological (Korotkov *et al.* 1993, Miller and Laikhtman 1994, Prengel *et al.* 1994, Bonilla *et al.* 1994*b*) or microscopic (Wacker 1998, Wacker and Jauho 1998*b*, Cao and Lei 1999) models, it has also been found in computer simulations that small amounts of imperfections and disorder, such as fluctuations of doping, or of well or barrier widths, sensitively influence the domain states (Schwarz 1995, Wacker *et al.* 1995*c*, Schöll *et al.* 1996, 1998*b*). If an ac driving bias is applied, moving domains exhibit particularly complex spatio-temporal dynamics, as simulations for Gunn diodes (Nakamura 1989, Mosekilde *et al.* 1990, 1993, Zongfu and Benkum 1991), modulation-doped heterostructures (Döttling and Schöll 1994), low-temperature impurity breakdown in p-Ge (Bergmann *et al.* 1996), and sequential resonant tunneling in superlattices (Bulashenko and Bonilla 1995, Bulashenko *et al.* 1996, Bonilla *et al.* 1996) show.

The formation of field domains in a sample with NNDC can be understood from a simple argument. By way of example, we consider the case in which the NDC results from negative differential mobility, as in the Gunn effect (Shaw *et al.* 1979), so that there exists a range of fields for which $dv/d\mathcal{E} < 0$, where v is the carrier drift velocity. To illustrate the response of an NNDC element to a charge fluctuation, consider a uniform field with a domain of increased field in the center of the sample as shown in Fig. 3.7(a). The charge distribution that produces this field is shown in Fig. 3.7(b). There is a net accumulation of charge on the left-hand side of the domain and a depletion layer on the right. If we consider *positively* charged carriers, the carriers and hence the domain will be moving to the right. Assuming that the field within the domain is within the NDC range and the field outside the domain is within the Ohmic range, but close to the field of peak velocity, then it is clear that the field fluctuation will initially grow with time. This happens because the higher upstream field in the center of the domain results in carriers moving more slowly than those at the edges, where the field is lower. Charge will therefore be depleted from the right-hand (leading) edge of the domain and accumulate at the left-hand (trailing) edge. This charge will add to what is already there, increasing the field in the domain. If the element is in a resistive circuit, the increasing voltage across the domain will decrease the current in the circuit and lower the field outside the domain. The field in the domain will continue to grow in the interior of the domain and drop outside. Thus an instability of the uniform field with respect to domain formation results.

The constitutive equations in one spatial dimension are the carrier continuity equation, neglecting recombination processes,

$$\frac{\partial n(x,t)}{\partial t} + \frac{\partial}{\partial x}\left(n(x,t)v(\mathcal{E}) - D\frac{\partial n}{\partial x}\right) = 0 \qquad (3.59)$$

and Gauss' Law,

$$\frac{\epsilon}{e}\frac{\partial \mathcal{E}(x,t)}{\partial x} = n(x,t) - N_0, \qquad (3.60)$$

where eN_0 is the negative uniform background charge density, and D is the diffusion constant. Equations (3.59) and (3.60) can be combined and integrated to give the current equation

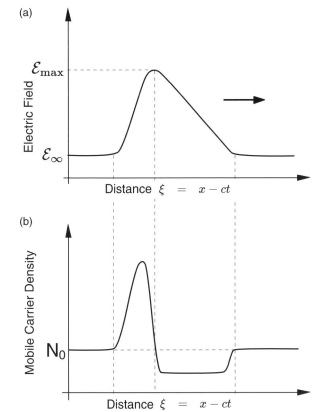

(a)

Electric Field

$\mathcal{E}_{max}$

$\mathcal{E}_\infty$

Distance $\xi = x - ct$

(b)

Mobile Carrier Density

N_0

Distance $\xi = x - ct$

Figure 3.7. (a) The electric-field profile, and (b) the carrier-density profile of a high-field Gunn domain. The domain is moving with velocity v in the positive x-direction (schematic).

$$\epsilon \frac{\partial \mathcal{E}(x,t)}{\partial t} + e\left(n(x,t)v(\mathcal{E}) - D\frac{\partial n}{\partial x}\right) = J(t) \tag{3.61}$$

with the external current density $J(t)$. Substituting eq. (3.60) into eq. (3.61) yields

$$\epsilon\left(\frac{\partial \mathcal{E}(x,t)}{\partial t} + v(\mathcal{E})\frac{\partial \mathcal{E}}{\partial x} - D\frac{\partial^2 \mathcal{E}}{\partial x^2}\right) + eN_0 v(\mathcal{E}) = J(t). \tag{3.62}$$

The stability of the uniform steady state $\mathcal{E}_0$ can be tested by linearizing (3.62) around $\mathcal{E}_0$ for small space- and time-dependent fluctuations

$$\delta\mathcal{E}(x,t) = e^{\lambda t}e^{ikx}. \tag{3.63}$$

Substituting this *Ansatz* into (3.62) for fixed external current density J yields the dispersion relation

$$\lambda = -\frac{1}{\tau_M} - Dk^2 - ikv(\mathcal{E}_0) \tag{3.64}$$

with the effective differential dielectric relaxation time

$$\tau_M \equiv \left(\frac{e}{\epsilon}N_0\frac{dv}{d\mathcal{E}}\right)^{-1}.$$

Hence for negative differential mobility $(dv/d\mathcal{E} < 0)$, τ_M is negative, and small fluctuations

$$\delta\mathcal{E}(x,t) = \exp\left[-t\left(\frac{1}{\tau_M} + Dk^2\right)\right]e^{ik(x-v_0t)} \tag{3.65}$$

grow in time for $k^2 < (D|\tau_M|)^{-1}$, i.e. for short wavelengths. Equation (3.65) describes an undamped traveling wave propagating in the positive x-direction with velocity $v_0 = v(\mathcal{E}_0)$. This instability leads to the bifurcation of moving field domains from the uniform steady state, which depends, however, sensitively upon the sample's length, the boundary conditions at the cathode contact, and the external circuit (Shaw *et al.* 1979, 1992). A simple criterion for the onset of a moving field domain can be derived from (3.65) by neglecting diffusion and assuming that the linearization remains valid throughout its transit across the element of length L. The growth at the end of the transit, i.e. after the transit time $t = L/v_0$, is then given by

$$G = \frac{\delta\mathcal{E}(L,t)}{\delta\mathcal{E}(0,0)} = \exp\left(\frac{L}{v_0|\tau_M|}\right) = \exp\left(\frac{N_0 Le|dv/d\mathcal{E}|}{\epsilon v_0}\right). \tag{3.66}$$

Substantial growth, i.e. a moving domain instability, occurs if the exponent in (3.66) is larger than unity, or

$$N_0 L > \frac{\epsilon v_0}{e|dv/d\mathcal{E}|}. \tag{3.67}$$

This is the N_0L-product stability criterion for the Gunn effect derived by McCumber and Chynoweth (1966) and Kroemer (1968).

The fully developed field domains must be calculated from the full nonlinear transport equation (3.62), subject to appropriate boundary conditions. A simple visualization of the moving-domain solution can be obtained by transforming (3.62) to the co-moving frame $\xi = x - v_D t$, where v_D is the domain velocity:

$$\epsilon \left((v(\mathcal{E}) - v_D) \frac{\partial \mathcal{E}}{\partial \xi} - D \frac{\partial^2 \mathcal{E}}{\partial \xi^2} \right) + e N_0 (v(\mathcal{E}) - v_\infty) = 0. \tag{3.68}$$

Here $J = e N_0 v_\infty$ with $v_\infty = v(\mathcal{E}_\infty)$ is the current density in the neutral material outside the domain. The result can be written in the form of a "dynamic system" of two first-order differential equations for $\mathcal{E}$ and

$$n = \frac{\epsilon}{e} \frac{\partial \mathcal{E}}{\partial x} + N_0,$$

where the parameter corresponding to time is now given by ξ. The topology of the solutions may be conveniently discussed in terms of a phase-portrait analysis (Shaw et al. 1992), which gives qualitative insight into the nature of the solutions and the effects of boundary conditions. Such methods were used extensively by Böer et al. (Böer and Quinn 1966, Böer and Voss 1968b, Böer and Döhler 1969) in the early investigations of Gunn domains and recombination domains. It turns out that the triangular shape of the Gunn domain depicted in Fig. 3.7 can be understood on the basis of an equal-areas rule (Butcher 1965, Knight and Peterson 1967) that can be derived from (3.68) by multiplying by $(1/n) \partial \mathcal{E}/\partial \xi$ and integrating from $\xi = \pm \infty$ to the position of the peak field in the domain $\mathcal{E}_{max}$ (Fig. 3.8). It states that the two hatched areas B_+ and B_- must be equal, and thereby determines, for a given domain velocity $v_D = v(\mathcal{E}_\infty)$, the peak field $\mathcal{E}_{max}$ of the domain. It should be noted that the low-field state $\mathcal{E}_\infty$ corresponds to a homogeneous steady state (fixed point), but the peak field $\mathcal{E}_{max}$ does not. Therefore the second rising branch of the NNDC characteristic is not necessary for the existence of the domain: The Gunn domain does not involve phase coexistence of two bistable homogeneous states. Note that more elaborate mathematical analyses, including realistic boundary conditions, and a proof of the existence and stability of traveling domain solutions, have also been given (Higuera and Bonilla 1992, Bonilla et al. 1994c, 1997a).

3.3.2 Filamentary instability

In the regime of SNDC, current filaments were observed in some early studies of electric and thermal breakdown (Böer 1961), for which the analogy between current filamentation in semiconductors and plasma discharges was pointed out, in p–i–n diodes (Barnett 1970, Jäger et al. 1986), in amorphous chalcogenide films (Bosnell and Thomas 1972, Petersen and Adler 1976, Adler et al. 1980), in low-temperature impurity breakdown (Aoki and Yamamoto 1983, Mayer et al. 1987, Brandl et al. 1989), and in p–n–p–i–n diodes (Niedernostheide et al. 1993). The analogy between semiconductors and gas discharges was later revived in investigations of pattern formation in gas-discharge systems (Purwins *et al.* 1987) and in detailed simulations of the spatio-temporal dynamics of current-filament formation (Gaa *et al.* 1996*a*, Schöll 1998*a*). Much theoretical and numerical work has been done following the seminal early papers by Ridley (1963), Volkov and Kogan (1967), Kogan (1968), Bass *et al.* (1970), Varlamov and Osipov (1970), and Kerner and Osipov (1976), for reviews see Schöll (1987), Shaw *et al.* (1992), Kerner and Osipov (1994), and Schöll *et al.* (1998*a*). Here we derive only a simple argument for current filamentation in a reaction–diffusion system of the type which we introduced in Chapter 1 and in Section 2.3.5, which was also the basis for the discussion of oscillatory instabilities in Section 3.2. More detailed results on current filamentation for specific model systems will be presented in Chapters 4, 5, and 7.

In contrast to the analysis in Section 3.2, now we have to consider the possibility that the internal variable $a(x, y, t)$ might be spatially dependent in the (x, y)-plane perpendicular to the direction of the applied voltage, due to some diffusive coupling, as was derived for various model systems in Sections 2.4.5–2.4.8. Then the dynamics of a is given by

$$\frac{\partial a(x, y, t)}{\partial t} = f(a, U) + D\left(\frac{\partial^2 a}{\partial x^2} + \frac{\partial^2 a}{\partial y^2}\right). \tag{3.69}$$

We restrict ourselves to a rectangular sample of size $L_x \times L_y$ and assume Neumann boundary conditions $\partial a/\partial x = 0$ at $x = 0, L_x$ and $\partial a/\partial y = 0$ at $y = 0, L_y$. Then the linear stability of the homogeneous state (a^*, U^*) is governed by the modes

$$\delta a(x, y, t) = \sum_{n,m=0}^{\infty} a_{n,m}(t) \cos\left(\frac{n\pi}{L_x}x\right) \cos\left(\frac{m\pi}{L_y}y\right) \tag{3.70}$$

with wave vectors $k_{x,n} = n\pi/L_x$ and $k_{y,n} = n\pi/L_y$. For constant voltage the amplitudes of the modes develop according to

$$\frac{da_{n,m}(t)}{dt} = \left[f_a(a^*, U^*) - D\left(\frac{(n\pi)^2}{L_x^2} + \frac{(m\pi)^2}{L_y^2}\right)\right]a_{n,m}. \tag{3.71}$$

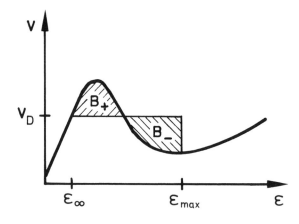

Figure 3.8. The equal-areas rule for Gunn domains. The hatched areas B_+ and B_- must be equal. v is the electron drift velocity, v_D is the domain velocity (schematic).

This leads to the dispersion relation

$$\lambda = f_a - D(k_{x,n}^2 + k_{y,m}^2). \tag{3.72}$$

A similar analysis holds for a cylindrical geometry, for which the linear modes are mth-order Bessel functions of the radial coordinate $J_m(kr)e^{im\phi}$, and the lowest mode $m = 0$ is cylindrically symmetric (Schöll 1987).

On branches that are stable for voltage-controlled conditions we have $\partial f/\partial a \equiv f_a < 0$ and all spatial fluctuations are damped out. For $f_a > 0$ (dotted lines in Fig. 3.2) spatial fluctuations with small wave vectors $k < \sqrt{f_a/D}$ grow in time. Since they do not change the total current $I = \int_A dx\, dy\, j(a, U)$ they are not impeded by the circuit condition (3.23), in contrast to homogeneous fluctuations. Thus, an instability occurs if at least one of the sample dimensions L_x and L_y is larger than $L_{crit} = \sqrt{D\pi^2/f_a}$, i.e. the $n = 1$ or $m = 1$ mode grows. (Note that the precise value of L_{crit} depends on the boundary conditions but the general dependency on the length is robust.) These inhomogeneous fluctuations can lead to the bifurcation of stable current filaments, especially if the operating point is stable against homogeneous fluctuations. More precisely, depending on the sample geometry, either plane current layers or cylindrical current filaments are formed. Such stable current filaments are characterized by the spatial coexistence of two stable homogeneous steady states (fixed points) at a given voltage: A high-conductivity state embedded in a low-conductivity state. The conditions of such phase coexistence can be formulated in terms of equal-areas rules, as discussed in Section 3.1, eq. (3.7), which determines the coexistence voltage $U = U_{co}$. Thus, unlike in case of the Gunn domains discussed in Section 3.3.1, the filament profile has a flat top (see Fig. 3.1), and requires a bistable system.

The filaments constitute another branch in the current–voltage characteristic of the sample which should exhibit *negative* differential conductance (Schöll 1987, Wacker 1993). If the capacitance is increased these filaments should exhibit an oscillatory instability similar to the homogeneous mode discussed in the previous section. If the bifurcation is supercritical (Schöll and Drasdo 1990), breathing boundaries of the filament are likely to occur. In the subcritical case, spiking filaments can appear (Wacker and Schöll 1994b). This indicates that current or voltage oscillations may occur even if there is no oscillatory instability of the homogeneous state. More complex spatio-temporal behavior can be found if two or more competing diffusive variables are involved (Niedernostheide *et al.* 1992b, Wacker and Schöll 1994b). Extensions of the linear mode analysis to the case of several internal variables and the coupling with electromagnetic modes have been given elsewhere (Schöll 1987).

3.4 Conditions for stability of current filaments

In this section we study the stability of stationary current filaments in a bistable semiconductor system in the presence of global coupling given by an external circuit

(Alekseev *et al.* 1998). The system is described by a reaction–diffusion model on a two-dimensional spatial domain with Neumann boundary conditions. We will prove generally for the voltage-driven regime that, in a convex domain, any filament has at least one unstable linear eigenmode. Introducing a global coupling may either eliminate the unstable mode with the largest increment or induce oscillatory instabilities. Filaments with negative differential conductance can be stabilized by strong global coupling. Stabilization of filaments with positive differential conductance can be achieved only by an active external circuit with negative resistance and capacitance. We present analytic arguments and numerical simulations suggesting that the boundary of the domain always attracts current filaments. Our numerical results also show that seed inhomogeneities may pin current filaments in the centers of sufficiently large domains. The competition between the attractive boundary and pinning by seed inhomogeneities will be illustrated by simulations for the specific reaction–diffusion model of Section 2.4.5.

Current filamentation in semiconductor systems with bistable current–voltage characteristics is one of the simplest nontrivial examples of pattern formation in active spatially extended active media. Current filaments are characterized by a current-density profile that varies in the plane perpendicular to the current flow, reflecting spatial coexistence of the two stable phases. According to the concept of pattern formation in activator–inhibitor systems, current filamentation may occur due to the competition between an internal mechanism of activation, which provides negative differential conductivity (NDC) of the semiconductor element (Ridley 1963), and an external mechanism of inhibition given by the constraint related to the external circuit. The theoretical description of stationary current filaments, which had originally been developed for semiconductors with an electron-overheating instability (Volkov and Kogan 1967, 1969, Bass *et al.* 1970), was later advanced for other semiconductor systems exhibiting S-shaped negative differential conductivity (SNDC) (Varlamov and Osipov 1970, Varlamov *et al.* 1970, Osipov and Kholodnov 1971, Bonch-Bruevich *et al.* 1975, Schöll 1987, Jäger *et al.* 1986, Niedernostheide *et al.* 1992*a*, Gorbatyuk and Rodin 1992*a*, 1992*b*, Shaw *et al.* 1992). It was also discovered that current filaments can exhibit temporal instabilities that lead to the formation of traveling or rocking filaments (Niedernostheide *et al.* 1992*b*), and small-amplitude or relaxation-type oscillations, known as breathing (Schöll and Drasdo 1990, Kunz and Schöll 1992) and spiking (Wacker and Schöll 1994*b*, Bose *et al.* 1994, Niedernostheide *et al.* 1996*b*) of a current filament, respectively. In many cases the problem can be treated in terms of a reduced one-dimensional reaction–diffusion equation, and an integro-differential equation corresponding to a global constraint given by the external circuit (Bass *et al.* 1970, Wacker and Schöll 1994*b*, Schimansky-Geier *et al.* 1991). Since this type of model equation takes into account only one lateral (transversal) degree of freedom for the current-density distribution, such a theory is adequate only for samples with effectively one-dimensional strip-like geometries such that the z-coordinate parallel to the current flow has been eliminated, and the second transversal dimension is so short that spatial instabilities can not develop. The description of real three-dimensional samples requires models on two-dimensional lateral spatial domains. Most studies of two-dimensional models have either assumed axial symmetry of the current distribution (Bonch-Bruevich *et al.* 1975, Schöll 1987, Schöll and Drasdo 1990, Kunz and Schöll 1992), or have considered the longitudinal (z) and one transversal (x) coordinate only (Gaa *et al.* 1996a, Kunihiro *et al.* 1997), which is appropriate for thin semiconductor films. Recent numerical simulations of filament dynamics on quadratic domains (Niedernostheide *et al.* 1997) have been performed for an activator–inhibitor model that takes into account both local inhibition processes inside the semiconductor structure and global inhibition due to the external circuit.

The purpose of this section is to develop analytic and numerical results for samples for which the two transversal dimensions are of comparable size. We concentrate on the stability of stationary current filaments in bistable semiconductor systems that do not experience local internal inhibition and present a general approach for arbitrary convex two-dimensional spatial domains with proper account taken for global coupling. A schematic sample geometry is shown in Fig. 3.9. An external circuit with load resistor R and capacitor C is attached to the top (K) and bottom (A) contacts of the sample.

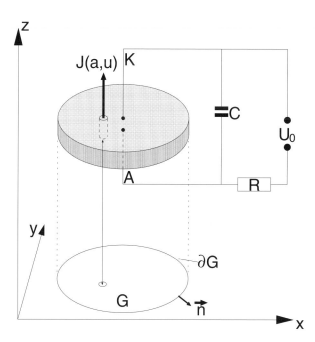

Figure 3.9. A schematic sketch of the semiconductor structure (shaded) and the external circuit attached to its cathode (K) and anode (A) contacts. The direction of current flow $j(a, u)$ and a cylindrical current filament are schematically indicated. The projection onto the (x, y)-plane shows the two-dimensional domain G under study; n denotes the normal vector at the boundary ∂G.

We assume that the internal state of the semiconductor system can be characterized by a single spatially distributed variable $a(x, y, t)$ that corresponds to the internal degree of freedom relevant to the bistability. As discussed in Section 2.4, the physical meaning of this variable might be the electron temperature (Volkov and Kogan 1967, 1969), concentration of excess carriers (Schöll 1987), bias of one of the p–n junctions in a thyristor (Varlamov and Osipov 1970, Gorbatyuk and Rodin 1992a), interface charge density of a heterostructure hot-electron diode (Wacker and Schöll 1994b), etc., depending upon the specific transport mechanism. The variable $a(x, y, t)$ and the voltage $u(t)$ across the sample determine the current density $j(a, u)$ in a cross-section of the device. For various semiconductor systems (Volkov and Kogan 1967, 1969, Bass et al. 1970, Varlamov and Osipov 1970, Varlamov et al. 1970, Jäger et al. 1986, Gorbatyuk and Rodin 1992a, Niedernostheide et al. 1992b, Wacker and Schöll 1994b, 1995, Meixner et al. 1997b) the spatio-temporal dynamics of $a(x, y, t)$ is described by a reaction–diffusion equation of the type (2.47):

$$\tau_a \frac{\partial a(x, y, t)}{\partial t} = f(a, u) + l_a^2 \, \Delta a(x, y, t). \tag{3.73}$$

Here τ_a and l_a characterize the relaxation time and transversal diffusion length, respectively. The local kinetic function $f(a, u)$ is a nonmonotonic function of a that, for fixed u in a certain range, has three zeros corresponding to high-conductivity, low-conductivity and negative-differential-conductivity (NDC) states (Fig. 3.10(a)). Since $\partial f / \partial a > 0$ for values of u in the NDC range, the variable a may be regarded as an activator. The functions $f(a, u)$ and $j(a, u)$ contain all necessary information about vertical transport along the z-direction in the structure shown in Fig. 3.9. Generally, in the steady state, the local dependency $a(u)$ is calculated from the null-isocline $f(a, u) = 0$ and inserted into $j(a, u)$ to find the local current-density–voltage characteristic $j(u) \equiv j(a(u), u)$, which can be S-shaped (Fig. 3.10(b)) or Z-shaped (Fig. 3.10(c)) (see Section 3.2 for a discussion of the general case and some examples). However, a particular type of nonlinearity is not crucial for the

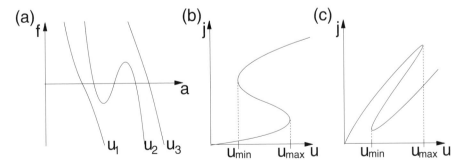

Figure 3.10. (a) The local kinetic function $f(a, u)$ for various values of u ($u_1 < u_{min}$, $u_{min} < u_2 < u_{max}$, and $u_3 > u_{max}$). In (b) and (c) S- and Z-shaped local current-density versus voltage characteristics $j(u)$, respectively (schematic).

present analysis. In our analytic considerations we will not specify the dependencies of $f(a, u)$ and $j(a, u)$, and will assume only the condition $\partial j / \partial u > 0$. In our numerical examples we will use the model functions that had originally been derived for the heterostructure hot-electron diode (HHED), see Section 2.4.5, but were shown to hold more generally for layered semiconductor structures (Wacker and Schöll 1994b):

$$f(a, u) = \frac{u - a}{(u - a)^2 + 1} - \mathcal{T}a, \qquad j(a, u) = u - a. \qquad (3.74)$$

These functions result in a Z-shaped $a(u)$-dependency and an S-shaped $j(u)$-dependency (Fig. 3.11). Note that dimensionless variables are used throughout.

The distribution $a(x, y, t)$ has to satisfy Neumann boundary conditions corresponding to a passive boundary with no flux:

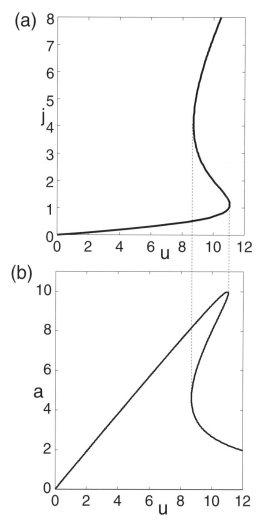

Figure 3.11. (a) The local current-density–voltage characteristic $j(u)$ and (b) the null-isocline $a(u)$ given by $f(a, u) = 0$ for the model functions eq. (3.74) with $\mathcal{T} = 0.05$ (after Alekseev *et al.* (1998)).

$$\left.\frac{\partial a(x, y, t)}{\partial n}\right|_{\partial G} = 0. \tag{3.75}$$

Here $\partial/\partial n$ is the normal derivative at the boundary ∂G of the domain G.

The temporal dynamics of the voltage $u(t)$ is described by Kirchhoff's equation for the external circuit (1.58):

$$\tau_u(u) \frac{du}{dt} = g(a, u), \qquad g(a, u) \equiv U_0 - u(t) - RA \langle J(a, u) \rangle,$$

$$\tau_u(u) \equiv RC(u), \qquad C(u) \equiv C_{\text{ext}}(u) + C_{\text{int}}(u), \tag{3.76}$$

where U_0 is the applied bias voltage, R is the load resistance, C_{int} is the internal differential capacitance of the sample, C_{ext} is the differential capacitance of the external circuit, A is the cross-section of the device, and the brackets $\langle \ \rangle$ denote the spatial average over the domain G. Since generally the derivatives $\partial g/\partial a$ and $\partial f/\partial u$ have different signs in the NDC range, the variable u acts as the inhibitor. (Specifically, $\partial g/\partial a > 0$ and $\partial f/\partial u < 0$ hold for the dependencies (3.74), whereas $\partial g/\partial a < 0$ and $\partial f/\partial u > 0$ hold for models of overheating instabilities (Volkov and Kogan 1967, 1969, Bass et al. 1970) and multilayered structures (Varlamov and Osipov 1970, Varlamov et al. 1970, Osipov and Kholodnov 1971, Gorbatyuk and Rodin 1992a, 1997).) Equation (3.76) represents a global constraint that is imposed on the dynamics of $a(x, y, t)$ and provides global coupling between distant parts of the cross-section.

Linearization of the equations (3.73) and (3.76) in the vicinity of a stationary solution $a_0(x, y)$, u_0 with respect to the perturbation

$$a(x, y, t) - a_0(x, y) = e^{\zeta t} \, \delta a(x, y), \tag{3.77}$$

$$u(t) - u_0 = e^{\zeta t} \, \delta u \tag{3.78}$$

with $\partial \delta a(x, y)/\partial n|_{\partial G} = 0$ yields

$$\tau_a \zeta \, \delta a = \widehat{H}_N \, \delta a + \frac{\partial f}{\partial u} \, \delta u, \qquad \widehat{H}_N \equiv l_a^2 \Delta + \Phi(x, y), \tag{3.79}$$

$$\Phi(x, y) \equiv \left.\frac{\partial f}{\partial a}\right|_{a_0, u_0},$$

$$\tau_u \zeta \, \delta u = -\left(1 + RA \left\langle \frac{\partial j}{\partial u} \right\rangle\right) \delta u - RA \left\langle \frac{\partial j}{\partial a} \, \delta a \right\rangle, \qquad \tau_u = \tau_u(u_0). \tag{3.80}$$

Here and further on the characteristic time τ_u and partial derivatives of f and j are computed for the steady state a_0, u_0. The stationary solution is stable if $\text{Re} \, \zeta < 0$ for all eigenvalues ζ.

The self-adjoint operator $\widehat{H}_N$ acts on the space of functions with Neumann boundary conditions. Its eigenfunctions Ψ_i and eigenvalues λ_i correspond to eigenmodes and eigenvalues of the voltage-driven system ($R = C = 0$): According to (3.79), (3.80) $\zeta = \lambda/\tau_a$ for $\delta u = 0$. So the voltage-driven system is stable if all $\lambda_i < 0$. In the presence of the global constraint, eq. (3.80) mixes the eigenmodes Ψ_i. In this case we should analyze both equations (3.79) and (3.80) simultaneously, but knowledge of the spectrum $\{\lambda_i\}$ remains the key to the stability problem.

Our aim is to establish a link between stability and such general features of a filamentary state as the location of the extremum and the sign of the differential conductivity.[1] First, in Section 3.4.1 we shall study the spectrum λ_i of the operator $\widehat{H}_N$ related to the spatial domain G of arbitrary shape, for voltage control. In Section 3.4.2 we concentrate on the effect of the global coupling on filament stability and establish a connection between the number of unstable modes (Re $\zeta > 0$) of the complete stability problem and the number of positive eigenvalues λ_i corresponding to the voltage-driven system. In Section 3.4.3 we use the results of the two previous subsections to formulate some general criteria for filament stability. Here we also present the results of numerical simulations and discuss the effect of embedded inhomogeneities of the semiconductor structure on filament stability.

3.4.1 Eigenvalues of a voltage-driven system

The eigenvalues of the operator $\widehat{H}_N$ may be ordered by decreasing size, $\lambda_1 > \lambda_2 > \cdots$. In this section we study the signs of the first two eigenvalues, λ_1 and λ_2. The first eigenmode Ψ_1 is strictly positive inside the domain G and hence represents the switching mode which leads to expansion or shrinking of a filament. The interpretation of the second eigenmode Ψ_2 depends on the particular solution $a_0(x, y)$. For filaments in the interior of the domain it corresponds to the shift of a filament along a certain direction.

To analyze the spectrum of the operator $\widehat{H}_N$ let us introduce the translation modes (these modes are also known as Goldstone modes (Schöll 1987))

$$\vartheta_G(x, y) \equiv \frac{da_0(x, y)}{d\xi}, \tag{3.81}$$

where $d/d\xi$ is the derivative along the direction ξ in the (x, y)-plane (Fig. 3.12). Boundary conditions for $\vartheta_G(x, y)$ can be obtained by direct calculation taking into account that $a_0(x, y)$ satisfies Neumann boundary conditions on ∂G:

$$\vartheta_G(l) = \frac{\partial a_0}{\partial l} \sin \alpha(l); \qquad \frac{\partial \vartheta_G(l)}{\partial n} = \frac{1}{R(l)} \frac{\partial a_0}{\partial l} \sin \alpha(l) + \frac{\partial^2 a_0}{\partial n^2} \cos \alpha(l). \tag{3.82}$$

Here l is a coordinate along the boundary, $\alpha(l)$ is an angle between the direction ξ and the external normal $\boldsymbol{n}$, and $R(l)$ is the radius of curvature of the domain boundary ∂G. The translation modes correspond to zero eigenvalue of the operator $\widehat{H}_{g(l)}$ which

---

[1] Whereas in a one-dimensional theory (Bonch-Bruevich *et al.* 1975, Schöll 1987) the complete classification of $a_0(x, y)$ follows from the phase-portrait analysis of eq. (3.73), in the two-dimensional case such a classification is not available even for domains of simple shape.

is defined by the same expression as that for $\widehat{H}_N$ (see eq. (3.79)) but acts on the space of functions with linear boundary conditions

$$\left(g(l)\Psi(x, y) + \frac{\partial \Psi(x, y)}{\partial n}\right)\Bigg|_{\partial G} = 0. \tag{3.83}$$

In particular, the function $g(l)$ can always be chosen in such a way that ϑ_G satisfies (3.83)[2]

$$g(l) = -\frac{1}{R(l)} - \left(\frac{\partial^2 a_0}{\partial n^2}\right)\left(\frac{\partial a_0}{\partial l}\right)^{-1} \cot \alpha(l). \tag{3.84}$$

The quadratic form $\widetilde{H}_{g(l)}$ of the operator $\widehat{H}_{g(l)}$ given by the scalar product $\langle \Psi \cdot \widehat{H}_{g(l)} \Psi \rangle$ can be represented as

$$\widetilde{H}_{g(l)}(\Psi) \equiv \int_G \Psi \widehat{H}_{g(l)} \Psi \, dx \, dy \tag{3.85}$$

$$= -l_a^2 \int_{\partial G} g(l)\Psi^2 \, dl + \int_G \left[-l_a^2 (\nabla \Psi)^2 + \Phi(x, y)\Psi^2\right] dx \, dy.$$

In order to determine the sign of the first eigenvalue of the operator $\widehat{H}_N$ we apply the variational principle (Reed and Simon 1972) which states that eigenfunctions of $\widehat{H}_N$ are extrema of its quadratic form

$$\widetilde{H}_N(\Psi) = \int_G \left[-l_a^2(\nabla \Psi)^2 + \Phi(x, y)\Psi^2\right] dx \, dy. \tag{3.86}$$

The eigenfunction Ψ_1 corresponding to the largest eigenvalue of $\widehat{H}_N$ provides the maximum of $\widetilde{H}_N$.

[2] Note that different directions ξ correspond to different operators $\widehat{H}_{g(l)}$.

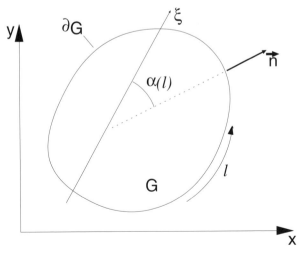

Figure 3.12. The domain G (schematic). The direction ξ corresponds to the translation mode ϑ_G, l is a coordinate along the boundary ∂G, and n is the external normal to the boundary.

Since the variation of the quadratic form (3.85) includes the boundary term $\int_{\partial G}(\partial \Psi/\partial n)\,\delta \Psi\,dl$ which vanishes only if $\partial \Psi/\partial n = 0$, $\tilde{H}_N(\Psi)$ automatically imposes Neumann boundary conditions on its extrema. Therefore, if we present any function Ψ for which $\tilde{H}_N(\Psi) > 0$, then the operator $\hat{H}_N$ has at least one positive eigenvalue. Since ϑ_G is a zero mode of $\hat{H}_{g(l)}$ ($\tilde{H}_{g(l)}(\vartheta_G) = 0$) and on account of (3.85) and (3.86), we obtain

$$\tilde{H}_N(\vartheta_G) = l_a^2 \int_{\partial G} \vartheta_G \frac{\partial \vartheta_G}{\partial n}\,dl. \tag{3.87}$$

Substituting (3.82) into (3.87) leads to

$$\tilde{H}_N(\vartheta_G) = l_a^2 \int_{\partial G} \left[\frac{1}{R}\left(\frac{\partial a_0}{\partial l}\right)^2 \sin^2 \alpha(l) + \frac{1}{2}\frac{\partial a_0}{\partial l}\frac{\partial^2 a_0}{\partial n^2}\sin 2\alpha(l)\right] dl. \tag{3.88}$$

Now let us choose two translation modes ϑ_G^1 and ϑ_G^2 corresponding to orthogonal directions. According to (3.88) the sum of their quadratic forms can be represented as

$$\Sigma \equiv \tilde{H}_N(\vartheta_G^1) + \tilde{H}_N(\vartheta_G^2) = l_a^2 \int_{\partial G} \frac{1}{R(l)}\left(\frac{\partial a_0}{\partial l}\right)^2 dl. \tag{3.89}$$

For any convex domain ($R(l) > 0$) we obtain $\Sigma \geq 0$. Let us prove that from $\Sigma \geq 0$ there follows $\lambda_1 > 0$. Indeed, in the case $\Sigma > 0$, at least one of the quadratic forms $\tilde{H}_N(\vartheta_G^1)$ and $\tilde{H}_N(\vartheta_G^2)$ is positive. Then, from the variational principle, it follows that $\hat{H}_N$ has at least one positive eigenvalue. The equalities $\Sigma = 0$ and $\lambda_1 = 0$ can not occur simultaneously because in this case either $\vartheta_G^1 = 0$ and $\vartheta_G^2 = 0$ or $\vartheta_G^1 = \Psi_1$ and $\vartheta_G^2 = 0$. In the first case $a_0 = $ constant; in the second case a_0 corresponds to a plane current layer and its translation mode can never coincide with Ψ_1 due to the boundary conditions. Therefore again $\lambda_1 > 0$. Thus, for the operator $\hat{H}_N$ defined on any convex domain, $\lambda_1 > 0$ and the ground-state eigenfunction Ψ_1 corresponds to an unstable mode.[3]

In order to determine the sign of the second eigenvalue we should establish some relation between $\hat{H}_N$ and $\hat{H}_{g(l)}$ and use the fact that the operator $\hat{H}_{g(l)}$ has a zero eigenmode ϑ_G. From the theory of operators (Reed and Simon 1972) it follows that $\hat{H}_N > \hat{H}_{g(l)}$ (i.e. $\lambda_i > \lambda_i^{g(l)}$ for any i) if $\tilde{H}_N > \tilde{H}_{g(l)}$ or if $\tilde{H}_N = \tilde{H}_{g(l)}$ and the domain $D(\tilde{H}_N)$ of $\hat{H}_N$ ($D(\hat{H}_N)$ is a set of functions on which $\hat{H}_N$ is defined) includes the domain $D(\tilde{H}_{g(l)})$ for $\tilde{H}_{g(l)}$: $D(\tilde{H}_N) \supset D(\tilde{H}_{g(l)})$. The difference between the quadratic forms $\tilde{H}_N$ (3.86) and $\tilde{H}_{g(l)}$ (3.85)

$$\tilde{H}_N(\Psi) - \tilde{H}_{g(l)}(\Psi) = l_a^2 \int_{\partial G} g(l)\Psi^2\,dl \tag{3.90}$$

is positive when $g(l) > 0$. However, the condition $g(l) > 0$ is violated for any solution a_0 that varies along the boundary ∂G. Indeed, at the boundary point where

---

[3] A similar but more cumbersome proof of this fact for n-dimensional domains was presented by Svirezhev (1987) for application to dissipative structures in ecological systems.

ξ and n are parallel, $\cot \alpha(l)$ changes sign whereas $\partial^2 a_0 / \partial n^2$ in general does not. Therefore $\widehat{H}_N$ and $\widehat{H}_{g(l)}$ can not be compared if $a_0(l) \neq$ constant. When $a_0(l) =$ constant eq. (3.84) implies that all translation modes satisfy the Dirichlet boundary conditions ($g(l) = \infty$). Then the operator $\widehat{H}_{g(l)}$ is a Dirichlet one: $\widehat{H}_{g(l)} = \widehat{H}_D$. In this case $\widetilde{H}_N = \widetilde{H}_D$. Since $\widetilde{H}_N$ can be considered on functions with arbitrary boundary conditions the domain of $\widetilde{H}_N$ contains the domain of $\widetilde{H}_D$. Thus, $\widehat{H}_N > \widehat{H}_D$.

Any inhomogeneous state a_0 that satisfies the condition $a_0(l) =$ constant has an extremum inside the domain G and its translation mode ϑ_G always has zeros. Since the ground-state eigenfunction of the operator $\widehat{H}_D$ is strictly positive, the eigenvalue $\lambda_i^D = 0$ corresponds to $i \geq 2$. Taking into account that $\widehat{H}_N > \widehat{H}_D$, we conclude that $\lambda_2 > 0$.

Thus we have established that, for any stationary inhomogeneous state $a_0(x, y)$ on a convex domain, corresponding to a current-density profile $j(x, y) = u_0 - a_0(x, y)$, $\lambda_1 > 0$ holds, and that, for any steady state that satisfies the condition $a_0(x, y)|_{\partial G} =$ constant on an arbitrary domain $\lambda_1 > 0$ and $\lambda_2 > 0$ hold. Note that the condition of convexity of the domain and the condition $a_0(x, y)|_{\partial G} =$ constant are sufficient but not necessary conditions for $\lambda_1 > 0$ and for $\lambda_1 > 0$, $\lambda_2 > 0$, respectively. Since for one-dimensional domains the translation mode ϑ_G always satisfies Dirichlet boundary conditions, it also follows from our analysis that $\lambda_1 > 0$ for any one-dimensional distribution $a_0(x)$ and that $\lambda_1 > 0$ and $\lambda_2 > 0$ for nonmonotonic profiles $a_0(x)$. Thus, under voltage control and Neumann boundary conditions, stationary current filaments are generally unstable.

3.4.2   The effect of the global constraint on filament stability

In this section we establish the relation between the number of unstable modes ($\text{Re}\, \zeta > 0$) in the presence of a global constraint and the number of positive eigenvalues λ_i corresponding to the voltage-driven regime. We advance the approach originally suggested for semiconductors with an electron-overheating instability (Bass *et al.* 1970). Stability of current filaments in controllable bistable systems with two external global constraints has been studied by Gorbatyuk and Rodin (1997).

Let us expand the eigenmode $\delta a(x, y)$ in the basis of the eigenfunctions Ψ_i of the operator $\widehat{H}_N$:

$$\delta a(x, y) = \sum_m \langle \delta a \cdot \Psi_m \rangle \Psi_m. \tag{3.91}$$

From eq. (3.79) we obtain the coefficients of this expansion

$$\delta a(x, y) = \sum_m \frac{\langle (\partial f / \partial u)\, \Psi_m \rangle}{\tau_a \zeta - \lambda_m} \delta u\, \Psi_m(x, y). \tag{3.92}$$

Substituting into eq. (3.80) leads to the characteristic equation

$$F(\zeta) = F_0(\zeta) + F_1(\zeta) = 0, \tag{3.93}$$

3.4 Conditions for stability of current filaments

$$F_0(\zeta) = 1 + RA\left\langle \frac{\partial j}{\partial u} \right\rangle + RA \sum_m \frac{\langle (\partial f/\partial u)\, \Psi_m \rangle \langle (\partial j/\partial a)\, \Psi_m \rangle}{\tau_a \zeta - \lambda_m}, \qquad F_1(\zeta) = \tau_u \zeta,$$

which determines the complex eigenvalues ζ of the linearized system (3.79) and (3.80). We separate the two parts $F_0(\zeta)$ and $F_1(\zeta)$ such that the characteristic time of the inhibitor τ_u enters $F_1(\zeta)$ only. Note that the ground-state eigenfunction Ψ_1 has no zeros whereas the Ψ_i oscillate in space for $i > 1$. Therefore the first term dominates the others in the sum in the last term of $F_0(\zeta)$.

Taking into account that $\zeta = 0$ for variations $\delta a(x, y)$ along the steady current–voltage characteristic and applying eq. (3.92), we can represent the differential conductance of the inhomogeneous state

$$\sigma_d \equiv A \frac{d\langle j(a, u)\rangle}{du} = A\left(\left\langle \frac{\partial j}{\partial u} \right\rangle + \left\langle \frac{\partial j}{\partial a} \frac{\delta a}{\delta u} \right\rangle \right) \tag{3.94}$$

as

$$\sigma_d = \sigma_u - A \sum_m \frac{\langle (\partial f/\partial u)\, \Psi_m \rangle \langle (\partial j/\partial a)\, \Psi_m \rangle}{\lambda_m}, \qquad \sigma_u \equiv A\left\langle \frac{\partial j}{\partial u} \right\rangle, \tag{3.95}$$

where σ_u denotes the differential conductance for fixed internal parameter $a(x)$. Note that $\sigma_u > 0$ for the form of $j(a, u)$ given in (3.74).

Zeros of the characteristic function $F(\zeta)$ are complex eigenvalues of the stability problem (3.79) and (3.80) and its poles are eigenvalues λ_i of the operator $\widehat{H}_N$. The number of poles P and the number of zeros N located inside any closed contour Γ in the complex ζ-plane are related by the argument principle

$$N = P + \frac{1}{2\pi} \Delta\mathrm{Arg}\, F(\zeta), \tag{3.96}$$

where $\Delta\mathrm{Arg}\, F(\zeta)$ is the variation of the argument of $F(\zeta)$ when ζ varies along the contour Γ (Derrick 1972). Since unstable modes correspond to $\mathrm{Re}\,\zeta > 0$, we choose the contour Γ encircling the right-hand half-plane (Fig. 3.13(a)) in the clockwise direction. The variation $\Delta\mathrm{Arg}\, F(\zeta)$ depends on how many times and in which direction the contour $F(\Gamma)$ turns around the origin. In the case of anticlockwise rotation the number of unstable eigenmodes is less then the number of positive eigenvalues λ_i by the number of turns; in the case of clockwise rotation it is larger by the same number. $N = P$ holds if $F(\Gamma)$ does not encircle the origin.

First, let us consider the case of fast inhibition $\tau_u \ll \tau_a$ when temporal instabilities are damped out in favor of spatial ones. Then we neglect the term F_1 in the characteristic equation. In this case

$$F(0) = 1 + R\sigma_d, \qquad F(\infty) = 1 + R\sigma_u. \tag{3.97}$$

Therefore the contour $F(\Gamma)$ encircles the origin once if $\mathrm{sign}\{F(\infty)\} = -\mathrm{sign}\{F(0)\}$ and the direction of rotation of $F(\Gamma)$ is anticlockwise (Fig. 3.13(b)). This leads to the following stability criterion:

$$-\sigma_{\mathrm{u}} < R^{-1} < -\sigma_{\mathrm{d}}. \tag{3.98}$$

When this criterion is satisfied the global constraint reduces the number of unstable modes related to the operator $\widehat{H}_{\mathrm{N}}$ by one:

$$N = P - 1; \tag{3.99}$$

otherwise

$$N = P. \tag{3.100}$$

In the last case the global constraint has no impact on stability, and the system's behavior is similar to that of the voltage-driven regime. In what follows we refer to the cases (3.99) and (3.100) as regimes of strong and weak global coupling, respectively.

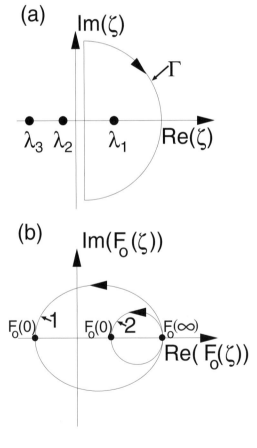

Figure 3.13. Stability analysis using the characteristic equation. (a) The contour Γ in the complex ζ-plane encircles the right half-plane which contains eigenvalues ζ with positive real parts. λ_1, λ_2, and λ_3 are poles corresponding to the eigenvalues for the voltage-driven case. (b) The mapping $F_0(\Gamma)$ of the contour Γ corresponding to the case of fast inhibition $\tau_a \gg \tau_u$. Contours 1 and 2 correspond to the cases of eq. (3.99) and (3.100), respectively (after Alekseev et al. (1998)).

Therefore, in the regime of weak coupling the system is stable if all eigenvalues λ_i of the operator $\widehat{H}_N$ are negative, whereas in the case of strong coupling the system remains stable even if the first eigenvalue λ_1 is positive. For filamentary states with NDC ($\sigma_d < 0$) the criterion (3.98) can be fullfilled with a linear passive external circuit and it takes the conventional form $R > |\sigma_d|^{-1}$. This occurs for a heavily loaded external circuit and corresponds to the regime of strong global coupling. For filamentary states with positive differential conductance (PDC) the criterion (3.98) demands $R < 0$. A negative external load resistance can be realized by active circuits: An experimental setup exhibiting an absolute negative resistance $R < 0$ and a negative (both absolute and differential) capacitance $C < 0$ has been reported by Martin $et\ al.$ (1994).

Since for $R < 0$ an instability can also be induced by the negative load itself, in a valid stability analysis one should then properly take into account also the action of the temporal degree of freedom related to the external circuit, i.e. one should treat the case $\tau_u \neq 0$. Since we have already shown that filamentary states with two or more positive eigenvalues $\lambda_1 > 0$ and $\lambda_2 > 0$ remain unstable if a global constraint is imposed, it is sufficient to consider the case of $\lambda_1 > 0$ and $\lambda_2 < 0$. Leaving only the first term corresponding to λ_1 in the sum in eq. (3.93), we write the characteristic equation in the form

$$C\tau_a\zeta^2 + \left[\tau_a(R^{-1} + \sigma_u) - C\lambda_1\right]\zeta - \lambda_1(R^{-1} + \sigma_d) = 0. \tag{3.101}$$

For $\lambda_1 > 0$ this equation leads to a modified version of the stability criterion (3.98):

$$-\sigma_u + \frac{C\lambda_1}{\tau_a} < R^{-1} < -\sigma_d, \qquad \text{if } C > 0, \tag{3.102}$$

$$-\sigma_d < R^{-1} < -\sigma_u + \frac{C\lambda_1}{\tau_a}, \qquad \text{if } C < 0. \tag{3.103}$$

For the cases $C > 0$ (3.102) and $C < 0$ (3.103) a Hopf bifurcation, whereby a pair of complex-conjugate eigenvalues ζ crosses the imaginary axis, occurs at the lower and upper bound, respectively. This condition shows that oscillatory instabilities are possible if the total capacitance $|C|$ exceeds a critical value C_{crit}.

The criteria (3.102) and (3.103) allow us to understand how to stabilize the filamentary states with PDC. Indeed, for $\sigma_d > 0$ the condition (3.102) can be satisfied only if $\sigma_d < \sigma_u$ (and $R < 0$). Since the main contribution to the differential conductance of a filamentary state usually comes from the shift of the filament wall (the second term in eq. (3.95), in most situations $|\sigma_d| > |\sigma_u|$. Therefore the condition (3.102) can not be satisfied, and filaments with PDC can be stable only in the case of $R < 0$ and $C < 0$ described by (3.103). The requirement of a negative capacitance has a simple qualitative explanation: For $R < 0$ and $C > 0$ the characteristic time $\tau_u = RC$ becomes negative, which indicates an instability induced by the external circuit.

Since we have not specified the type of spatial distribution $a_0(x, y)$ being studied, the analysis performed in this section is also valid for homogeneous states. For instance, for uniform states on the middle branch of an S- or Z-shaped characteristic $\lambda_1 = \partial f / \partial a > 0$ and $\lambda_2 \approx \partial f / \partial a - (\pi l_a / L)^2$, where L is the largest transverse dimension of the domain G, cf. Section 3.3. For the case of a small domain with $\lambda_1 > 0$ and $\lambda_2 < 0$ our analysis leads to the results obtained for space-clamped elements in Section 3.2. For a sufficiently large domain the second eigenvalue is also positive ($\lambda_1 > 0$ and $\lambda_2 > 0$) and a homogeneous state can not be stabilized by a global constraint.

An oscillatory instability of current filaments might result either in homogeneous oscillations or in breathing (Schöll and Drasdo 1990) or spiking (Bose *et al.* 1994) filaments. It is known from one-dimensional analyses that hysteresis between stationary current filaments and the spiking mode takes place for the major part of the parameter space (Niedernostheide *et al.* 1996*b*). Therefore, if a stationary filament is already unstable, the spiking mode is unlikely to appear. As for inhomogeneous states $\lambda_1 < \partial f / \partial a$, according to eq. (3.102) a filamentary state is more stable with respect to oscillatory instabilities than are homogeneous states with NDC. For this reason, in the simplest case of S-type NDC a limit cycle corresponding to homogeneous oscillations already exists for system parameters at which a Hopf bifurcation of the filamentary state occurs, and hence homogeneous self-generated oscillations rather than breathing filaments emerge as the most probable result of the instability. Obviously, the global behavior after the bifurcation also essentially depends on the form and position of the null-isoclines $f(a, u) = 0$ and $g(a, u) = 0$ (or the local current-density–voltage characteristic and the load line, which give an equivalent representation in the (j, u)-plane) and, in particular, on the number of their intersections. This is especially important for systems with Z-shaped current–voltage characteristics, for which several points of intersection usually exist.

3.4.3   Current filaments on two-dimensional domains

General results of the stability analysis performed in Sections 3.4.1 and 3.4.2 are summarized in Table 3.1. In the regime of weak global coupling (a weakly loaded external circuit) any stationary filamentary state $a_0(x, y)$ on a convex domain is unstable. The switching mode dominates the growth of perturbations. This leads to either expansion or shrinking of the current filament. In the case $R > 0$ the stability criterion (3.102) is never satisfied for filaments with PDC, and therefore the presence of a global constraint and NDC of the filamentary state are necessary conditions of stability. Stability of a filament with PDC requires a negative external load $R < 0$ in the case $\sigma_d < \sigma_u$ (3.102), and a negative external load $R < 0$ together with negative differential capacitance $C < 0$ in the case $\sigma_d > \sigma_u$ (3.103).

For domains that are not convex the sign of λ_1 is unknown in general. It has been shown (Svirezhev 1987) that $\lambda_1 < 0$ holds for a dumbbell-shaped domain that represents two circular domains connected by a sufficiently narrow crosspiece. This domain corresponds to two weakly coupled SNDC elements. Because $\lambda_1 < 0$, coexistence of high-conductivity and low-conductivity states is possible in such a system even in the voltage-driven regime.

In the case of strong global coupling and NDC of the filamentary state the number of unstable modes $\lambda_i > 0$ associated with the voltage-driven system is reduced by one. Therefore the filament stability depends on the sign of the second eigenvalue λ_2. We have proved that $\lambda_2 > 0$ holds for any distribution $a_0(x, y)$ that has a maximum inside the domain and does not vary along its boundary ($a_0|_{\partial G} = $ constant). Here we do not require convexity of the domain. According to this conclusion, a filament located in the interior of the domain is unstable and is attracted by the boundary of the domain.

Let us illustrate this statement by the example of radially symmetric filaments in circular domains. Assuming radial symmetry, we write eq. (3.73) in the steady state in polar coordinates (r, ϕ) as in Bonch-Bruevich et al. (1975) and Schöll (1987):

Table 3.1. *Results of the stability analysis of current filaments*

Type of global coupling	Type of steady current density distribution	
	Arbitrary nonuniform distribution, convex domain G	Distribution with an extremum inside G, $a_0|_{\partial G} = $ constant, arbitrary domain G
No coupling: (voltage-driven regime) $R = 0$	Unstable: $\lambda_1 > 0$; sign λ_2 undetermined	Unstable: $\lambda_1 > 0$, $\lambda_2 > 0$
Weak global coupling: $R^{-1} \notin \left[-\sigma_u + \dfrac{C\lambda_1}{\tau_a}; -\sigma_d \right]$ for $C > 0$ $R^{-1} \notin \left[-\sigma_d; -\sigma_u + \dfrac{C\lambda_1}{\tau_a} \right]$ for $C < 0$	Unstable: at least one unstable mode	Unstable: at least two unstable modes
Strong global coupling: $R^{-1} \in \left[-\sigma_u + \dfrac{C\lambda_1}{\tau_a}; -\sigma_d \right]$ for $C > 0$ $R^{-1} \in \left[-\sigma_d; -\sigma_u + \dfrac{C\lambda_1}{\tau_a} \right]$ for $C < 0$	Stable if and only if $\lambda_2 < 0$	Always unstable against translation mode

$$l_a^2 \left(\frac{\partial^2 a}{\partial r^2} + \frac{1}{r} \frac{\partial a}{\partial r} \right) + f(a, u) = 0. \tag{3.104}$$

According to the phase-portrait analysis of eq. (3.104) (Bonch-Bruevich *et al.* 1975, Schöll 1987) there are two types of current distributions: Filaments in the center

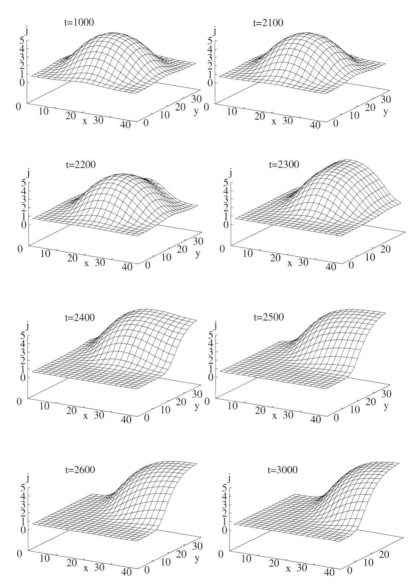

Figure 3.14. Instability of a central filament in a square domain. The temporal evolution of the current density $j(x, y)$ is shown for the following dimensionless numerical parameters: $j_0 \equiv U_0/(RA) = 1.94$, $C^{-1} = 0.035$, $\mathcal{T} = 0.05$, $\tau_a = l_a = 1$, $L = 40$, and $L_{cr} = 11.5$, where $L_{cr} \equiv \pi l_a (\partial f/\partial a)^{-1/2}$ is the minimum system size necessary for filament formation (after Alekseev *et al.* (1998)).

of the domain and annular current layers attached to the boundary. With increasing total current a "hot" central filament (i.e. a filament with high current density in the center) expands over the domain and becomes a "cold" (i.e. low-current) annulus attached to the boundary. In a similar way a cold central filament experiences a transformation into a hot annulus with decreasing total current. Because the condition $a_0|_{\partial G} = \text{constant}$ is satisfied, we conclude that all filamentary and annular states described above are unstable against translation, which breaks the radial symmetry.

On the basis of our analytic considerations we conjecture that all current distributions that have a maximum in the interior of the domain are unstable against translation regardless of the condition $a_0|_{\partial G} = \text{constant}$. This is corroborated by numerical simulations for the model of eqs. (3.74) and (3.76) for rectangular domains in the current-driven regime ($R \to \infty, U_0 \to \infty$ and $j_0 \equiv U_0/(RA) = \text{constant}$) which, obviously, corresponds to the case of strong global coupling. Figure 3.14 shows the temporal evolution of a center filament in a square domain. The initial configuration corresponds to the stationary solution; the condition $a_0|_{\partial G} = \text{constant}$ is slightly violated. This stationary solution remains stable when perturbations orthogonal to the translation mode are introduced into the numerical simulation (therefore $\lambda_3 < 0$). However, the central filament becomes unstable for random perturbations of arbitrarily small amplitude when the condition of orthogonality is not satisfied ($\lambda_2 > 0$). This instability results in the eventual formation of a corner filament that has its maximum at the boundary. For other parameters the transient process may also lead to an edge current layer. Hot and cold corner filaments and plane edge current layers parallel to the boundary (Fig. 3.15) are the only structures which arise from random initial conditions in the regime of strong global coupling and represent stable steady states for rectangular domains. Since $\lambda_1 > 0$ and $\lambda_2 < 0$ their stability depends on the sign of their differential conductance; stability requires $\sigma_d < 0$.

Current–voltage characteristics for domains of two different sizes are presented in Fig. 3.16. Generally, with increasing current the differential conductance changes sign

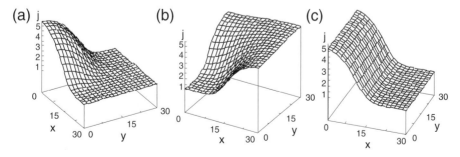

Figure 3.15. Stable filament configurations on a square domain: (a) a hot corner filament, (b) a cold corner filament, and (c) an edge-current layer. The values of the average current $j_0 = U_0/(RA)$ are 1.5, 3.5, and 2.0, respectively; $L = 30$, other numerical parameters are the same as those in Fig. 3.14 (after Alekseev *et al.* (1998)).

when the filament wall reaches the boundary of the domain: Starting from this point the main contribution to the differential conductance is given not by the shift of the filament wall (the second term in eq. (3.95)) but by the change of current density in the homogeneous part of the current distribution (the first term in eq. (3.95). Therefore the differential conductance becomes positive. For edge current layers such a transition occurs for sufficiently thin hot and cold layers. For hot corner filaments the bottom point of the characteristic where the differential conductance changes sign corresponds to a sufficiently small filament; the change at the top of the characteristic takes place when the filament has expanded such that the filament boundary reaches the adjacent corners of the domain (at this point a hot filament covers approximately three quarters

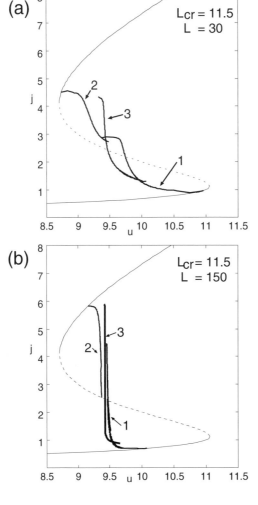

Figure 3.16. Current–voltage characteristics for (a) small ($L = 30 \approx 3L_{cr}$) and (b) large ($L = 150 \approx 13L_{cr}$) square domains. The thin line represents homogeneous states, the thick lines depict the spatially averaged current density of inhomogeneous steady states. Curves 1, 2, and 3 correspond to a hot corner filament, a cold corner filament, and an edge-current layer, respectively. These branches are stable only in the regime of strong global coupling. Numerical parameters are the same as those in Fig. 3.14 (after Alekseev *et al.* (1998)).

($\approx \pi/4$) of the domain area). For cold corner filaments the top and bottom points are interchanged. Sweeping up and down the current–voltage characteristic (Fig. 3.16) in the current-controlled regime one observes multistability and multiple hysteresis cycles between the edge current layer and corner filaments. The particular scenario of the transitions essentially depends on the ratio L/L_{cr}, where L_{cr} is the minimum system size necessary for filament formation.

Our overall conclusion is that a boundary attracts filaments from the interior of the domain. However, the growth increment of the unstable mode decreases with increasing system size. Since therefore, in sufficiently large domains, a filament in the interior of the domain has nearly neutral stability, it may be pinned by a small embedded inhomogeneity of the semiconductor structure. To study the competition between the attracting boundary and pinning at inhomogeneities, we present the following numerical simulations. Inhomogeneities sensitively influence the transport parameter T (in the HHED, for example, this parameter corresponds to a tunneling rate). We assume that T becomes a function of space, and $T(x, y)$ has a localized peak of amplitude δT in the center of the square domain. Then, for a sufficiently large value of δT, the central filament becomes stabilized, i.e. it is pinned at the inhomogeneity. Figure 3.17 shows the stability in the $(\delta T, L)$ diagram: The minimum amplitude $\delta T/T$ of the seed inhomogeneity which is necessary to stabilize a central filament decreases approximately exponentially with the system's size L. This results from the asymptotically exponential decrease of the eigenvalue $\lambda_2(L)$ of the central filament. Thus, in large domains seed inhomogeneities dominate over boundary effects.

In conclusion, in bistable semiconductor systems on two-dimensional spatial domains with convex passive boundaries (the Neumann boundary condition), any stationary nonuniform current-density distribution has at least one unstable mode $\lambda_1 > 0$ when the system is operated in the voltage-driven regime. If the second eigenvalue is negative ($\lambda_2 < 0$), such a distribution can be stabilized by global coupling provided by the external circuit if the stability criteria (3.102) and (3.103) are satisfied. For filaments with negative differential conductance, stability can be achieved with a passive ($R > 0$) heavily loaded external circuit. Filaments with positive differential conductance σ_d can be stabilized only by an active external circuit with negative resistance $R < 0$ when $\sigma_d > \sigma_u$. If $\sigma_d > \sigma_u$, stabilization of filaments with positive differential conductance requires also a negative total differential capacitance of the device and external circuit $C = C_{int} + C_{ext} < 0$. We have proved analytically that $\lambda_2 > 0$ holds for current distributions that have their maxima in the interior of the spatial domain and a constant value at its boundary. Such distributions, corresponding to central current filaments, remain unstable against translation even if the conditions (3.102) and (3.103) are satisfied. Our numerical simulations suggest that the condition of a constant boundary value may even be relaxed and that Neumann boundaries generally attract current filaments. This is in agreement with numerical results obtained for other reaction–diffusion systems (Niedernostheide *et al.* 1997). In structurally imperfect systems, seed inhomogeneities tend to pin current filaments in the center; thus there is competition between attractive inhomogeneities and the boundary. In large systems, imperfections dominate over boundary effects: The amplitude of the inhomogeneity which is sufficient for pinning current filaments decreases exponentially with increasing transverse dimension of the system.

3.5   Lateral current-density fronts in globally coupled bistable systems

Stationary current filaments exist in bistable systems only if the condition of coexistence of the two stable fixed points, usually given in terms of an equal-areas rule, is satisfied. Otherwise a front between the two phases propagates, as discussed in Section 3.1 for a simple reaction–diffusion system. In this section we study the propagation of a lateral current-density front in a bistable semiconductor system with S- or Z-shaped current–voltage characteristics (Meixner *et al.* 2000*a*). Such trigger fronts can switch a bistable system from one uniform steady state to another. An external load circuit introduces a global coupling that leads to positive or negative

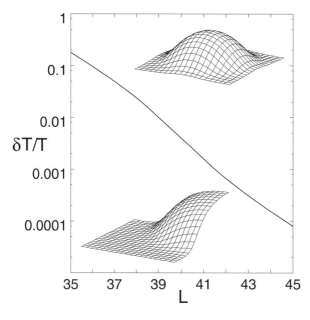

Figure 3.17. The effect of structural imperfections on filament stability. The two different regimes are shown in the diagram of the relative amplitude $\delta T/T$ of seed inhomogeneities versus the system's size L. In the upper regime a stable central filament is pinned by a seed inhomogeneity; in the lower regime the attractive influence of the boundary dominates. Other numerical parameters are the same as those in Fig. 3.14 (after Alekseev *et al.* (1998)).

feedback upon the front dynamics in Z- and S-type systems, respectively. This results in accelerated or decelerated front motion. The type of feedback can be reversed if the system is operated in an active external circuit with negative load resistance. Double-barrier resonant-tunneling diodes (DBRTs) and heterostructure hot-electron diodes (HHEDs) are used as examples of Z- and S-type systems, respectively, but similar results can be obtained for gate-driven thyristor structures, which also exhibit a Z-shaped current–voltage characteristic, as shown in Section 2.4.8 (Meixner *et al.* 1998*a*, 1998*b*).

3.5.1   Trigger fronts

Traveling front patterns in bistable and excitable active media (Mikhailov 1994) are a universal phenomenon that occurs in nonlinear spatially extended systems of diverse natures, ranging from physics (Zel'dovich and Frank-Kamenetskij 1938, Gunn 1964, Loeb 1965, Bonch-Bruevich *et al.* 1975, Pelcé 1988, Shaw *et al.* 1992) and chemistry (Kuramoto 1988) to biology and ecology (Pelcé 1988, Svirezhev 1987). In many cases, in addition to a local diffusive coupling, such systems also experience a *global coupling*. Generally, global coupling is related to external constraints imposed upon the system's dynamics. In the presence of global coupling, some dynamic variables of the active media (e.g. the global excitation level) depend on the spatially averaged parameters of the traveling pattern.

In *bistable semiconductor systems* the global coupling is an inherent feature of the spatio-temporal dynamics of the current-density patterns (e.g. current filaments, fronts, and pulses). The mechanism of this coupling has been described in Section 1.5.2. For any evolution of the current-density pattern that is accompanied by a variation of the total current through the device, the voltage drop at an external load and/or internal series resistance changes. That causes a variation of the voltage u dropping across the device which usually characterizes the global excitation level of the bistable semiconductor system. This type of feedback is well known with respect to stationary, breathing, and spiking current filaments.

In this section we study the effect of global coupling on the propagation of *lateral current-density fronts*. Excitation of such fronts, propagating in the direction perpendicular to the direction of the current flow, is possible for systems with bistable current–voltage characteristics if the low-current state and the high-current state coexist in a certain interval of the voltage u. Such bistability may be associated either with an S- or with a Z-shaped current-density–voltage characteristic (Fig. 3.18). As discussed in Section 2.4, an S-shaped characteristic represents a classical example of bistability both in bulk semiconductors (e.g. due to an overheating instability (Volkov and Kogan 1969) or impact-ionization breakdown (Schöll 1987)) and in layered semiconductor structures (in p–i–n diodes (Jäger *et al.* 1986), avalanche transistors (Osipov and Kholodnov 1973), heterostructure hot-electron diodes (HHEDs) (Hess *et al.* 1986, Belyantsev *et al.* 1986, Wacker and Schöll 1994*a*), p–n–p–n (Gorbatyuk

and Rodin 1992*a*) and p–n–p–i–n (Niedernostheide *et al.* 1994*a*, Gorbatyuk and
Niedernostheide 1996) multilayered structures). The S-shaped characteristic also
includes the whole family of switching devices of modern electronics (thyristors,
MOSFETs, etc.) (Sze 1998). The *Z-shaped bistability* has only more recently
received attention; it occurs, for example, in double-barrier resonant-tunneling structures (DBRTs) (Goldman *et al.* 1987) and gate-driven p–n–p–n structures (Meixner
et al. 1998*a*). Transverse current-density fronts in DBRTs have recently been
discussed (Glavin *et al.* 1997, Mel'nikov and Podlivaev 1998, Meixner *et al.* 2000*a*).
S- and Z-type systems constitute the major classes of bistable semiconductors. We
shall point out similarities and differences between the globally coupled dynamics of
lateral current-density patterns for these two classes of bistable systems and illustrate
these findings with two significant examples, viz. the HHED and the DBRT.

We consider one-dimensional fronts propagating along the x-axis in long, narrow
samples as sketched in Fig. 3.19, such that the second transverse dimension L_y (along
the y-axis) is so short that pattern formation can not develop. For a wide class of
semiconductor systems the internal state can be characterized by a single variable
$a(x, t)$ that corresponds to the internal degree of freedom relevant for the bistability.
The reduced dynamic description in terms of this variable can be derived from a full
three-dimensional transport model, the Poisson equation, and continuity equations via
adiabatic elimination of fast variables, as was done, for instance, by Volkov and Kogan
(1969), Osipov and Kholodnov (1973), Wacker and Schöll (1994*b*), Niedernostheide
et al. (1994*a*), Gorbatyuk and Rodin (1992*a*), Glavin *et al.* (1997), and Mel'nikov and
Podlivaev (1998) for various semiconductor systems. This equation takes the form of
a reaction–diffusion equation (2.47):

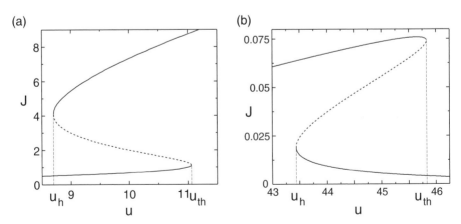

Figure 3.18. (a) S-shaped and (b) Z-shaped current–voltage characteristics $j(u)$.
The threshold voltage u_{th} and the holding voltage u_h mark the boundaries of the
bistability domain. The characteristics are calculated (a) for the heterostructure
hot-electron diode (HHED) (see eq. (3.110), $T = 0.05$) and (b) for the double-barrier
resonant-tunneling structure (DBRT) (see eq. (3.111)) (after Meixner *et al.* (2000*a*)).

$$\tau_a \frac{\partial a(x, t)}{\partial t} = f(a(x, t), u(t)) + l^2 \frac{\partial^2 a(x, t)}{\partial x^2}. \tag{3.105}$$

Here τ_a and l are the relaxation time and diffusion length of the variable a, respectively. Generally, the combination of lateral diffusion and drift (due to the lateral electric field induced by an inhomogeneous charge distribution) effectively results in a term with a concentration-dependent diffusion coefficient $D(a) \equiv l^2(a)/\tau_a$ (D'yakonov and Levinstein 1978, Wacker and Schöll 1995, Glavin *et al.* 1997, Mel'nikov and Podlivaev 1998, Cheianov *et al.* 2000) (see also Sections 2.4.5 and 2.4.6). Here for simplicity we set $l = $ constant. We assume passive (no-flux) boundaries described by Neumann boundary conditions for $a(x, t)$

$$\frac{\partial a(0, t)}{\partial x} = \frac{\partial a(L_x, t)}{\partial x} = 0. \tag{3.106}$$

The temporal dynamics of the voltage $u(t)$ across the device is described by Kirchhoff's equation for the external circuit:

$$\tau_u \frac{du}{dt} = U_0 - u(t) - RL_y \int_0^{L_x} j(a, u) \, dx,$$
$$\tau_u \equiv RC, \qquad C \equiv C_{\text{ext}} + C_{\text{int}}, \tag{3.107}$$

where $j(a, u)$ is the current density in a cross-section of the system, which is determined locally by the internal variable a and the voltage drop u, U_0 is the applied

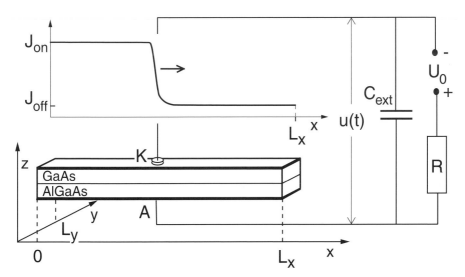

Figure 3.19. A sketch of the bistable semiconductor element and the external circuit (bias voltage U_0, load resistance R, and capacitance C_{ext}). The GaAs/AlGaAs heterostructure hot-electron diode (HHED) is shown as an example. The cathode (K) and anode (A) contacts are equipotential surfaces, and therefore the voltage $u(t)$ across the device is independent from x and y. An inhomogeneous current-density distribution $j(x)$ corresponding to a front propagating in the lateral x-direction is shown as an inset.

bias voltage, R is the load resistance, C_{int} is the internal differential capacitance of the sample, and C_{ext} is the capacitance in the external circuit (see Fig. 3.19). Equation (3.107) describes the *global constraint* imposed upon the internal dynamics of the current-density pattern by the external circuit. Note that, in contrast to the case considered by Bass *et al.* (1970), Elmer (1990, 1992), Gorbatyuk and Rodin (1997), and Hempel *et al.* (1998), this constraint has the form of a *dynamic* equation for the global excitation parameter u.

The local kinetic function $f(a, u)$ is a nonmonotonic function of a that, for $u_h < u < u_{th}$, has three zeros $a_1 < a_2 < a_3$ reflecting the bistability (Fig. 3.20). For the homogeneous steady state, the local dependency $a(u)$ is calculated from the null-isocline $f(a, u) = 0$ and inserted into $j(a, u)$ in order to determine the local current density as a function of the voltage $j(u) \equiv j(a(u), u)$. Here we restrict ourselves to the case in which the resulting current–voltage characteristic is S- or Z-shaped. We denote the current densities in the high-conductivity and low-conductivity states by J_{on} and J_{off}, respectively. The value a_2 corresponds to the state on the intermediate branch of the current–voltage characteristic. The higher value of a does not necessarily correspond to the higher value of the current density j and both situations ($a_1 \rightarrow J_{off}$ and $a_3 \rightarrow J_{on}$, and $a_1 \rightarrow J_{on}$ and $a_3 \rightarrow J_{off}$) are possible in various systems.

In the following we assume that t and x are measured in units of τ_a and l, respectively ($t \rightarrow t/\tau_a$ and $x \rightarrow x/l$), and rewrite eqs. (3.105) and (3.107) as

$$\frac{\partial a}{\partial t} = f(a, u) + \frac{\partial^2 a}{\partial x^2}, \tag{3.108}$$

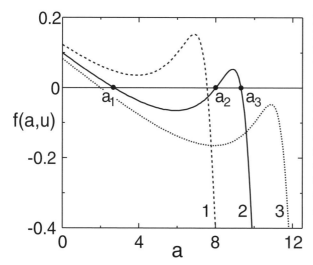

Figure 3.20. The local kinetic function $f(a, u)$ as a function of the internal variable a for $u < u_h$ (curve 1), $u_h < u < u_{th}$ (curve 2), and $u > u_{th}$ (curve 3). The model of the heterostructure hot-electron diode (see eq. (3.110)), in which large a corresponds to small values of the current density j, is used as a numerical example ($\mathcal{T} = 0.05$) (after Meixner *et al.* (2000a)).

$$\varepsilon \frac{du}{dt} = U_0 - u - r\langle j(a, u)\rangle, \qquad \varepsilon \equiv \frac{\tau_u}{\tau_a}, \tag{3.109}$$

$$r \equiv RL_x L_y, \qquad \langle j\rangle \equiv \frac{1}{L_x} \int_0^{L_x} j \, dx.$$

We aim to describe *the universal features* of the globally coupled dynamics of current-density fronts in systems with S- and Z-shaped current–voltage characteristics and generally do not specify the dependencies $f(a, u)$ and $j(a, u)$ which underlie these characteristics. In the figures we use two important semiconductor devices – the heterostructure hot-electron diode (HHED) and the double-barrier resonant-tunneling structure (DBRT) – as examples of S- and Z-type bistabilities, respectively. For the HHED the internal variable a has the meaning of an interface charge density; and the local kinetic function $f(a, u)$ and the current density $j(a, u)$ are given by (Section 2.4.5)

$$f(a, u) = \frac{u - a}{(u - a)^2 + 1} - Ta, \qquad j(a, u) = u - a. \tag{3.110}$$

These dependencies result in an S-shaped current–voltage characteristic, as shown in Fig. 3.18(a). For the DBRT the variable a is the electron concentration in the well. The local kinetic function $f(a, u)$ and the local current density $j(a, u)$ to be used in Eqs. (3.105) and (3.107) have been derived in Section 2.4.6:

$$f(a, u) = \frac{1}{e}(J_{ew}(a, u) - J_{wc}(a)),$$

$$j(a, u) = \frac{1}{2}(J_{ew}(a, u) + J_{wc}(a)). \tag{3.111}$$

The emitter–well and well–collector current densities J_{ew} and J_{wc}, respectively, are given by (2.74) and (2.75). The corresponding Z-shaped current–voltage characteristic is shown in Fig. 3.18(b). For both models we use dimensionless variables throughout the text. In our simulations of the DBRT we use the following typical structural parameters: $E_F = 5$ meV, $E_W = 40$ meV, $d = 20$ nm, $\Gamma_{scatt} = 1$ meV, $\Gamma_L = \Gamma_R = 0.5$ meV, $\epsilon = 12\epsilon_0$, and $m = 0.067$ (for GaAs). Consequently, in the DBRT model the units of time, voltage, and current density are $\tau_a = 3.3$ ps, $\Gamma/e = 2$ mV, and $(e\rho E_F\Gamma)/\hbar \approx 70.2$ kA cm^{-2}, respectively. The length scale $l = \sqrt{(D\tau_a)}$ scales with the square root of the effective transverse diffusion constant D; $D = 100$ cm^2 s^{-1}, for instance, yields $l = 180$ nm. The stationary current–voltage characteristic $j = J_{ew} = J_{wc}$ which results from (2.74), (2.75), and (2.78) is shown in Fig. 3.18(b).

First, in Section 3.5.2, we consider self-similar front propagation as it occurs in the voltage-controlled regime and focus on the dependency of the front velocity v upon the applied voltage u. Section 3.5.3 is devoted to the globally coupled dynamics. Here we consider the regimes corresponding to a conventional positive load resistor in the external circuit as well as the situation when the system is operated via an active external circuit.

3.5.2   The front velocity as a function of the applied voltage

Let us consider first the case of the voltage-driven circuit $u = $ constant. In this case the front between two stationary homogeneous states $J_{on}(u)$ and $J_{off}(u)$ propagates in a self-similar way with constant velocity v:

$$j(x, t) = j_0(x - vt), \qquad j_0 \to J_{on}, J_{off} \qquad \text{for } x \to 0, L_x. \qquad (3.112)$$

Here $v > 0$ holds for *hot fronts* corresponding to the propagation of the high-current-density state into the low-current-density state, and $v < 0$ holds for *cold fronts* corresponding to the propagation of the low-current-density state into the high-current-density state. Since the current density $j(x, t)$ itself is not an independent variable it should be expressed through the order parameter $a(x, t)$ and the applied voltage $u(t)$ according to $j(x, t) = j(a(x, t), u(t))$. For $u = $ constant the equivalent description in terms of a is given by

$$a(x, t) = a_0(x - vt), \qquad a_0 \to a_L, a_R \qquad \text{for } x \to 0, L_x, \qquad (3.113)$$

where the asymptotic values a_L and a_R should be chosen as a_1 or a_3 according to the asymptotic values for j:

$$J_{on} = j(a_L, u), \qquad J_{off} = j(a_R, u). \qquad (3.114)$$

The front is described by eq. (3.108) and corresponds to the propagation of a stable state into a metastable state. Relaxation of the initial profile to the asymptotic solution $a_0(x - vt)$ occurs exponentially fast (Ebert and van Saarloos 1998). It follows from eq. (3.108) that the front width W_f is close to $W_f \approx (\partial f / \partial a)^{-1}$. We assume that $W_f \ll L_x$, which justifies the definition of the boundary conditions (3.112) and (3.113) on the interval $[0, L_x]$ instead of on $(-\infty, +\infty)$. In the co-moving frame eq. (3.108) takes the form

$$\frac{d^2 a_0(x)}{dx^2} + v \frac{d a_0(x)}{dx} + f(a_0, u) = 0. \qquad (3.115)$$

On multiplying eq. (3.115) by $d a_0/dx$ and integrating over $\int_0^{L_x} dx$ we obtain the general expression for the front velocity (3.9):

$$v(u) = \frac{A(u)}{B(u)}, \qquad (3.116)$$

$$A(u) \equiv \int_{a_R}^{a_L} f(a, u) \, da,$$

$$B(u) \equiv \int_0^{L_x} (d a_0/dx)^2 \, dx.$$

Since $B > 0$ the direction of the front propagation is determined by the sign of $A(u)$: We have hot fronts with $v > 0$ for $A(u) > 0$ and cold fronts with $v < 0$ for $A(u) < 0$.

It can readily be seen that, at the bifurcation points $u = u_h$ and $u = u_{th}$ where a new state emerges with an increase or decrease of the control parameter u, respectively, the function $f(a, u)$ has a fixed sign on the whole interval $[a_1, a_3]$. The sign of $A(u)$ depends on which of the two states a_1 and a_3 experiences the bifurcation but is always such that the "old" state propagates into the "new" state. Therefore, for Z-type systems, we predict hot and cold fronts for $u = u_h$ and $u = u_{th}$, respectively, whereas for S-type systems we expect cold and hot fronts for $u = u_h$ and $u = u_{th}$, respectively. For a certain value u_{co} between these points $A(u_{co}) = 0$ and the front has zero velocity. Assuming monotonicity, we conclude that the $v(u)$-dependencies obtained for the HHED and DBRT by direct numerical simulations (Figs. 3.21(a) and (b)) are qualitatively the same for all bistable systems with S- and Z-shaped characteristics. Note that these dependencies look similar but the direction of the front propagation is inverted.

At $u = u_h, u_{th}$ the intermediate value a_2 coincides with a_1 or a_3 and the local kinetic function is tangent to the line $f = 0$ for $a = a_2$. The numerical simulations

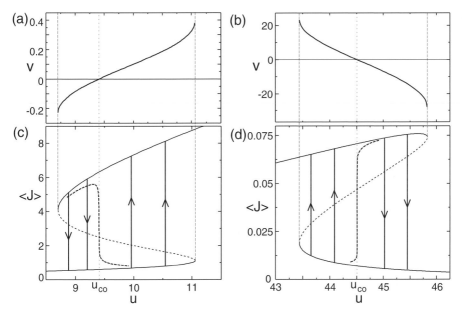

Figure 3.21. Self-similar front propagation in the voltage-controlled regime $u = \text{constant}$ for S-shaped ((a) and (c)) and Z-shaped ((b) and (d)) bistabilities. The front velocity v is shown in (a) and (b) as a function of the applied voltage u. The phase flow in the $(\langle j \rangle, u)$-plane for self-similar front propagation in the voltage-controlled regime is shown in (c) and (d). The thick solid line depicts the local current–voltage characteristic of uniform states. The thin dotted line at $u = u_{co}$ corresponds to a stationary front for $L_x \gg W_f$, the thick dashed line corresponds to the stationary fronts for finite system size $L_x \approx 5W_f$. The numerical parameters of the HHED ($\mathcal{T} = 0.05$) and the DBRT are used in (a) and (c) and (b) and (d), respectively (after Meixner *et al.* (2000a)).

show that $|dv/du| = \infty$ at these points (see also Meixner *et al.* (1998a)). We propose that this is a universal feature of the $v(u)$-dependency caused by the fact that $|dB/du| = \infty$ due to the divergence $|da_{L,R}/du| = \infty$ at the bifurcation point.

In the $(\langle j \rangle, u)$-phase plane the trajectories corresponding to self-similar front propagation are represented by straight vertical arrows (Figs. 3.21(c) and (d)). The phase flow is directed up and down for hot and cold fronts, respectively. In a large system ($L_x \gg W_f$) the line $u = u_{co}$ exactly corresponds to stationary fronts, or kinks. In a finite-sized system the branch of stationary kinks slightly deviates from the vertical line in such a way that the sign of its differential conductance coincides with the differential conductivity of the intermediate branch of uniform states, e.g. we have negative differential conductance (NDC) for S-shaped bistability and positive differential conductance (PDC) for Z-shaped bistability (see Fig. 3.21(c) and (d)). Let us support this observation by the following analytic arguments. The differential conductance of both homogeneous and inhomogeneous states $a_0(x)$

$$\sigma_d \equiv L_x L_y \frac{d\langle j(a, u)\rangle}{du} = L_x L_y \left(\left\langle \frac{\partial j}{\partial u} \right\rangle + \left\langle \frac{\partial j}{\partial a} \frac{\delta a}{\delta u} \right\rangle \right) \tag{3.117}$$

can be represented as (see Section 3.4)

$$\sigma_d = \sigma_u - L_x L_y \sum_m \frac{\langle (\partial f/\partial u) \Psi_m \rangle \langle (\partial j/\partial a) \Psi_m \rangle}{\lambda_m}, \tag{3.118}$$

$$\sigma_u \equiv L_x L_y \left\langle \frac{\partial j}{\partial u} \right\rangle > 0,$$

where σ_u denotes the differential conductance for the fixed internal parameter a, Ψ_m and λ_m are eigenmodes and eigenvalues of the corresponding stationary state, and $\partial f/\partial u$ and $\partial j/\partial a$ are calculated at the stationary solution $a_0(x)$. In nonlinear systems the main contribution to the differential conductance is due to the response δa of the internal parameter to the variation δu of the applied bias, and therefore generally $|\sigma_d| \gg \sigma_u$. The stationary front experiences a weak translation instability: This implies that only the first eigenvalue λ_1 is positive, $|\lambda_1| \ll |\lambda_m|$ for $m > 1$, and the "ground-state" eigenmode $\Psi_1 \geq 0$ corresponds to translation (Bass *et al.* 1970, Schöll 1987, Alekseev *et al.* 1998), cf. Section 3.4. Therefore σ_d is determined solely by the first term in the sum in eq. (3.118). For the homogeneous state $a_0 = a_2$ on the intermediate branch of the current–voltage characteristic the eigenmodes Ψ_m are given by $\Psi_m(x) = \cos[\pi(m-1)x/L_x]$, $\lambda_1 = \partial f/\partial a > 0$, and $\partial f/\partial u, \partial j/\partial a = $ constant. Therefore, for $m > 1$, all terms in the sum in eq. (3.118) are equal to zero. Thus, both for stationary fronts $a_0(x)$ and for the homogeneous states $a_0 = a_2 = $ constant we have

$$\sigma_d \approx L_x L_y \frac{\langle (\partial f/\partial u)\Psi_1 \rangle \langle (\partial j/\partial a)\Psi_1 \rangle}{\lambda_1}, \qquad \Psi_1 \geq 0, \qquad \lambda_1 > 0. \tag{3.119}$$

(Note that eq. (3.119) is applicable neither to narrow filaments nor in the vicinity of the turning points $u = u_h$ and u_{th}.) According to eq. (3.119) $\text{sgn}\{\sigma_d\}$ is determined by sgn$\{\partial f/\partial u\}$ and sgn$\{\partial j/\partial a\}$. Assuming that the functions $\partial f/\partial u$ and $\partial j/\partial a$ are of fixed sign, which is usually the case, we conclude that sgn$\{\sigma_d\}$ for stationary fronts (or, more generally, for wide filaments) and for homogeneous states on the intermediate branch is the same.

The deviation of the thick dashed curve from the vertical line at $u = u_{co}$ reflects the attraction of the system's boundaries in a finite-sized system which causes a translation instability of a stationary front ($\lambda_1 > 0$) in the voltage-controlled regime. However, this instability can hardly be detected in numerical simulations even for $L_x > 10W_f$ due to the pinning of fronts (Mitkov *et al.* 1998), which is caused in this case by the discreteness of the grid. In realistic systems of large transverse dimensions the front stability in the voltage-driven regime is determined by structural imperfections rather than by boundary effects (Alekseev *et al.* 1998). However, the sign of the differential conductance will be shown to have a crucial impact on the stability in presence of global coupling and will be discussed at length in the next sections.

3.5.3 Globally coupled front dynamics

Reduced equations of motion

The total current through a semiconductor element depends on the front position w, and, in presence of a global constraint, the control parameter u changes as the front propagates. Since the local kinetic function $f(a, u)$ and the asymptotic values a_L and a_R in the globally coupled regimes are not constant, the self-similarity of front propagation is broken. By parametrizing the front by its position $w(t)$ and the values in the two phases $a_L(t)$ and $a_R(t)$ we can describe the front dynamics by writing ordinary differential equations:

$$\dot{w} = v\{a(x, t), u(t)\}, \tag{3.120}$$

$$\dot{a}_L(t) = f(a_L(t), u(t)), \tag{3.121}$$

$$\dot{a}_R = f(a_R(t), u(t)), \tag{3.122}$$

$$\varepsilon\dot{u} = U_0 - u - r\left[j(a_L(t), u(t))\frac{w(t)}{L_x} + j(a_R(t), u(t))\left(1 - \frac{w(t)}{L_x}\right)\right]. \tag{3.123}$$

Generally, the front velocity has a functional dependency on the instantaneous front shape and the instantaneous local kinetic profile (3.120), which implies that the front *Ansatz* given by (3.120)–(3.122) is not complete. This difficulty can be overcome if the time hierarchy of the three following relaxation processes is properly taken into account. First, the front shape relaxes to the quasistationary shape corresponding to the local kinetic function $f(a, u)$ and instantaneous asymptotic values a_L and a_R. Second, a_L and a_R relax to the quasistationary values defined by $f(a, u) = 0$. Third, the applied voltage u tends to the value $U_0 - r\langle j(t)\rangle$ given by the load line. It has been argued (Elmer 1990, 1992, Meixner *et al.* 1998a) that the two first relaxation processes are fast with respect to the characteristic time W_f/v it takes for the front to advance by its own width W_f. In this case the actual velocity v is determined by the instantaneous value of u and coincides with the velocity of self-similar propagation at this value of u. Then eqs. (3.120)–(3.123) can be considerably simplified:

$$\dot{w} = v(u), \tag{3.124}$$

$$\varepsilon\dot{u} = U_0 - u - r\left[J_{\text{on}}(u)\frac{w(t)}{L_x} + J_{\text{off}}(u)\left(1 - \frac{w(t)}{L_x}\right)\right]. \tag{3.125}$$

The relaxation time $\varepsilon = \tau_u/\tau_a$ of the variable u is controlled by the external circuit ($\tau_u = RC$). For $C \to 0$ the feedback upon the front dynamics is instantaneous, for $C \to \infty$ the feedback vanishes and the front propagates as in the voltage-controlled case.

The validity of the reduction (3.124) and (3.125) has been confirmed by direct numerical simulations of the full equations (3.105) and (3.107) in comparison with the predictions of the reduced model (3.124) and (3.125) (Meixner *et al.* 1998a, 2000a). The deviation of the front velocity has always been found to be less than 0.5%.

Global coupling via passive external circuit ($R > 0$)

Qualitatively, the type of feedback on front dynamics is determined by the slope of the $v(u)$-dependency and the sign of the load resistance R in the external circuit. The propagation of hot fronts is accompanied by an increase of the total current and, according to eq. (3.107), for $R > 0$ the voltage u decreases. Taking into account that $dv/du > 0$ for S-type systems and $dv/du < 0$ for Z-type systems (Figs. 3.21(a) and (b)), we conclude that the front velocity decreases and increases, respectively. Similar reasoning holds for cold fronts. This results in negative feedback on front dynamics for S-type systems and positive feedback for Z-type systems.

In Figs. 3.22 and 3.23, we present numerical solutions of eqs. (3.108) and (3.109) for $R > 0$ and sufficiently small $\varepsilon = \tau_u/\tau_a$. Since in this case the relaxation of u is fast, the trajectories in the $(\langle j\rangle, u)$-plane exactly follow the load lines. The front propagation is decelerated for the S-system and accelerated for the Z-system. For the load line which intersects the line $u = u_{\text{co}}$ the system possesses a fixed point corresponding to a stationary front at a certain position $w = w_0$ (Figs. 3.22 and 3.23, trajectories 1 and 4). This point is a stable node for the S-system (Fig. 3.22) and a saddle-point for the Z-system (Fig. 3.23). A general stability analysis of stationary transverse current patterns has been given in Section 3.4. The stability criterion is

$$-\sigma_u + \frac{C\lambda_1}{\tau_a} < R^{-1} < -\sigma_d, \qquad C > 0, \tag{3.126}$$

where $\lambda_1 > 0$ is the eigenvalue of the unstable mode corresponding to translation. According to (3.126) only fronts with $\sigma_d < 0$ can be stabilized for $R > 0$. That implies that stationary fronts in Z-systems driven via an ordinary passive external circuit with $R > 0$ are never stable. The upper bound of the criterion (3.126) corresponds to the saddle-node bifurcation where the system has one real positive eigenvalue. That indicates that the stationary front has lost stability with respect to the translation mode, and monotonic front propagation will switch the system to the homogeneous state. The lower bound corresponds to an oscillatory instability where a pair of complex-conjugate eigenvalues crosses the imaginary axis. Note that the main contribution to the differential conductivity of the stationary front comes from the shift of the front and therefore generally $|\sigma_d| \gg \sigma_u$.

With increasing $\varepsilon \equiv \tau_u/\tau_a$ the system experiences a transition from instantaneous feedback to delayed feedback. Now the slow relaxation of u results in deviations of the trajectories from the load line. For S-shaped characteristics the fixed point evolves from a stable node to a stable focus (Fig. 3.24). This corresponds to oscillatory transient processes leading to the steady state (Fig. 3.24, trajectories 1 and 4). For sufficiently slow feedback the front propagation becomes self-similar as in the case of the voltage-controlled conditions. For the stationary pattern the increase of ε eventually leads to a violation of the left-hand inequality in (3.126) and the fixed

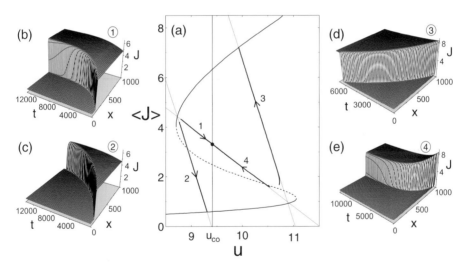

Figure 3.22. Globally coupled dynamics of lateral current-density fronts in an S-system (HHED) for positive load $R > 0$ and instantaneous feedback. (a) A phase portrait including the S-shaped current–voltage characteristic (solid line) with various load lines (dotted). The front dynamics is indicated by the trajectories 1, 2, 3, and 4. The fixed point at $u = u_{co}$ is a stable node, and corresponds to a stationary front. In (b) and (c) are shown decelerated cold fronts evolving either to a stationary front ((b); trajectory 1) or to the uniform off state ((c); trajectory 2). In (d) and (e) are shown decelerated hot fronts evolving either to a stationary front ((e); trajectory 4), or to the uniform on state ((d); trajectory 3). Parameters of the load lines are $U_0 = 11.5$ and $r = 0.42$ for trajectories 1 and 4, $U_0 = 9.4$ and $r = 0.15$ for trajectory 2, and $U_0 = 11.0$ and $r = 0.15$ for trajectory 3; $\varepsilon = 100$ and $L_x = 1000$ (after Meixner *et al.* (2000a)).

point becomes an unstable focus reflecting an oscillatory instability of the stationary front (see Meixner *et al.* (1998*a*) for an example of such behavior). However, since this instability is caused by the interaction of the front wall with the boundaries, it

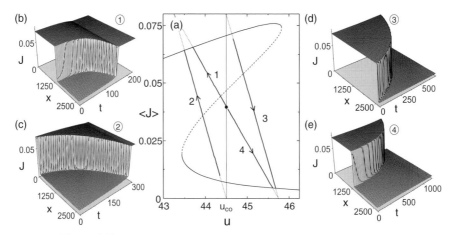

Figure 3.23. Globally coupled dynamics of lateral fronts in a Z-system (DBRT) for positive load $R > 0$ and instantaneous feedback. (a) A phase portrait including the Z-shaped current–voltage characteristic (solid line) with various load lines (dotted). The front dynamics is indicated by the trajectories 1, 2, 3, and 4. The fixed point at $u = u_{co}$ is a saddle-point. The regimes corresponding to the trajectories 1 and 4 can be realized only for special initial conditions. In (b) and (c) are shown accelerated hot fronts for trajectories 1 (b) and 2 (c). In (d) and (e) are shown accelerated cold fronts for trajectories 3 (d) and 4 (e). Parameters of the load lines are $U_0 = 96$ and $r = 45$ for trajectories 1 and 4, $U_0 = 92$ and $r = 25$ for trajectory 2, and $U_0 = 96$ and $r = 25$ for trajectory 3; $\varepsilon = 10^{-12}$ and $L_x = 500$ (after Meixner *et al.* (2000*a*)).

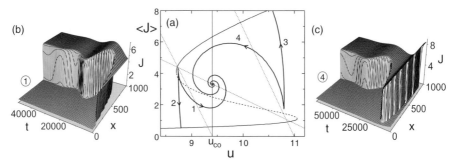

Figure 3.24. Oscillatory dynamics of globally coupled fronts in an S-system (HHED) in the case of delayed feedback ($R > 0$). (a) A phase portrait including the S-shaped current–voltage characteristic (solid line). The load lines (dotted) and the initial conditions are the same as those in Fig. 3.22, $\varepsilon = 10^8$. In contrast to Fig. 3.22, the trajectories 1, 2, 3, and 4 deviate from the corresponding load lines due to the delay in the feedback. The fixed point at $u = u_{co}$ is a stable focus. In (b) and (c) is shown oscillatory slowing down of the cold (b) and hot (c) fronts corresponding to trajectories 1 and 4, respectively.

represents essentially *a boundary effect* that is not relevant for the front dynamics in a large system ($W_f \ll L_x$). For Z-systems the phase portrait does not undergo qualitative changes with increasing ε and the stationary point always remains a saddle-point.

Global coupling via active external circuit ($R < 0$)

The type of feedback can be changed if the system is operated by an active circuit simulating a negative load resistance $R < 0$. Such a circuit has been implemented (Martin *et al.* 1994) in order to stabilize the uniform states corresponding to the intermediate branch of the DBRT current–voltage characteristic and can also be used in studies of lateral patterns. Note that, to secure $\tau_u = RC > 0$, such a circuit should also provide $C < 0$.

For $R < 0$ the propagation of a hot front is accompanied by an increase of u. That results in positive feedback for S-systems and negative feedback for Z-systems. The last case is most interesting here since efficient control over front propagation and stabilization of lateral patterns becomes possible. The front dynamics in a Z-system for $R < 0$ and instantaneous feedback is illustrated on Fig. 3.25. Now the front propagation in the Z-system is decelerated. For $C < 0$ the stability criterion takes the form (Section 3.4)

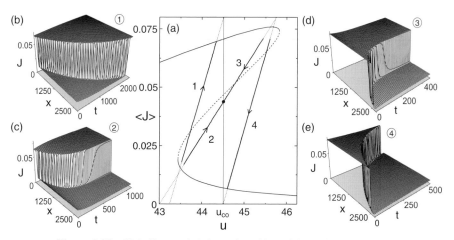

Figure 3.25. Globally coupled dynamics of lateral fronts in a Z-system (DBRT) operated via an active external circuit with $r < 0$ and $\varepsilon > 0$ ($R < 0$ and $C < 0$). The relaxation time $\varepsilon = 10^{-12}$ corresponds to instantaneous feedback. (a) A phase portrait including the Z-shaped current–voltage characteristic (solid line) and load lines (dotted). The fixed point at $u = u_{co}$ is a stable node. In (b) and (c) are shown decelerated hot fronts evolving either to the uniform on state ((b); trajectory 1) or to a stationary front state ((c); trajectory 2). In (d) and (e) are shown decelerated cold fronts evolving either to a stationary front state ((d); trajectory 3) or to the uniform off state ((e); trajectory 4). Parameters of the load lines are $U_0 = 88$ and $r = -20$ for trajectory 1; $U_0 = 87$ and $r = -40$ for trajectories 2 and 3, and $U_0 = 91.25$ and $r = -20$ for trajectory 4 (after Meixner *et al.* (2000a)).

$$-\sigma_{\rm d} < R^{-1} < -\sigma_{\rm u} + \frac{C\lambda_1}{\tau_a}, \qquad C < 0. \tag{3.127}$$

In eq. (3.127) the lower and upper bounds correspond to a saddle-node bifurcation and oscillatory instabilities, respectively. According to (3.127) a stationary pattern with $\sigma_{\rm d} > 0$ can be stable for $R < 0$ and $C < 0$. For sufficiently small C the fixed point is a stable node as in Fig. 3.25. An increase of $\varepsilon = \tau_u/\tau_a$ leads to oscillatory motion as shown in Fig. 3.26, and the fixed point becomes a stable focus. It is readily seen that the front behavior in Z-type systems for $R < 0$ is qualitatively the same as that in S-type systems for $R > 0$.

However, the implementation of the active external circuit leads also to some new regimes that have no analog for $R > 0$. First of all, if a spatially extended bistable element is operated via an active external circuit, temporal instabilities can be caused by the negative differential conductivity of this circuit itself. Indeed, for $R^{-1} > -\sigma_{\rm u}$ the condition for an oscillatory instability is fullfilled even for $C = 0$. This instability is not related to front oscillations but represents circuit-induced oscillations. These oscillations can be excluded by choosing a negative relaxation time in the external circuit.

For $R^{-1} > -\sigma_{\rm u}$, $\tau_u < 0$, $R < 0$, and $C > 0$ we can also meet a new situation if the slopes of the load line is smaller than the slopes of the on and off branches of the current–voltage characteristic. Generally, these regimes are of limited interest since: (i) being close to the current-controlled condition they are not favorable for front propagation since the load line does not connect on and off branches of the current–voltage characteristic within the bistability regime; and (ii) the global dynamics becomes unbounded in this case as an effect of the active circuit. Let us note that, for $R^{-1} > -\sigma_{\rm u}$, the propagation of a hot or cold front can be accompanied by a decrease or increase, respectively, of $\langle j \rangle$. Indeed, on eliminating u adiabatically from (3.124) and (3.125), by direct calculation we obtain

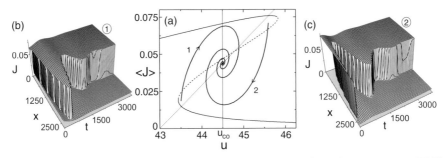

Figure 3.26. Oscillatory dynamics of current-density fronts in a Z-system (DBRT) for $R < 0$ and $C < 0$ in the case of delayed feedback, $\varepsilon = 10^{-9}$. (a) A phase portrait including the Z-shaped current–voltage characteristic (solid line). The load line (dotted) corresponds to $U_0 = 87$ and $r = -40$. In (b) and (c) is shown oscillatory slowing down of the cold (b) and hot (c) fronts corresponding to the trajectories 1 and 2, respectively (after Meixner *et al.* (2000a)).

$$\frac{dj}{dt} = \frac{1}{L_x} \frac{J_{\text{on}}(u) - J_{\text{off}}(u)}{1 + R\sigma_u} v(t). \tag{3.128}$$

Therefore, if $(1 + R\sigma_u) < 0$, we have $dj/dt < 0$ for $v > 0$. This regime, in which the type of feedback may become *indeterminate*, has been studied for the model of a gate-driven p–n–p–n thyristor structure (Meixner *et al.* 1998a).

Conclusions

Let us now summarize the main results. We have studied the dynamics of lateral current-density fronts in bistable semiconductor systems with S- and Z-shaped current–voltage characteristics (Figs. 3.18 and 3.19). Both hot fronts ($v > 0$, high-current state propagates into low-current state) and cold fronts ($v < 0$, low-current state propagates into high-current state) can be excited in these systems for $u_h < u < u_{th}$. In the voltage-controlled case the front propagates in a self-similar way and the direction as well as the absolute value of the velocity can be controlled by varying u (Figs. 3.21(a) and (b)). The front has zero velocity for a certain voltage u_{co} determined by the equal-areas rule $A = 0$ (see eq. (3.116)). The directions of the front propagation for S- and Z-type systems are different, which is reflected by the different signs of the slopes of the $v(u)$-dependency.

In the presence of global coupling via an external circuit the voltage u applied to the structure varies as the front propagates due to the dependency of the total current on the front position. This results in a feedback upon the front dynamics. The type of this feedback is determined by the slope of the $v(u)$-dependency and the type of external load. For positive external load $R > 0$ the feedback is negative for S-systems and positive for Z-systems (Figs. 3.22 and 3.23). This results in deceleration and acceleration of the front propagation, respectively. The type of the feedback can be reversed if the system is operated via an active external circuit simulating negative load $R < 0$ and capacitance $C < 0$ (the latter is required in order to keep the relaxation time RC positive). The case of negative feedback (for Z-systems) is of most interest since deceleration and efficient control over fronts in Z-systems becomes possible (Fig. 3.25). The propagation of globally coupled fronts is monotonic as long as the relaxation time of the external circuit $\tau_u = RC$ is sufficiently small and the feedback upon the front dynamics is instantaneous (Figs. 3.22 and 3.23). Delayed negative feedback can result in oscillatory slowing down both in S- and in Z-systems, as shown in Fig. 3.24 and Fig. 3.26, respectively. Generally, we conclude that S- and Z-systems are dual and the behavior of S-type systems for $R > 0$ versus $R < 0$ is qualitatively the same as the behavior of Z-type systems for $R < 0$ versus $R > 0$, respectively. The heterostructure hot-electron diode (S-system) and the double-barrier resonant-tunneling structure (Z-system) which we have used to illustrate our findings serve as prominent examples of semiconductor heterostructures falling into these two classes.

Despite the fact that globally coupled front propagation is not self-similar, due to the fast relaxation of the front profile the front velocity is determined by the instantaneous value of the voltage u with good accuracy. Therefore the equations of motion can be reduced to (3.124) and (3.125).

The type of the feedback upon propagating fronts is closely related to the stability of stationary lateral patterns considered elsewhere (Bass *et al.* 1970, Alekseev *et al.* 1998). It is known that, in the presence of global coupling, the stability of an inhomogeneous pattern (a stationary lateral current-density front or filament) that has one unstable mode in the voltage-controlled regime is determined by the sign of its differential conductance. Figures 3.21(c) and (d) help us to understand the stability criterion qualitatively. In the globally coupled regime the stability of the steady state is determined by the projection of the phase flow onto the load line. For $\sigma_d < 0$ this projection is directed toward the steady point for $R > 0$, indicating stability, as happens in S-systems; for $\sigma_d > 0$ this holds for $R < 0$, as happens in Z-systems.

Operating the system via an active external circuit with $R < 0$ results in new effects that do not occur for conventional circuits. If a spatially extended nonlinear element is operated with a load $R < 0$, a temporal instability can be induced by the active circuit itself. Therefore such circuits should provide $C < 0$ in order to exclude circuit-induced relaxation-type oscillations in the regimes considered above.

The globally coupled dynamics of traveling pulses in *excitable media* has recently been studied by Hempel *et al.* (1998). Global coupling influences the velocity of a traveling pulse but the self-similar character of the motion remains. As has been shown here, in *bistable media* the global coupling completely destroys the self-similarity, resulting in accelerated, decelerated, or oscillatory motion. These findings have important consequences for gate-driven bistable systems in which the front dynamics experiences a feedback both from the gate and from the main external circuits. Generally, these systems should be considered as bistable media with two global constraints (Gorbatyuk and Rodin 1997). Gate-driven p–n–p–n structures possess a Z-shaped and an S-shaped current–voltage characteristic in the gate (cathode–gate) and the main (cathode–anode) circuit, respectively. The interplay between the positive feedback via the gate circuit (an *activatory global constraint*) and negative feedback via the main circuit may result in *large-amplitude self-sustained oscillations* of the front (Meixner *et al.* 1998b).

It is important to note that, in semiconductor switching devices with S-shaped current–voltage characteristics, such as in thyristors, turn-on and -off can often be triggered locally via the propagation of current-density fronts (D'yakonov and Levinstein 1978). Since the device operation implies switching not only from a low-current state to a high-current state but also from a high-voltage state to a low-voltage state, the semiconductor device essentially interacts with the external circuit during this transient process and the propagation of the current-density front experiences a strong global coupling. It is not sufficient to treat only self-similar propagating fronts in this case. Additionally, nonuniform attractors (stationary filamentary states) of the transient process should be properly taken into account (Gorbatyuk and Rodin 1992a). The concepts of globally coupled front dynamics which we have developed here provide an adequate description of inhomogeneous switching processes in these devices.

Chapter 4

Impact-ionization-induced impurity breakdown

In this chapter we discuss a model system that has been studied thoroughly both experimentally and theoretically within the last decade: Impurity impact-ionization breakdown at low temperatures. This system exhibits a variety of temporal and spatio-temporal instabilities ranging from first- and second-order nonequilibrium phase transitions between insulating and highly conducting states via current filamentation and traveling waves to various chaotic scenarios. There are several models that can account for periodic and chaotic current self-oscillations and spatio-temporal instabilities. Here we focus on a model for low-temperature impurity breakdown that combines Monte Carlo simulations of the microscopic scattering and generation–recombination processes with macroscopic nonlinear spatio-temporal dynamics in the framework of continuity equations for the carrier densities coupled with Poisson's equation for the electric field. A period-doubling route to chaos, traveling-wave instabilities, and the dynamics of nascent and fully developed current filaments are discussed including two-dimensional simulations for thin-film samples with various contact geometries.

4.1   Introduction

Impact ionization of charge carriers is a widespread phenomenon in semiconductors under strong carrier heating. It is a process in which a charge carrier with high kinetic energy collides with a second charge carrier, transferring its kinetic energy to the latter, which is thereby lifted to a higher energy level. Impact-ionization processes may be classified as band–band processes or band–trap processes depending on whether the second carrier is initially in the valence band and makes a transition from the valence band to the conduction band, or initially at a localized level (trap, donor, or acceptor) and makes a transition to a band state (Landsberg 1991). Furthermore, impact-ionization processes are classified as electron or hole processes if the ionizing hot carrier is a conduction-band electron or a hole in the valence band, respectively. The result of the process is carrier multiplication, which may induce electric instabilities at sufficiently high electric fields (Shaw *et al.* 1992). Impact ionization from shallow donors or acceptors is responsible for impurity breakdown at low temperatures (Stillman *et al.* 1977). Being an autocatalytic process, it induces a nonequilibrium phase transition (Schöll 1987) between a low-conductivity and a high-conductivity state and may lead to a menagery of spatio-temporal instabilities, including self-oscillations and chaos (Schöll 1992). Band-to-band impact ionization in MOSFETs is the reason for the parasitic substrate current and influences the tail of the carrier distribution function. It eventually induces the avalanche effect, limiting the bias voltage which can be safely applied to a device.

Impact ionization has been modeled on various levels of theoretical description ranging from Shockley's (1961) simple lucky-electron model via the early work of Robbins and Landsberg (1980) and Ridley's (1983, 1987) lucky-drift model to most recent quantum-kinetic treatments, e.g. by Bude *et al.* (1992), Quade *et al.* (1994*b*), and Redmer *et al.* (2000).

In this chapter we shall focus on nonlinear spatio-temporal dynamics and chaos induced by impact ionization of shallow impurities. It has been known for a long time that current filaments or high-field domains are formed in semiconductors with S-shaped and N-shaped current-density–field characteristics, respectively (Ridley 1963). However, it is only recently that research has focussed upon the secondary bifurcation of these elementary spatial structures, leading to complex spatio-temporal dynamics (Niedernostheide 1995). Coherent spatio-temporal structures such as breathing, traveling, or rocking filaments, spatio-temporal spiking of the current density, and oscillating domains have been identified in experiments and computer simulations. In particular, the stationary structure and the possible nonlinear dynamic behavior of current filaments have been investigated widely both experimentally (Aoki *et al.* 1988, Aoki and Kondo 1991, Spangler *et al.* 1992, 1994, Peinke *et al.* 1992, Spangler and Prettl 1994, Eberle *et al.* 1996, Hirschinger *et al.* 1997*a*, Aoki 2000) and theoretically (Schöll 1987, Wacker and Schöll 1994*b*, Niedernostheide *et al.* 1994*a*, Kerner and Osipov 1994, Kozhevnikov *et al.* 1995, Novák *et al.* 1995, 1998*b*, Sablikov *et al.* 1996, Gaa *et al.* 1996*a*, Aoki 1997, Schöll *et al.* 1998*a*) in a variety of semiconducting materials, e.g. p-Ge, n-GaAs, n-InSb, and n-Si at liquid-helium temperatures.

Breathing filaments (Schöll and Drasdo 1990, Kunz and Schöll 1992) and the intermittent and chaotic behavior of laterally traveling filaments (Hüpper *et al.* 1993*b*, 1993*a*) have been studied in one-dimensional simulations in which only the transverse spatial coordinate perpendicular to the current flow was taken into account. On the one hand, there has been recent progress in the microscopic analysis of low-temperature impurity breakdown in terms of single-particle (Kuhn *et al.* 1993) and many-particle

Monte Carlo (MC) simulations (Quade *et al.* 1994*a*) for p-Ge and for n-GaAs (Kehrer *et al.* 1995*a*, Kostial *et al.* 1995*a*), on the other hand the spatio-temporal modes of the breakdown process have been investigated in detail. For instance, the dynamics of stochastically induced and spatially inhomogeneous switching has been studied in a one-dimensional longitudinal model for p-Ge, which neglects the transverse spatial degree of freedom and therefore can not explain filamentation (Kunz and Schöll 1996). Longitudinally traveling charge-density waves have been found in simulations of n-GaAs (Gaa and Schöll 1996). In order to study the nascence of current filaments the transverse and the longitudinal degrees of freedom have been combined for a realistic two-dimensional sample geometry with appropriately modeled contacts, and detailed microscopic information on the generation–recombination kinetics obtained from MC simulations has been included (Gaa *et al.* 1996*a*, 1996*b*, Schwarz *et al.* 2000*a*).

4.2   A model for impact-ionization breakdown

4.2.1   Constitutive equations

We consider a doped semiconductor at low temperatures at which the donors and acceptors are not thermally ionized, i.e. the carriers are frozen out at the impurities. In the following we shall assume that we have n-type material with donor density N_D, partially compensated by acceptors of density $N_A < N_D$; p-type semiconductors can be treated analogously. Extensions to ambipolar conduction with two types of carriers (electrons and holes) have been discussed elsewhere (Schöll 1987). The carrier density in the conduction band, and hence the current density, is determined by the generation–recombination (GR) processes of carriers between the conduction band and the donor levels. The experimentally observed S-shaped current-density–field relation in the regime of impurity breakdown can be explained in terms of standard GR kinetics only if impact ionization from at least two impurity levels is taken into account (Schöll 1982*a*). Therefore we model the infinite hydrogen-like energy spectrum of the shallow donors by the ground state and an "effective" excited state close to the band edge. In this case the state of the system can be characterized by the spatial distribution of the carrier densities in the conduction band $n(\mathbf{r}, t)$ as well as in the impurity ground state and excited state, $n_1(\mathbf{r}, t)$ and $n_2(\mathbf{r}, t)$, respectively, where $\mathbf{r}$ is the position vector and t denotes time.

In the following we shall adopt an approach that combines Monte Carlo (MC) simulations of the microscopic scattering processes and GR kinetics with rate equations for the macroscopic spatio-temporal dynamics of the carrier densities which occurs on a much slower time scale than the microscopic dynamics. First, we will list the macroscopic constitutive equations.

The temporal evolution of n is governed by the continuity equation

$$\dot{n} = \frac{1}{e} \nabla \cdot \boldsymbol{j} + \phi(n, n_1, n_2, |\mathcal{E}|), \tag{4.1}$$

where the overdot denotes the partial derivative with respect to time, e is the electron charge, and $\mathcal{E}$ is the local electric field within the sample. Within the drift–diffusion approximation the current density $\boldsymbol{j}$ can be expressed as

$$\boldsymbol{j} = e(n\mu\mathcal{E} + D\nabla n) \tag{4.2}$$

with the field-dependent mobility μ and diffusion coefficient D. We assume the validity of the Einstein relation $D = \mu k_B T_L/e$, where k_B is Boltzmann's constant and T_L is the lattice temperature.

The rate ϕ of GR processes depends on the local values of the carrier densities in the conduction band and at the impurities, given by n, n_1, and n_2, respectively, and the strength of the electric field $\mathcal{E} = |\mathcal{E}|$. Analogously, the rates ϕ_1 and ϕ_2 determining the temporal evolutions of n_1 and n_2 can be defined as

$$\dot{n}_i = \phi_i(n, n_1, n_2, \mathcal{E}) \tag{4.3}$$

with $i = 1, 2$.

In an explicit model (Schöll 1982a, 1987) for n-type GaAs at 4.2 K, the GR rates are given by

$$\phi = X_1^S n_2 - T_1^S n p_t + X_1 n n_1 + X_1^* n n_2, \tag{4.4}$$

$$\phi_1 = T^* n_2 - X^* n_1 - X_1 n n_1, \tag{4.5}$$

$$\phi_2 = -\phi - \phi_1, \tag{4.6}$$

where $p_t = N_D - n_1 - n_2$ is the density of ionized donors, X_1^S is the thermal ionization coefficient of the excited level, T_1^S is its capture coefficient, X_1 and X_1^* are the impact-ionization coefficients for ionization from the ground and excited level, respectively, and X^* and T^* denote the transition coefficients for transition from the ground level to the excited level and vice versa, respectively (Fig. 4.1).

The electric field is coupled to the carrier densities via Gauss' law

$$\epsilon \nabla \cdot \mathcal{E} = \rho \equiv e(N_D^* - n_1 - n_2 - n), \tag{4.7}$$

where $\epsilon \equiv \epsilon_0 \epsilon_r$ with the absolute and relative dielectric constants ϵ_0 and ϵ_r, respectively, ρ is the charge density, and $N_D^* \equiv N_D - N_A$ is the effective doping density. For the numerical treatment of time-dependent problems in the drift–diffusion approximation it is often advantageous (Selberherr 1984, Gajewski 1985, Kunz *et al.* 1996) to substitute Gauss' law (4.7) by the charge-conservation equation, given by

$$\nabla \cdot \boldsymbol{J} = 0 \tag{4.8}$$

with

$$\boldsymbol{J} = \epsilon \dot{\mathcal{E}} + \boldsymbol{j}, \tag{4.9}$$

where $\boldsymbol{J}$ is the total current density composed of displacement-current and conduction-current densities. Equation (4.8) can be obtained from Gauss' law (4.7) by differentiating with respect to time, and subsequent elimination of $\dot{n}$, $\dot{n}_1$, and $\dot{n}_2$ using (4.1) and (4.3), taking into account the carrier conserving character of the GR processes, i.e. $\phi + \phi_1 + \phi_2 = 0$. If the initial values of n, n_1, n_2, and $\mathcal{E}$ satisfy Gauss' law (4.7), eq. (4.7) and eqs. (4.8) and (4.9) are equivalent. One of the advantages of the above substitution is the treatment of the boundary conditions as natural boundary conditions for J (Gajewski 1985, Gajewski and Gärtner 1992).

4.2.2   The current-density–field characteristic

The spatially homogeneous steady states are given by the simultaneous solution of (4.4)–(4.6) and the local charge-neutrality condition following from (4.7)

$$N_{\mathrm{D}}^* - n_1 - n_2 - n = 0. \tag{4.10}$$

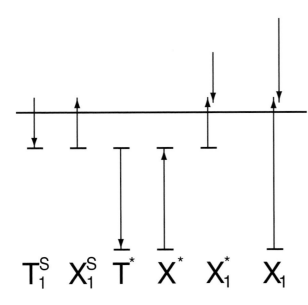

Figure 4.1. The generation–recombination (GR) model for low-temperature impurity breakdown, involving the conduction band, the donor ground state, and its first excited state. The GR processes are denoted by T_1^{S} (capture), X_1^{S} (thermal ionization), T^* (relaxation), X^* (excitation), and X_1^* and X_1 (impact ionization).

Solving (4.5) and (4.6) for $n_1(n)$ and $n_2(n)$ and substituting into (4.10) yields a polynomial equation in n of order three, which may have up to three positive solutions depending upon the values of the GR coefficients. A detailed analysis (Schöll 1987) shows that there are in general three positive solutions (two of which are stable) in a range of fields $\mathcal{E}$ with $\mathcal{E}_h < \mathcal{E} < \mathcal{E}_{th}$, leading to an S-shaped $n(\mathcal{E})$ relation. For electric fields smaller than the holding field $\mathcal{E}_h$ the system is monostable, and, if $\mathcal{E}$ exceeds the threshold field $\mathcal{E}_{th}$, again only one spatially homogeneous steady state exists. The resulting bistability can be physically understood as follows. For low fields almost all carriers are trapped in the ground state of the impurities, and n is small. When the electric field increases, the few carriers in the conduction band gain energy in the field, until they are able to impact ionize carriers from the ground level, while the excited level is practically not populated at the low temperatures considered. Breakdown (threshold switching from the high- to the low-resistivity branch) occurs at the threshold field $\mathcal{E}_{th}$ that corresponds to the (large) energy needed to ionize carriers from the ground level. On the high-n branch, when the electric field is decreased below $\mathcal{E}_{th}$, some of the carriers in the conduction band will be trapped in the excited level. Since the time for transition to the ground state is very slow, the GR cycle in the steady state runs essentially between the excited level and the conduction band. Hence the population of the excited level remains high until the energy of the free carriers is too small to impact ionize the carriers in the excited level. So the freeze-out (i.e. switching back to the high resistivity branch) occurs only at fields as low as $\mathcal{E}_h$ corresponding to the (lower) energy needed to ionize carriers from the excited level. This results in hysteresis under voltage control (see Fig. 4.2).

Under spatially homogeneous conditions, the local $j(\mathcal{E})$ characteristic follows from the $n(\mathcal{E})$ relation by

$$j(\mathcal{E}) = en(\mathcal{E})\mu(\mathcal{E})\mathcal{E}, \tag{4.11}$$

where $\mu(\mathcal{E})$ may introduce additional nonlinearities. These may lead to a modification of the S-shaped current-density–field relation resulting, for example, in an additional N-shaped postbreakdown regime as observed in ultrapure p-Ge (Teitsworth *et al.* 1983).

The two-level model is representative of a larger class of GR mechanisms involving $M \geq 2$ localized levels (the ground state and excited states), which, in the steady state, give a nonmonotonic ρ versus n relation due to depletion of the impurity ground state by impact ionization (Crandall 1970, Kastalsky 1973, Zabrodskij and Shlimak 1975, Pickin 1978). This has been elaborated upon in detail elsewhere (Schöll 1987).

4.2.3 Generation–recombination coefficients

The essential nonlinearities of the constitutive model equations (4.1)–(4.8) in the regime of low-temperature impurity breakdown are contained in the dependencies of the GR coefficients upon n, n_1, n_2, and $\mathcal{E}$. In a simple phenomenological description the strong, threshold-like dependencies of the impact-ionization coefficients upon the local field can be modeled by Shockley's (1961) formula:

$$X_1(\mathcal{E}) = X_1^0 \exp(-E_1/\mathcal{E}), \tag{4.12}$$

$$X_1^*(\mathcal{E}) = X_1^{*0} \exp(-E_1^*/\mathcal{E}). \tag{4.13}$$

A much more detailed microscopic understanding, however, can be achieved by deriving the GR coefficients from Monte Carlo simulations of the individual scattering and generation–recombination processes. For p-Ge this has been carried out by single-particle (Hüpper *et al.* 1989, Kuhn *et al.* 1993) and many-particle Monte Carlo (MC) simulations (Quade *et al.* 1994*a*). Here we focus on n-GaAs (Kehrer *et al.* 1995*a*, Kostial *et al.* 1995*a*), in which – besides ensemble Monte Carlo simulations for inhomogeneously (delta-)doped samples – single-particle MC simulations for the spatially homogeneous steady state have been performed. Thermal

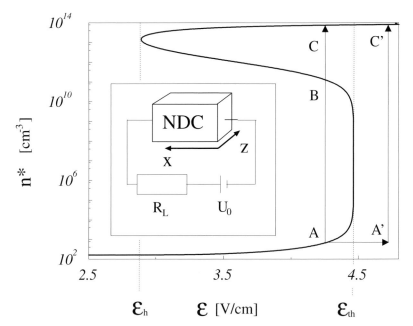

Figure 4.2. Low-temperature impurity breakdown. The S-shaped carrier-density versus electric-field relation $n(\mathcal{E})$ has been calculated from the model in Fig. 4.1 for p-Ge at 4 K. Bistability between the holding field $\mathcal{E}_h$ and the threshold field $\mathcal{E}_{th}$ occurs. Two cases of switching from the lower to the upper branch of the characteristic can be distinguished: If the field is increased to $\mathcal{E} > \mathcal{E}_{th}$ the initial low-conductivity state A becomes unstable (A'): The system switches to the upper branch (state C'); for $\mathcal{E} < \mathcal{E}_{th}$, however, sufficiently strong fluctuations may drive the system from the metastable steady state A to the highly conducting absolutely stable state C. The inset shows the typical experimental setup: The NDC device is connected to a bias voltage U_0 and a load resistor of resistance R_L (after Kunz *et al.* (1996)).

ionization of the excited donor level, acoustic-phonon-assisted recombination into the excited level (Lax 1960, Abakumov *et al.* 1978), and impact ionization both from the ground and from the excited donor level are included as band–impurity processes. The relevant intraband scattering processes are elastic ionized-impurity scattering (the Conwell–Weisskopf approximation) and inelastic acoustic-deformation-potential scattering. (Optical-phonon scattering is neglected because of the low lattice temperature, although optical-phonon emission becomes relevant for energies above 36 meV (Kostial *et al.* 1993*b*). However, those states are not frequently populated except at the highest fields (Kostial *et al.* 1995*a*).)

The microscopic rates of all band–impurity processes depend upon the carrier densities in the band and impurity states, which in turn depend upon the nonequilibrium carrier distribution function. To obtain these carrier densities, the MC method has to be combined self-consistently with the rate equations (4.1)–(4.3) in the homogeneous steady state, whereby the GR coefficients X_1, X_1^*, and T_1^S are calculated by averaging the microscopic transition probabilities (P_{ii}^1, P_{ii}^2, and P_{rec} for impact ionization from the ground state and the excited state, and capture, respectively) over the nonequilibrium distribution function $f(k)$, which is extracted from the MC simulation at each step:

$$X_1(n, n_1, n_2, \mathcal{E}) = \frac{1}{nn_1} \int d^3k \, f(k; n, n_1, n_2, \mathcal{E}) P_{ii}^1(k, n_1),$$

$$X_1^*(n, n_1, n_2, \mathcal{E}) = \frac{1}{nn_2} \int d^3k \, f(k; n, n_1, n_2, \mathcal{E}) P_{ii}^2(k, n_2), \qquad (4.14)$$

$$T_1^S(n, n_1, n_2, \mathcal{E}) = \frac{1}{np_t} \int d^3k \, f(k; n, n_1, n_2, \mathcal{E}) P_{rec}(k, p_t).$$

Note that f, and hence X_1, X_1^*, and T_1^S, in turn depend parametrically on n, n_1, n_2, and $\mathcal{E}$. An iteration procedure, whereby n_1 and n_2 are expressed in terms of their steady-state dependencies on n and $\mathcal{E}$, is used to solve the above problem self-consistently (Kehrer *et al.* 1995*a*). The microscopic scattering rates which are used in the MC simulation and the resulting nonequilibrium distribution function for two states on the S-shaped current density-field characteristic are shown in Fig. 4.3 (Kostial *et al.* 1995*a*). As expected, states of higher energy are populated more strongly in the high-conductivity state (Fig. 4.3(b)) than they are in the low-conductivity state (Fig. 4.3(a)).

As a result, the impact-ionization coefficients X_1 and X_1^* as well as the capture coefficient T_1^S depend not only on the local electric field $\mathcal{E}$, but also on the electron concentration n. This dependency on n is associated with a higher electron temperature T_e^{up} on the upper branch of the S-shaped $n(\mathcal{E})$ characteristic than the values T_e^{lo} on the lower and the middle branch, as shown in Fig. 4.4 (Gaa *et al.* 1996*b*). Here we use the notion of the electron temperature T_e in the usual sense of the mean electron energy $\frac{3}{2}k_B T_e = \langle E \rangle = [\hbar^2/(2m)]\langle k^2 \rangle$, i.e. the electron temperature is extracted from the MC data essentially as the second moment of the nonequilibrium distribution function $f(k)$. This should not be confused with the concept of a heated Maxwellian distribution function, in which the electron temperature appears as a parameter. The latter concept is much more restrictive since it assumes strong electron–electron scattering, which is not included in the present MC simulation. Note that the generalized definition of the electron temperature reduces to the conventional one in the special case of a heated Maxwellian.

The MC data for the electron temperature (dots in Fig. 4.4) can be represented by smooth fit functions (lines in Fig. 4.4). A strong increase of T_e with rising electric field is clearly visible on the high-conductivity branch as opposed to only a slight increase on the low-conductivity branch. Physically the strong increase is associated with a population inversion between the donor ground and excited states on the upper branch. For $\mathcal{E} > 9$ V cm^{-1} impact ionization no longer contributes to energy relaxation in a significant way since the donor ground states are almost completely ionized and impact ionization from the excited state, which is now more strongly populated than the ground state, dominates. The GR cycle now runs between the donor excited state and the conduction band. Because of the much smaller energies involved, this process cools less efficiently.

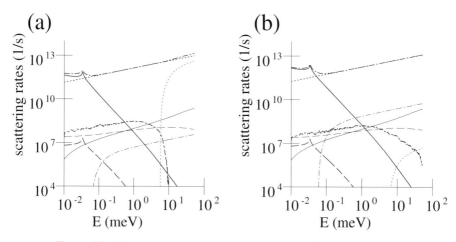

Figure 4.3. Microscopic scattering rates as a function of the carrier energy used in a Monte Carlo simulation of low-temperature impurity breakdown in n-GaAs; showing emission (thin solid line) and absorption (thin broken line) of acoustic phonons, impact ionization from the donor ground level (thin dotted line) and excited level (thin dash–dotted line), capture (thick solid line) and thermal ionization of donors (thick broken line), and ionized impurity scattering (thick dotted line). The total scattering rate is indicated by the thick dash-dotted line. These rates are calculated at 4.2 K and $\mathcal{E} = 11$ V cm^{-1} with the parameters of Table 4.2 on p. 190 for (a) the low-conductivity state and (b) the high-conductivity state of the S-shaped current-density–field characteristic. The calculated electron-distribution function $f(E)$ multiplied by the density of states $D(E)$ is shown by the ragged thick broken line (after Kostial *et al.* (1995*a*)).

It is possible to express the dependency of the GR coefficients upon $\mathcal{E}$ and n through the electron temperature $T_e(\mathcal{E}, n)$ as shown in Fig. 4.5 (Gaa *et al.* 1996*b*). X_1 exhibits a sharp increase between 30 and 50 K, while X_1^* shows a much weaker dependency on T_e because of the much smaller ionization energy.

In the following sections our strategy will be to insert the fitted analytic representations of the MC data into the macroscopic equations (4.1)–(4.8). We use this approach in order to take into account as much detailed information as possible about the microscopic scattering processes, while still retaining manageable expressions. As a check of consistency, the spatially uniform stationary $n(\mathcal{E})$ characteristic obtained with the fit functions from (4.4)–(4.6) in the steady state using the condition of charge neutrality $N_D^* = n(\mathcal{E}) + n_1(\mathcal{E}, n) + n_2(\mathcal{E}, n)$ can be compared with the characteristic obtained by direct MC simulation shown in the inset

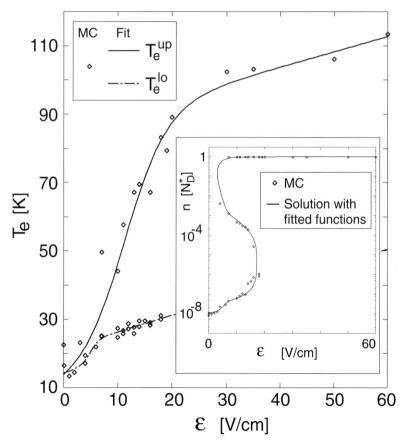

Figure 4.4. The electron temperature T_e as a function of the electric field $\mathcal{E}$ for n-GaAs at 4.2 K: Monte Carlo (MC) data and a fitted analytic representation. The inset shows the spatially homogeneous steady-state electron concentration n as a function of $\mathcal{E}$ calculated with the parameters of Table 4.2 (after Gaa *et al.* (1996*b*)).

of Fig. 4.4. The S-shaped negative differential conductivity (SNDC) is successfully reproduced.

Similar results on the impact-ionization breakdown in p-type Ge have also been obtained. The results of an ensemble MC simulation for p-Ge (Quade *et al.* 1994a) are presented in Table 4.1 and Fig. 4.6.

4.3   Chaotic dielectric relaxation oscillations

In this section we shall demonstrate that already in its simplest version the above model for impurity breakdown exhibits complex nonlinear dynamic behavior including self-generated voltage oscillations and a period-doubling route to chaos. To this end we will use the simple phenomenological expressions (4.12) and (4.13) for the

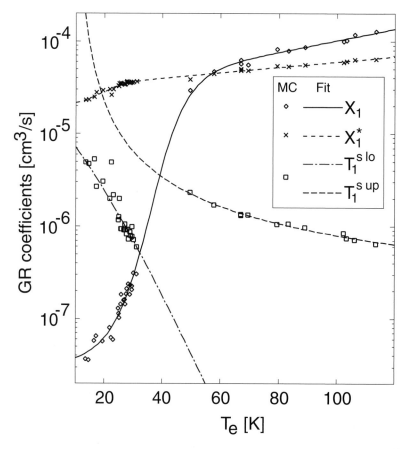

Figure 4.5. Impact-ionization coefficients X_1 and X_1^* and the capture coefficient T_1^S as functions of the electric field $\mathcal{E}$ for n-GaAs at 4.2 K: Monte Carlo (MC) data and a fitted analytic representation (after Gaa *et al.* (1996b)).

impact-ionization coefficients and confine attention to spatially homogeneous states. Current control will be assumed.

Table 4.1. *Material parameters for p-Ge at 4 K*

$N_A = 10^{14}$ cm^{-3}, $N_D = 5 \times 10^{12}$ cm^{-3}, $\epsilon_r = 16$, $T_L = 4$ K, $\mu = \mu_0 = 10^5$ cm^2 V^{-1} s^{-1}
$\tau_M = 10^{-12}$ s, $L_D = 5.6 \times 10^{-6}$ cm, $\mathcal{E}_0 = 60.8$ V cm^{-1}
$X^* = 10^{-15}$, $T^* = 7.21 \times 10^{-5}$, $X_1^S = 1.4 \times 10^{-6}$

$$X_1(\mathcal{E}) = x_1 \exp\left[-x_2(\alpha\mathcal{E})^{x_3}\right]$$
$$X_1^*(\mathcal{E}) = x_1^* \exp\left[-x_2^*(\alpha\mathcal{E})^{x_3^*}\right]$$
$$T_1^S(\mathcal{E}) = t_1 \exp\left[-t_2(t_3 + \alpha\mathcal{E})^2\right] + t_4(t_5 + \alpha\mathcal{E})^{t_6}$$

with

$x_1 = 7.85 \times 10^{-4}$	$x_1^* = 4.18 \times 10^{-2}$	$t_1 = -1.2 \times 10^{-3}$
$x_2 = 11.3$	$x_2^* = 3.72$	$t_2 = -0.254$
$x_3 = -0.745$	$x_3^* = -0.66$	$t_3 = 0.2$
$\alpha = 60.8$		$t_4 = 1.73 \times 10^{-3}$
		$t_5 = 0.421$
		$t_6 = -0.887$

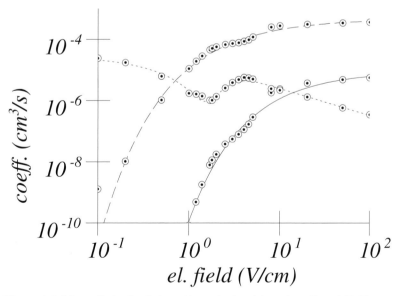

Figure 4.6. Monte Carlo simulation of impurity breakdown in p-Ge at 4 K. The simulated GR coefficients are shown together with analytic fit functions $X_1(\mathcal{E})$ (full lines), $X_1^*(\mathcal{E})$ (dashed line), and $T_1^S(\mathcal{E})$ (dotted line) (after Quade *et al.* (1994a)).

4.3.1   Oscillatory instability

Chaotic current or voltage oscillations in high-purity n-type GaAs (Aoki *et al.* 1981, Maracas *et al.* 1985, Brandl *et al.* 1987, Knap *et al.* 1988), p-type Ge (Teitsworth *et al.* 1983, Peinke *et al.* 1985, Bumeliene *et al.* 1985, Spinnewyn *et al.* 1989), n-type InSb (Seiler *et al.* 1985, Fujii *et al.* 1989), n-type Si (Yamada *et al.* 1988), n-type GaInAs (Godlewski *et al.* 1992) and in heterostructures such as GaAs MESFETs (Yano *et al.* 1992) have been observed under a variety of experimental conditions, ranging from 4 K to room temperature and including weak infrared or visible illumination as well as complete shielding against external light. In the majority of these experiments the oscillations occurred in the regime of impurity-impact ionization, indicating that GR processes between localized states and extended band states played a crucial role. The experiments can be divided into two classes: (i) driven chaos, which is induced by a periodically modulated bias voltage; and (ii) self-generated chaos, which is observed under dc conditions and is broadly independent of external circuit conditions.

The onset of chaos is usually preceded by typical bifurcation scenarios. If the first bifurcation, an oscillatory instability of the steady state, is a Hopf bifurcation, analytic conditions can easily be derived by linearizing the dynamic equations for small fluctuations around the steady state. As we have seen in Section 3.2.3, a Hopf bifurcation is not possible if only one impurity level and positive differential mobility are considered. Self-sustained oscillations are, however, obtained if the external current is modulated with a suitable driving frequency. A theoretical model of this type was first proposed by Teitsworth and Westervelt (1984). A period-doubling route to chaos upon increasing the driving amplitude was found under the assumption of an impact-ionization coefficient that increases with $\mathcal{E}$ up to a maximum, after which it decreases.

Here we shall describe a model for self-generated oscillations under dc bias that was suggested on the basis of the two-level GR mechanism depicted in Fig. 4.1 (Schöll 1985, 1986*b*). Under the condition of spatial uniformity, the constitutive equations (4.1)–(4.9) reduce to

$$\dot{n} = X_1^S n_2 - T_1^S n p_t + X_1 n n_1 + X_1^* n n_2, \tag{4.15}$$

$$\dot{n}_1 = T^* n_2 - X^* n_1 - X_1 n n_1, \tag{4.16}$$

$$\epsilon \dot{\mathcal{E}} = J - e n \mu(\mathcal{E}) \mathcal{E}, \tag{4.17}$$

where $n_2 = N_D^* - n_1 - n$ and $p_t = N_A + n$ due to the charge-neutrality condition (4.10). J is the time-independent external current density used as a control parameter, and $\mathcal{E}$ is the uniform field across the sample. The phase space of the model consists of $(\mathcal{E}, n, n_1)$ and thus has the minimum dimension which is necessary to allow chaos in an autonomous dynamic system. As discussed in Section 3.2.3, an explicit condition for a Hopf bifurcation of a limit cycle can be derived by linearizing the system (4.15)–(4.17) around the steady state for small fluctuations:

$$\delta\mathcal{E}, \delta n, \delta n_1, \delta n_2 \sim \exp(\lambda t) \tag{4.18}$$

with complex eigenvalue λ and requiring that a pair of complex conjugate eigenvalues becomes unstable. A necessary but not sufficient condition (see Section 4.4) is that the differential dielectric relaxation time

$$\tau_\mathrm{M} = \epsilon\left(en\frac{\partial(\mu\mathcal{E})}{\partial\mathcal{E}}\right)^{-1}$$

is longer than the effective impact-ionization time $1/\lambda_1$, which singles out relaxation semiconductors, and that NDC occurs.

4.3.2   The period-doubling route to chaos

For typical numerical parameters, the integration of the full nonlinear system (4.15)–(4.17) yields a sequence of period-doubling bifurcations leading to chaos (Fig. 4.7). Here $\mathcal{E}^0$, defined by the steady-state current-density–field characteristic $J = en(\mathcal{E}^0)\mu(\mathcal{E}^0)\mathcal{E}^0$, has been chosen as a control parameter, rather than J. As $\mathcal{E}^0$ is increased, first a limit cycle of period T is generated by a supercritical Hopf bifurcation (Fig. 4.7(a)). Upon further increase, oscillations of period $2T$ (b), $4T$ (c), $8T$ (d), and chaotic oscillations (e) are successively displayed. Figure 4.7(e) shows the chaotic attractor beyond the accumulation point of the period-doublings. It is characterized by nonperiodic spiral-type motion in phase space, depending sensitively upon initial conditions. A bifurcation diagram can be obtained by plotting the local maxima $\mathcal{E}_n$ of the field $\mathcal{E}(t)$ versus the control parameter $\mathcal{E}^0$ (Fig. 4.8). It displays a period-doubling cascade, followed by chaotic bands and periodic windows as $\mathcal{E}^0$ is increased. Successive local maxima can be used to construct the first-return map $f : \mathcal{E}_n \rightarrow \mathcal{E}_{n+1}$. Its graph of $\mathcal{E}_{n+1}$ versus $\mathcal{E}_n$ resembles a parabola and belongs to Feigenbaum's (1978) universality class of one-dimensional noninvertible maps with quadratic extrema. Figures 4.7 and 4.8 reveal a close similarity to the experimentally observed behavior in high-purity semiconductors. The physical origin of the oscillatory instability may be visualized as follows. Injected carriers are trapped in donors (or, in case of p-type material, acceptors), building up a space-charge field. When this field reaches a certain threshold, impact ionization of the trapped carriers sets in, increasing the free-carrier density, which results in enhanced dielectric relaxation of the field. Subsequently, the field drops below the impact-ionization threshold, and the carriers are retrapped, which completes the oscillation cycle. Thus an autocatalytic, i.e. activating, step provided by impact ionization is coupled to a restoring force, i.e. a negative feedback, provided by delayed dielectric relaxation of the field. This represents a typical activator–inhibitor mechanism. Chaos is generated by the mixing of orbits which is provided by the redistribution of trapped carriers between the ground and the excited state due to the competing impact ionization of these two levels. This elucidates the role of the additional degree of freedom furnished by the two-level model versus a single-level GR model.

Variants of this model were subsequently used to explain complex routes to chaos including period-doubling, Hopf bifurcations to a torus, torus doubling, torus wrinkling into chaos and crisis-induced intermittent chaos (Aoki *et al.* 1986, Aoki and Yamamoto 1989, Tominaga and Mori 1994). The model was also applied to n-InSb at low temperature under a strong transverse magnetic field that splits degenerate donor states into several distinct levels (Abe 1988).

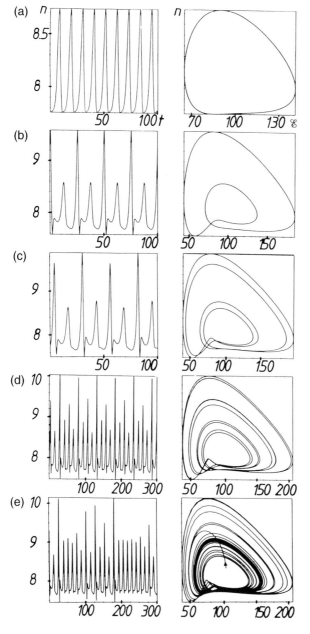

Figure 4.7. The period-doubling route to chaos in a two-level GR model for low-temperature impurity breakdown. The normalized carrier density n versus normalized time t (left-hand column) and phase portraits of n versus the normalized electric field $\mathcal{E}$ (right-hand column) are shown for increasing values of the control parameter $\mathcal{E}_0$: (a) 102 (period one), (b) 105 (period two), (c) 105.3 (period four), (d) 105.42 (period eight), and (e) 105.5 (chaos). Note the different time scales in (d) and (e) (after Schöll (1986*b*)).

4.4   **Current filamentation**

Semiconductors are spatially extended nonlinear dynamic systems that give rise to
self-organized pattern formation far from thermodynamic equilibrium. It has been
known for a long time (Ridley 1963) that current filaments are generally formed in
semiconductors with S-type negative differential conductivity (SNDC). For the model
of low-temperature impurity breakdown this can be shown rigorously by analyzing
small space- and time-dependent fluctuations around the homogeneous steady state.
Such an analysis has been carried out in detail for planar and cylindrical geometries
(Schöll 1987). It turns out that the uniform state of negative differential conductivity
is unstable against the formation of filamentary current-density distributions. In this
section we present a simple theory of the resulting stationary and oscillating filament
patterns taking into account only the transverse spatial coordinate. Throughout this
section current control will be assumed.

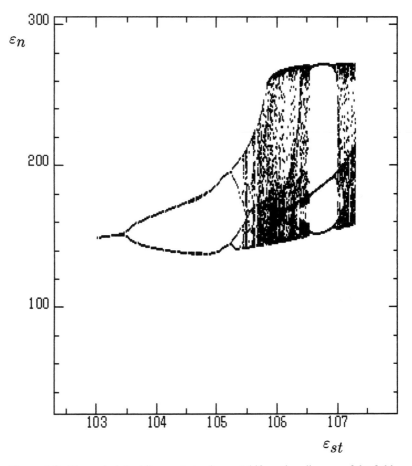

Figure 4.8. The period-doubling route to chaos. A bifurcation diagram of the field
maxima $\mathcal{E}_n$ versus the control parameter $\mathcal{E}_0$ (after Schöll (1989)).

4.4.1   Filamentary instability

The fully developed spatial structures must be calculated from the nonlinear transport equations (4.1)–(4.9). Although a full numerical solution of these equations provides the most detailed information about the spatial distributions of the current density and the electric field, it does not give immediate physical insight.

Therefore there is a need for approximations. In particular, it is desirable to derive simple analytic relations that provide some relevant information about the filaments, such as the filament radius as a function of the material parameters and the applied bias, without explicitly solving the differential equations for the spatial profiles. This is the purpose of *equal-areas* rules, which may be visualized geometrically in analogy with the Maxwell rule in thermodynamics which describes the coexistence of a gas and a liquid with a planar boundary (Schöll and Landsberg 1988). Mathematically, equal-areas rules are definite first integrals of a dynamic system, connecting control parameters with certain boundary values of the profiles of the self-organized structure (Butcher 1965). Conditions for planar or cylindrical current filaments associated with SNDC were developed in the form of equal-areas rules both for unipolar and for bipolar GR mechanisms by Schöll (1982b, 1986a).

The dominant structure formation in the system considered occurs perpendicular to the current flow, in the form of current filaments, cf. Section 3.3.2. Therefore, in the simplest approximation, in this section we assume spatial inhomogeneity in this direction only. As we will see later on in Section 4.6, contact effects may lead to spatial distortions also in the longitudinal direction, but in the present section these will be neglected, which is a reasonable approximation for a sufficiently large distance between the contacts.

The electric field can be decomposed into $\mathcal{E} = \mathcal{E}_\| + \mathcal{E}_\perp$, where $\|$ and $\perp$ denote components parallel and perpendicular to the current flow. It will turn out in the following discussion that the perpendicular internal electric field $\mathcal{E}_\perp$ is generated by the space charges in the filament wall; it is necessary in order to stabilize the filament. We consider two different geometries.

- Planar geometry: Let us choose the coordinate system such that the drift current and the drift field $\mathcal{E}_\| = \mathcal{E}_x$ are in the x-direction, and the internal space-charge field $\mathcal{E}_\perp = \mathcal{E}_z$ is in the z-direction. The sample is considered to be sufficiently small in the y-direction that we can assume homogeneity in this direction. This allows *planar filaments*. Since all occurring velocities are much smaller than the velocity of light, the electric field is treated as purely longitudinal, i.e. free from vortices ($\nabla \times \mathcal{E} = 0$). Hence $\partial \mathcal{E}_x / \partial z = 0$, and the dynamic variables are given by $n = n(z, t)$, $n_i = n_i(z, t)$, $\mathcal{E}_x = \mathcal{E}_x(t)$, and $\mathcal{E}_z = \mathcal{E}_z(z, t)$.

- Cylindrical geometry: Assuming rotational symmetry around the axis of current flow and planar Ohmic contacts gives $n = n(R, t)$, $\mathcal{E}_\| = \mathcal{E}_\|(t)$, and $\mathcal{E}_\perp = \mathcal{E}_\perp(R, t)$, where R is the radial coordinate. This describes *cylindrical filaments*.

For a planar geometry equations (4.1), (4.9), and (4.7) reduce to

$$\dot{n} = \frac{\partial}{\partial z}\left(n\mu\mathcal{E}_z + D\frac{\partial n}{\partial z}\right) + \phi(n, n_1, n_2, |\mathcal{E}|), \tag{4.19}$$

$$\dot{n}_i = \phi_i(n, n_1, n_2, \mathcal{E}) \quad \text{with } i = 1, 2, \tag{4.20}$$

$$\epsilon\frac{\partial\mathcal{E}_z}{\partial z} = \rho \equiv e(N_D^* - n_1 - n_2 - n), \tag{4.21}$$

with $|\mathcal{E}| = (\mathcal{E}_x^2 + \mathcal{E}_z^2)^{1/2}$. The temporal evolution of the electric field due to dielectric relaxation is given by

$$\epsilon\dot{\mathcal{E}}_x = J - e\langle n\rangle\mu\mathcal{E}_x, \tag{4.22}$$

$$\epsilon\dot{\mathcal{E}}_z = -e\left(n\mu\mathcal{E}_z + D\frac{\partial n}{\partial z}\right). \tag{4.23}$$

For the current balance (4.22) we have integrated the total current density $J = \epsilon\partial\mathcal{E}/\partial t + j$ over a closed surface including the cross-section of the sample, taking into account $\text{div }J = 0$, as discussed in Section 1.5.2. The brackets denote the spatial average over the sample width W: $\langle n\rangle \equiv (1/W)\int_0^W n\,dz$. Equation (4.23) can be obtained by adding (4.19) to (4.20) and combining the result with eq. (4.21).

The stability of the homogeneous steady states of the system $(n^0, n_1^0, n_2^0, \mathcal{E}_x^0, \mathcal{E}_z^0)$ against small fluctuations (Hüpper *et al.* 1993*a*)

$$(\delta n(z, t), \delta n_1(z, t), \delta n_2(z, t), \delta\mathcal{E}_x(z, t), \delta\mathcal{E}_z(z, t)) \sim e^{ikz}e^{\lambda t} \tag{4.24}$$

can be determined by linearizing (4.19)–(4.23) around this steady state. A detailed exposition of the linear stability analysis is available elsewhere (Schöll 1987), so we give only a brief summary.

Since the electric field is vortex free, $k\,\delta\mathcal{E}_x = 0$ holds. This implies that only homogeneous modes ($k = 0$) allow $\delta\mathcal{E}_x \neq 0$. This motivates the separate treatment of homogeneous and inhomogeneous modes.

For the case $k \neq 0$, $\delta\mathcal{E}_x = 0$ holds so that the linearized eq. (4.22) can be neglected. In this case a Fourier transform of the linearized system leads to three branches of the dispersion relation $\lambda(k)$, since the dynamics of the five dynamic variables n, n_1, n_2, $\mathcal{E}_x$ and $\mathcal{E}_z$ is restricted through $\delta\mathcal{E}_x = 0$ and the Poisson equation (4.21).

For convenience we introduce a vector notation $n_t = (n_1, n_2)$ and $\phi_t = (\phi_1, \phi_2)$. In the following linear stability analysis all variables are rendered dimensionless by normalizing all concentrations by the effective doping density $N_D^* = N_D - N_A$, and time and space by the effective dielectric relaxation time $\tau_M \equiv \epsilon_r\epsilon_0/(eN_D^*\mu^*)$ and the effective Debye length $L_D \equiv (D^*\tau_M)^{1/2}$, respectively. The electric field $\mathcal{E}$ is normalized by $\mathcal{E}^* \equiv k_BT_L/(eL_D)$, where μ^* and D^* are the low-field mobility and diffusion constant, respectively.

Since for normal GR kinetics the rates are linear in n_t, (4.20) leads to

$$\lambda \, \delta n_t = B(n) \, \delta n_t + d \, \delta n, \tag{4.25}$$

where the components of the matrix B and the vector d are given by $B_{ij} = \partial \phi_i / \partial n_j$ and $d_i = \partial \phi_i / \partial n$. Thus the fluctuations δn_t can be expressed in terms of δn only:

$$\delta n_t = \frac{-\mathrm{adj}(B - \lambda)}{G(\lambda)} d \, \delta n, \tag{4.26}$$

where $G(\lambda) = \det(B - \lambda)$, and $(\mathrm{adj} \, B)_{i,j}$ is $(-1)^{i+j}$ times the determinant of the matrix obtained by deleting the jth row and the ith column of B (i.e. the adjunct of B). The zeros of $G(\lambda)$ describe fluctuations of the trapped-electron concentrations with $\delta n = 0$ and $\delta \mathcal{E}_z = 0$. Now $\delta \rho$ in the linearized version of Gauss' law (4.21) can be expressed as

$$\delta \rho = -\frac{H(\lambda)}{G(\lambda)} \, \delta n. \tag{4.27}$$

Linearizing and differentiating (4.23) with respect to z, observing $\mathcal{E}_z^0 = 0$, and substituting into (4.21) yields the final eigenmode equation

$$V(\lambda) \equiv (\lambda + \mu n^0) \frac{H(\lambda)}{G(\lambda)} = -\mu k^2, \tag{4.28}$$

where $H(\lambda) = G(\lambda) - \sum_{i,j} \mathrm{adj}(B - \lambda)_{ij} \, d_j = \det(A_{\mathrm{GR}} - \lambda)$. A_{GR} is the Jacobian matrix of the subsystem describing the charge-neutral fluctuations of the carrier densities with $\delta \mathcal{E} = 0$ and $\delta \rho = 0$.

The time scales introduced in the model are

- the time associated with the transport of the carriers, i.e. the dielectric relaxation time τ_{M}, and
- the GR lifetime τ_{GR}.

Depending on the ratios of these time scales, three major regimes can be identified:

- $\tau_{\mathrm{M}} \gg \tau_{\mathrm{GR}}$ (a relaxation semiconductor, corresponding to low conductivity),
- $\tau_{\mathrm{M}} \ll \tau_{\mathrm{GR}}$ (a lifetime semiconductor, corresponding to high conductivity), and
- $\tau_{\mathrm{M}} \approx \tau_{\mathrm{GR}}$ (the intermediate regime).

For *lifetime* semiconductors we can drop the left-hand side of (4.23) and obtain the approximate form of (4.28):

$$n^0 \frac{H(\lambda)}{G(\lambda)} = -k^2. \tag{4.29}$$

For *relaxation* semiconductors we can drop the left-hand side of (4.20), thus

$$\delta \rho = -\frac{H(0)}{G(0)} \, \delta n, \tag{4.30}$$

where $H(0) = \det A_{\mathrm{GR}}$ and $G(0) = \det B$. This adiabatic elimination of the fast carrier dynamics is valid only if the quasistationary state obtained is stable against fluctuations of the fast variables (van Kampen 1985). We obtain the linearized continuity equation by summing up (4.19) and (4.20):

$$\lambda \, \delta\rho = -\mu \left(n^0 + k^2 \frac{G(0)}{H(0)} \right) \delta\rho. \tag{4.31}$$

For $\delta\rho = 0$ it follows from (4.30) and (4.23) that

$$\lambda \, \delta\mathcal{E}_z = -\mu n^0 \, \delta\mathcal{E}_z. \tag{4.32}$$

In conclusion, the approximate eigenmodes in case of a relaxation semiconductor are given by

$$\lambda_1 = -\mu n^0, \qquad \lambda_2 = -\mu \left(n^0 + k^2 \frac{G(0)}{H(0)} \right). \tag{4.33}$$

The last term can be connected with the stationary homogeneous current-density–field relation, as was discussed in a more general context in Chapter 3, Section 3.2. Since, under steady-state conditions

$$\phi_t(n, n_t^0(n, \mathcal{E}_z), \mathcal{E}_z) = 0 \tag{4.34}$$

holds, the total derivative of ϕ_t with respect to X (where X is n or $\mathcal{E}_z$) vanishes:

$$\frac{d\phi_t}{dX} = \frac{\partial \phi_t}{\partial X} + B \frac{dn_t}{dX} = 0, \tag{4.35}$$

thus

$$\frac{dn_t}{dX} = -B^{-1} \frac{\partial \phi_t}{\partial X}. \tag{4.36}$$

With $X = n$ we get

$$-\left(\frac{\partial \rho}{\partial n} \right)_{\mathcal{E}} = \frac{H(0)}{G(0)}. \tag{4.37}$$

With this the eigenvalues (4.33) read

$$\lambda_1 = -\mu n^0, \qquad \lambda_2 = -\mu \left[n^0 - k^2 \left(\frac{\partial \rho}{\partial n} \right)_{\mathcal{E}}^{-1} \right], \tag{4.38}$$

where λ_1 corresponds to a pure damped dielectric relaxation mode and λ_2 to a coupled relaxation GR mode, which may be unstable on the NDC branch of the current-density–field characteristic because $(\partial \rho / \partial n)_{\mathcal{E}} > 0$ holds there.

Equation (4.37) implies an important general theorem about the stability of the homogeneous steady states (Schöll 1987), cf. Section 3.2, eq. (3.46):

$$\prod_{i=1}^{2} \lambda_i(k = 0) = H(0) = -G(0) \left(\frac{\partial \rho}{\partial n} \right)_{\mathcal{E}}. \tag{4.39}$$

The eigenvalues associated with $G(\lambda)$ are usually all negative (which can be proven for the specific GR rates in (4.4) and (4.5)). Since ρ is a continuous function of n, the right-hand side of (4.39) changes sign between consecutive zeros. Since these zeros correspond to the homogeneous steady states, this implies that stable and unstable steady states alternate for a fixed electric field. In particular, for the middle (NDC) branch of the S-shaped current-density–field characteristic $\lambda(k = 0) > 0$ holds, corresponding to a *long-wavelength filamentary instability*.

For homogeneous modes ($k = 0$) a necessary condition for an *oscillatory* instability (Hopf bifurcation) of the special model (4.4) and (4.5) is $n\, dv/d\mathcal{E} < \lambda_1$, where λ_1 is the largest eigenvalue of A_{GR} (Schöll 1987). In dimensional units, this is exactly the condition that the differential dielectric relaxation time

$$\tau_M = \epsilon \left(en \frac{\partial(\mu\mathcal{E})}{\partial\mathcal{E}} \right)^{-1}$$

is longer than the effective impact-ionization time $1/\lambda_1$, which was already mentioned in Section 4.3. This condition can be achieved only on the NDC branch of the static current–field characteristic if the concentration n is low and the differential mobility $dv/d\mathcal{E}$ is small and positive. For the material parameters used in this section (see Table 4.1 on p. 163) the system is stable against homogeneous fluctuations.

As a result, we obtain a long-wavelength transversal instability on the NDC branch of the spatially homogeneous current-density–field characteristic. A plot of the dispersion relation resulting from (4.28) for p-type Ge at 4.2 K according to Table 4.1 (Hüpper *et al.* 1993*a*) reveals that $\lambda > 0$, i.e. an instability, occurs for a range of k-vectors. Fluctuations with wave vectors $k < k_c$, where k_c is the marginal wave vector with $\lambda(k_c) = 0$, grow in time. For a sufficiently high current density, the eigenvalue spectrum according to (4.28) matches perfectly with the approximate form (4.29), which allows one to neglect the fast transversal dielectric relaxation in this regime. For lower current densities significant differences between the exact solution and the lifetime approximation are seen.

The undamped inhomogeneous transversal fluctuations trigger the formation of current filaments. The physical origin of this instability is the autocatalytic impact-ionization process. As the fluctuations grow, the nonlinearities of the dynamic system come into play, and the full nonlinear system must be considered. In the next section we shall show that these nonlinear equations do indeed allow stationary solutions in the form of current filaments.

4.4.2   Stationary filaments

As we have shown in the preceding section by a linear mode analysis, the spatially homogeneous steady state of negative differential conductivity (SNDC) is unstable against transversal spatial fluctuations that lead to the bifurcation of current filaments. The fully developed stationary current filaments will now be discussed within a simple one-dimensional theory that neglects longitudinal spatial variations, but takes the nonlinearities fully into account.

In the steady state, the constitutive equations (4.21) and (4.23) reduce to

$$\epsilon \frac{\partial \mathcal{E}_z}{\partial z} = \rho(n, \mathcal{E}), \tag{4.40}$$

$$D \frac{\partial n}{\partial z} = -n\mu \mathcal{E}_z, \tag{4.41}$$

while (4.22) yields the integrated current–field relation

$$J = e\langle n \rangle \mu \mathcal{E}_x. \tag{4.42}$$

The solutions $n(z)$ and $\mathcal{E}_z(z)$ can be conveniently discussed in the $(n, \mathcal{E}_z)$-plane. We briefly summarize the discussion given in detail by Schöll (1987). The solutions include filamentary profiles that describe the spatial coexistence of the two stable homogeneous phases n_l (low conductivity) and n_h (high conductivity). The condition for the existence of such a kink-shaped profile $n(z)$ with the boundary conditions

$$n(-\infty) = n_l, \quad n(\infty) = n_h \tag{4.43}$$

(or, equivalently, with n_l and n_h interchanged) with a thin planar interfacial layer (filament wall) can be cast into the form of a simple geometrical construction under the assumption $|\mathcal{E}_z| \ll |\mathcal{E}_x|$: An *equal-areas rule*.

Solving eq. (4.41) for $\mathcal{E}_z$, differentiating, and substituting into (4.40) gives

$$\frac{\partial}{\partial z} \left(\frac{1}{n} \frac{\partial n}{\partial z} \right) + \frac{\mu}{\epsilon D} \rho(n, \mathcal{E}_x) = 0, \tag{4.44}$$

which can be cast into the form

$$\frac{\partial}{\partial z} \left(\frac{\partial \ln n}{\partial z} \right) + \frac{\mu}{\epsilon D} \rho(n, \mathcal{E}_x) = 0. \tag{4.45}$$

Equation (4.44) can be multiplied by $(1/n)\, \partial n/\partial z$ and integrated over $\int_{-\infty}^{+\infty} dz$, yielding

$$\frac{\mu}{\epsilon D} \int_{n_l}^{n_h} \rho(n, \mathcal{E}_x) \frac{dn}{n} = 0, \tag{4.46}$$

where the boundary conditions $n(-\infty) = n_l, n(\infty) = n_h$, and $(\partial n/\partial z)(-\infty) = (\partial n/\partial z)(\infty) = 0$ have been used.

Condition (4.46) singles out a specific value of the electric field $\mathcal{E}_x$: The *coexistence* field $\mathcal{E}_{co}$ with $\mathcal{E}_h < \mathcal{E}_{co} < \mathcal{E}_{th}$. In terms of $(n, \mathcal{E}_z)$ phase portraits, the coexistence solutions correspond to heteroclinic orbits (saddle-to-saddle separatrices), whereas for all other values of $\mathcal{E}_x$ only homoclinic orbits (closed saddle-to-saddle-loops) exist.

The coexistence condition (4.46) may be visualized as an equal-areas rule requiring that the two hatched areas in Fig. 4.9 be equal. This is analogous to Maxwell's rule in thermodynamics that determines the pressure under which the liquid and gas phases can coexist with a planar boundary; this is characteristic of first-order phase transitions, cf. the discussion in Section 3.1.

The current–voltage characteristic expected from this simple model which neglects inhomogeneities in the x-direction (hence $\mathcal{E}_x$ is proportional to the sample voltage) is given by (4.42). It is schematically shown in Fig. 4.10(a). The full and dashed lines correspond to homogeneous steady states with positive and negative differential conductivities, respectively. The dash–dotted branch is associated with a current filament. Its almost vertical part corresponds to the lateral widening of the filament by moving fronts with strongly increasing current whereas the nonmonotonic parts at its lower and upper ends describe the nucleation of a current filament from a homogeneous low-conductivity state (I), and its transition to a homogeneous high-conductivity state (III), respectively. For sufficiently large sample widths these branches are only slightly above, or slightly below, the respective homogeneous branches. Owing to the larger conductivity, i.e. larger slope, in the high-conductivity state (III), the current differences $\Delta I = I_4 - I_3$ of the turning points in the II–III transition region are much larger than those ($\Delta I = I_2 - I_1$) in the I–II transition region. In the latter region the filamentary branch may even become monotonic, as shown in Fig. 4.10(b) for samples with smaller widths. This explains why hysteresis is more likely to be experimentally observed in the II–III transition region than it is in the I–II transition region (Kostial et al. 1995a). Since the latter refers to nucleation phenomena – which are generally very susceptible to perturbations and fluctuations – it is plausible that the details of these transitions, e.g. jumps or steps, depend sensitively upon inhomogeneities. As

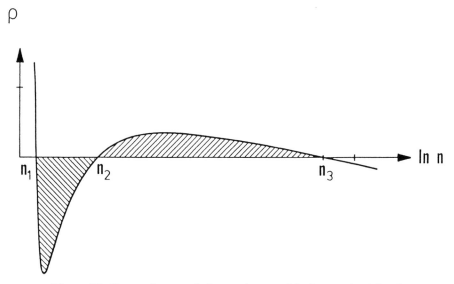

Figure 4.9. The equal-areas rule for coexistence of the low-conductivity phase n_1 and the high-conductivity phase n_3 with a planar interface. The two hatched areas must be equal. The charge density ρ is plotted versus the logarithm of the carrier density n.

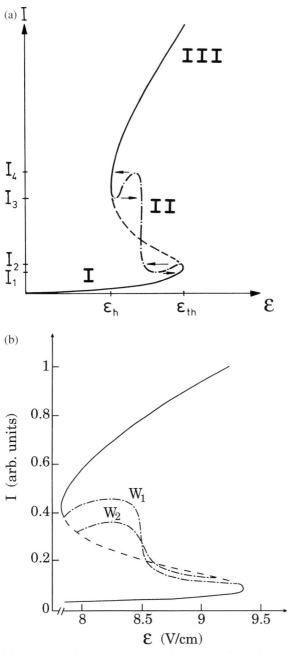

Figure 4.10. The current–voltage characteristic including filamentary states during impurity-impact-ionization breakdown (planar geometry). (a) A wide sample. Stable and unstable homogeneous states are represented by full and dashed lines, respectively. Inhomogeneous filamentary states are shown as dash–dotted lines. Nonequilibrium phase transitions between homogeneous states and current filaments under current control are indicated by arrows. (b) Narrow samples (widths $W_1 = 100 L_D$ and $W_2 = 75 L_D$, where L_D is an effective Debye length.); calculated for n-GaAs at 4.2 K with Neumann boundary conditions (after Schöll (1987)).

discussed elsewhere (Schöll 1987, Schimansky-Geier *et al.* 1991), there are profound analogies with nucleation phenomena such as metastability and supersaturation that also occur in equilibrium phase transitions.

Under approximately current-controlled conditions, the system switches along a horizontal load line between the branches, as indicated by arrows in Fig. 4.10(a). Note that there are small regions of NDC just before the switches occur, as has been observed experimentally (Kostial *et al.* 1995*a*). With decreasing sample width W, the filamentary branch shrinks as shown in Fig. 4.10(b), such that the hysteresis loop becomes smaller and finally vanishes. Physically, this occurs when the width becomes so small that it can not accommodate two filament walls, which have an intrinsic length scale of the order of a few Debye lengths (Schöll 1987).

4.4.3 Breathing filaments

The stationary current filaments may become unstable under certain conditions, resulting in oscillatory spatio-temporal dynamics. "Breathing" current filaments were suggested as a possible mechanism for self-generated current oscillations (Schöll 1987, 1988, Schöll and Drasdo 1990). Space- and time-resolved experimental investigations have indeed demonstrated that, in a certain regime of the S-shaped current–voltage characteristic, there arise spontaneous nonlinear oscillations that are localized at the filament boundary, which represent a breathing motion of the filament walls (Brandl *et al.* 1989, Rau *et al.* 1991).

In this section we shall discuss breathing current filaments under the following simplifying approximations (Schöll and Drasdo 1990).

■ There is cylindrical symmetry, and $n = n(R, t)$, $\mathcal{E}_\parallel = \mathcal{E}_\parallel(t)$, and $\mathcal{E}_\perp = \mathcal{E}_\perp(R, t)$ with the radial coordinate R.

■ The occupation of the various impurity levels relaxes on a faster time scale than do $\mathcal{E}$ and n, such that the concentration of trapped carriers distributed over the ground state and the excited states of the impurities can be expressed as $n_1(n, \mathcal{E})$ and $n_2(n, \mathcal{E})$, respectively, in eq. (4.7), using (4.3) in the steady state. This corresponds to the approximation of the relaxation semiconductor discussed in the previous section.

■ $|\mathcal{E}_\parallel| \gg |\mathcal{E}_\perp|$ holds.

■ Simple phenomenological expressions will be used for the GR coefficients rather than the Monte Carlo data described in Section 4.2.2.

First, we shall restate the constitutive equations (4.1) and (4.7) for a cylindrical geometry:

$$\dot{n} = D\left(n'' + \frac{n'}{R}\right) + \frac{1}{R}\frac{\partial}{\partial R}(Rn\mu\mathcal{E}_\perp), \tag{4.47}$$

$$\epsilon \left(\mathcal{E}'_\perp + \frac{\mathcal{E}_\perp}{R} \right) = \rho(n, \mathcal{E}_\parallel) \equiv e(N_D^* - n_1 - n_2 - n). \tag{4.48}$$

The prime denotes the derivative with respect to the radial coordinate. A quick check of consistency shows that eqs. (4.47) and (4.48) and (4.22) and (4.23) contain the homogeneous steady state given by $\rho(n, \mathcal{E}_\parallel) = 0$, $\mathcal{E}_\perp = 0$, and $J = en\mu(\mathcal{E}_\parallel)\mathcal{E}_\parallel$. The essential nonlinearity is contained in $\rho(n, \mathcal{E}_\parallel)$ due to the GR processes implicit in $n_1(n, \mathcal{E}_\parallel)$ and $n_2(n, \mathcal{E}_\parallel)$. Impact ionization from at least two impurity levels may induce a nonmonotonic dependency of ρ on n in some range of applied fields $\mathcal{E}_\parallel$, resulting in bistability of the uniform steady states (low conductivity n_l and high conductivity n_h), as discussed in Section 4.2.3.

In the time-independent case eqs. (4.47) and (4.48) and (4.22) and (4.23) reduce to

$$\mathcal{E}'_\perp + \frac{\mathcal{E}_\perp}{R} = \rho(n, \mathcal{E}_\parallel)/\epsilon, \tag{4.49}$$

$$n\mu\mathcal{E}_\perp + Dn' = 0. \tag{4.50}$$

The solutions $n(R)$ and $\mathcal{E}_\perp(R)$ include filamentary profiles that are singled out by the boundary conditions

$$n(0) \approx n_h, \quad n(\infty) = n_l, \quad \mathcal{E}_\perp(0) = 0. \tag{4.51}$$

The filaments have a radius $R_0^*(\mathcal{E}_\parallel)$ and a thin wall (transition layer) of thickness $\Delta R \ll R_0^*$. Solving eq. (4.50) for $\mathcal{E}_\perp$, differentiating, and substituting into (4.49) gives

$$\left(\frac{n'}{n} \right)' + \frac{n'}{Rn} + \frac{\mu}{\epsilon D}\rho(n, \mathcal{E}_\parallel) = 0 \tag{4.52}$$

in analogy with the result (4.44) for a planar geometry. Similarly, eq. (4.52) can be integrated over $\int_0^\infty (n'/n)\, dR$, yielding

$$\Sigma + \frac{\mu}{\epsilon D} \int_{n_h}^{n_l} \rho(n, \mathcal{E}_\parallel)\, \frac{dn}{n} = 0, \tag{4.53}$$

where

$$\Sigma \equiv \int_0^\infty \frac{1}{R}\left(\frac{n'}{n} \right)^2 dR \approx \frac{1}{R_0^*} \int_0^\infty \left(\frac{n'}{n} \right)^2 dR \tag{4.54}$$

can be evaluated self-consistently for $R_0^* \gg \Delta R$ (Schöll 1986a); it is analogous to a surface-tension term in thermodynamics. Equations (4.53) and (4.54) determine $R_0^*(\mathcal{E}_\parallel)$ as a monotonically decreasing function of the applied field $\mathcal{E}_\parallel$. At the coexistence value $\mathcal{E}_\parallel = \mathcal{E}_{co}$ introduced in the preceding section for planar filaments, $R_0^* \to \infty$, hence $\Sigma \to 0$, and the equal-areas rule for a planar geometry (4.46) is recovered.

The stationary filamentary state $R_0^*(\mathcal{E}_\parallel)$ leads to an additional, monotonically decreasing branch in the stationary current–voltage characteristic. A detailed analysis of the stability of filaments and of the influence of finite lateral boundaries shows that different nonequilibrium phase transitions associated with the nucleation (at a minimum radius) and growth of current filaments can be induced (Schöll 1987). These phenomena are analogous to droplet formation in a first-order gas–liquid phase transition.

Breathing filaments are characterized by a rigid shift of the filament wall with negligible change in the profile. In order to model such dynamics, one can identify as the most relevant dynamic variables the radius of the current filament $R_0(t)$ and the longitudinal field $\mathcal{E}_\parallel(t)$. For stationary filaments $R_0 = R_0^*$ holds. We use the *Ansatz*

$$n(R, t) = n^0(R - R_0(t) + R_0^*). \tag{4.55}$$

The spatial degrees of freedom in eqs. (4.47) and (4.23) may be projected out by multiplying with a suitable weight function G and integrating over $\int_0^\infty dR$ in a similar way to that in the derivation of the static equal-areas rule (4.53). The choice of G is guided by the requirement that it should be sharply peaked at the filament wall in order to project out the relevant contribution of the profile. With $G = (\partial/\partial R)n^0(R - R_0 + R_0^*)$ we obtain from eq. (4.47) approximately

$$\frac{\partial R_0(t)}{\partial t} = D(\mathcal{E}_\parallel)\left(\frac{1}{R_0^*(\mathcal{E}_\parallel)} - \frac{1}{R_0}\right). \tag{4.56}$$

Equation (4.56) holds only if the filament radius is larger than the width of the filament wall. The dynamics of the field $\mathcal{E}_\parallel$ is given by eq. (4.22), which has already been integrated over the cross-section of the current flow. With the help of the approximation

$$\langle n\rangle(R_0) \equiv \frac{1}{A}\int_A n(R, t)\, df \approx n_l + (n_h - h_l)\frac{\pi R_0^2}{A} \tag{4.57}$$

eq. (4.22) may be formulated in terms of the dynamic variables $R_0(t)$ and $\mathcal{E}_\parallel$ only:

$$\epsilon\dot{\mathcal{E}}_\parallel = J - e\mu\mathcal{E}_\parallel\langle n\rangle(R_0). \tag{4.58}$$

Equations (4.56) and (4.58) represent a nonlinear dynamic system, which can be regarded as an expansion of the partial differential equations (4.19) and (4.22) in terms of the *breathing* mode.

Self-generated oscillations induced by periodically breathing current filaments correspond to limit cycles in the $(R_0, \mathcal{E}_\parallel)$ phase space. The condition for a Hopf bifurcation of such a limit cycle can be obtained analytically by linearizing eqs. (4.56) and (4.58) around the steady state. It is a special case of the general conditions discussed in Chapter 3, Section 3.2.3. A Hopf bifurcation occurs for sufficiently large load resistance and negative differential conductance upon decreasing the current density J (Schöll and Drasdo 1990).

The oscillatory mechanism is based upon two features.

■ An instability of the filament radius $R_0(t)$, such that R_0 tends to increase further if R_0 is above a critical radius $R_0^*(\mathcal{E}_\parallel)$, and to decrease for $R_0 < R_0^*(\mathcal{E}_\parallel)$. Thus R_0^* is analogous to the critical droplet radius (*critical nucleus*) in equilibrium or nonequilibrium phase transitions. The microscopic mechanism in our case is impact ionization of impurities, but this enters only implicitly into the dynamics through the function $R_0^*(\mathcal{E}_\parallel)$. Any other autocatalytic mechanism that yields a *decreasing* function $R_0^*(\mathcal{E}_\parallel)$ will furnish similar results. This is the *activator* variable.

■ A restoring force, which is provided by dielectric relaxation of the longitudinal electric field $\mathcal{E}_\parallel$. It is essentially controlled by the average carrier density $\langle n \rangle (R_0)$ which forces $\mathcal{E}_\parallel$ to decrease with increasing R_0.

The physical mechanism underlying the breathing oscillations is thus similar to that of the spatially homogeneous dielectric relaxation oscillations discussed in Section 4.3. The two models are simple approximations of two different modes of dielectric relaxation oscillations: *breathing* and *bulk-dominated*, respectively. Both types of oscillations are predicted to occur on the negative conductance branch of the (filamentary or uniform, respectively) current–voltage characteristic, generated by Hopf bifurcations with decreasing applied current. This agrees well with experiments on p-Ge at 4 K by Rau *et al.* (1991), who found two different oscillation modes in different regimes of the falling current–voltage characteristic, associated with large-amplitude bulk-dominated relaxation oscillations and small-amplitude structure-limited (breathing) oscillations, respectively.

4.4.4   Ostwald ripening under global constraints

The formation of stable stationary filaments can be understood as a result of global circuit interactions in analogy with Ostwald ripening in a van der Waals gas (Schimansky-Geier *et al.* 1991, 1992, Kunz and Schöll 1992), if a load resistance $R_L \neq 0$ is considered. The reduced model for the dynamics of cylindrical filaments derived in Section 4.4.3 thus furnishes an example of the principle of global coupling that was introduced in Section 3.1 for the simple cubic reaction–diffusion model. If the dielectric relaxation occurs fast, the transverse and parallel electric fields, $\mathcal{E}_\perp$ and $\mathcal{E}_\parallel$, respectively, can be eliminated adiabatically. The latter gives with $J \equiv U_0/(R_L A)$

$$J = \left(\frac{L}{R_L A} + e \langle n \rangle \mu \right) \mathcal{E}_\parallel. \tag{4.59}$$

Substituting

$$\mathcal{E}_\parallel = \frac{J}{L/(R_L A) + e \langle n \rangle \mu} \tag{4.60}$$

into eq. (4.47) introduces a global interaction into eq. (4.47) via the laterally averaged carrier density $\langle n \rangle \equiv \int_A n(R, t)\,df/A$. Performing now a similar procedure to that in Section 4.4.3 yields the reduced dynamic equation (4.56) for $R_0(t)$ with a time-dependent

$$R_0^* = \frac{\int_0^\infty dR\,(n')^2}{\int_0^\infty dR\,\frac{1}{n}(n')^3 + \frac{\mu}{\epsilon D}\int_{n_1}^{n_3} dn\,n\rho(n, \mathcal{E}_\parallel(\langle n \rangle))}. \tag{4.61}$$

The denominator in this equation corresponds to the supersaturation discussed in Chapter 3, eq. (3.19). Its dependency upon $\mathcal{E}_\parallel$ may be approximated by $\alpha(\mathcal{E}_\parallel - \mathcal{E}_{co})^{1/2}$ with $\alpha > 0$ for $\mathcal{E}_\parallel \approx \mathcal{E}_{co}$. Owing to the global coupling (4.60) the denominator decreases with increasing $\langle n \rangle \sim R_0^2$, whence the critical radius R_0^* increases with increasing $R_0(t)$. Hence, by (4.56), the growth of R_0 is reduced, thus stabilizing the filament at a finite radius.

Physically, the formation of a stable filament can be understood as follows. As the filament radius R_0 increases, $\langle n \rangle$ and the current through the sample increase. Thus, the voltage drop across the load resistor increases, and the sample voltage $\mathcal{E}_\parallel L$ decreases. This results in an increase of R_0^* to a stable, finite value.

If several current filaments are coupled via this global circuit interaction, the initially largest filament will eventually survive as a stable, stationary filament, i.e. the set of globally coupled equations (4.56) for the radii of the individual filaments has several fixed point attractors, each corresponding to a *single* filament (Schimansky-Geier *et al.* 1991). This is the result of the winner-takes-all dynamics introduced by the global constraint. If we now consider the case that the dielectric relaxation time of $\mathcal{E}_\parallel$ is finite, breathing filaments occur as stable periodic solutions of eqs. (4.56) and (4.58). (Note that the same behavior can be induced by a global constraint with a discrete time delay (Schimansky-Geier *et al.* 1992).) Again, the coupling of several sets of equations (4.56) with (4.58) leads to a selection mechanism allowing the survival of the initially largest filament only. As before, a single stationary filament is asymptotically approached. Thus the breathing limit cycles are suppressed by the global coupling, and the system again has a number of fixed point attractors corresponding to single stable filaments, albeit transient symmetric breathing of the filaments is possible due to a periodic repeller that separates the different basins of attraction in phase space (Kunz and Schöll 1992).

4.4.5   The complex dynamics of current filaments

In order to obtain more insight into the potentially complex nonlinear dynamic behavior of the current filaments, we shall now analyze the full nonlinear equations (4.19)–(4.23) for a planar one-dimensional geometry (Hüpper *et al.* 1993a). In the numerical simulations we shall employ a more elaborate model for the GR processes than that in the preceding section, using Monte Carlo data for p-type Ge (Quade *et al.* 1994a) as given in Table 4.1. As throughout this chapter, all formulae are written for n-type material. In order to apply these formulae to p-type materials, one should replace electron densities by hole densities, N_D by N_A and vice versa, and the signs of the electric fields and charge density should be reversed. This has to be borne in mind when interpreting the figures given below.

Depending on the actual ratio of the relevant time scales, different strategies have to be applied for this purpose. For higher current densities, for which the dielectric relaxation is very fast compared with the GR processes, we can eliminate the electric field adiabatically. This implies for the longitudinal field

$$\mathcal{E}_x = \frac{J}{e\mu\langle n\rangle}. \tag{4.62}$$

The transverse displacement current density can also be neglected, which implies that the drift current density $e\mu n\mathcal{E}_z$ is compensated by the diffusion current density $eD\,\partial n/\partial z$. Physically, this means that the transport of the free carriers due to drift and diffusion is much faster than GR processes. The carriers rearrange themselves instantaneously (on the faster time scale τ_M) in such a manner that

$$D\frac{\partial n}{\partial z} = -\mu n\mathcal{E}_z \tag{4.63}$$

is satisfied. There remains a *local* dynamics

$$\frac{\partial n}{\partial t} = \phi(n, n_1, n_2, \mathcal{E}) \tag{4.64}$$

together with (4.20) and the additional constraints (4.21), (4.62), and (4.63).

In the intermediate-time-scale regime, the full dynamic system has to be taken into account. For the simulation a particle-in-cell (PIC) algorithm (Hockney and Eastwood 1981, Cenys *et al.* 1992) is used. We divide the cross-section of the sample into N equal cells. In each cell C_i ($i = 1, \ldots, N$) we assume that we have homogeneously distributed carrier densities $n^{(i)}$ and $n_t^{(i)}$, which are considered as "macroparticles". This allows us to write the charge density as

$$\rho(z, t) = \sum_{i=1}^{N} \rho^{(i)}(t)\Theta_i(z, t) \tag{4.65}$$

with

$$\rho^{(i)}(t) = e\left(N_D^* - n^{(i)}(t) - \sum_{j=1}^{2} n_j^{(i)}(t)\right), \tag{4.66}$$

where

$$\Theta_i(z, t) = \begin{cases} 1 & \text{for } z \in [z^{(i)} - h/2, z^{(i)} + h/2], \\ 0 & \text{otherwise.} \end{cases} \tag{4.67}$$

$z^{(i)} = (i - \frac{1}{2})W/N$ is the center of the macroparticle i. We obtain the transverse electric field in the center of each cell $\mathcal{E}_z^{(i)}(t) = \mathcal{E}_z(z^{(i)}, t)$ by integration of (4.21):

$$\mathcal{E}_z^{(i)} = \mathcal{E}_z^{(0)} + \frac{h}{\epsilon}\left(\frac{\rho^{(i)}}{2} + \sum_{j=1}^{i-1} \rho^{(j)}\right), \tag{4.68}$$

where $h = W/N$ is the length of each cell and $\mathcal{E}_z^{(0)}$ is the boundary condition $\mathcal{E}_z(0, t)$. The carrier densities $n^{(i)}$ can vary in the time interval Δt due to (i) drift, (ii) diffusion, and (iii) GR processes. Δt is assumed to be so short that we need consider the linearized processes only.

The center $z^{(i)}$ of each macroparticle will move in Δt due to the drift according to

$$z^{(i)} \mapsto z^{(i)} + \mu \mathcal{E}_z^{(i)} \Delta t. \tag{4.69}$$

The weight of each macroparticle varies due to GR processes:

$$n^{(i)} \mapsto n^{(i)} + \phi(n^{(i)}, \boldsymbol{n}_t^{(i)}, \mathcal{E}_z^{(i)}) \Delta t. \tag{4.70}$$

The trapped carriers are treated in a similar manner.

The diffusion tends to expand a macroparticle and decrease its weight by

$$h \mapsto h + \Delta h = h + \frac{2D \Delta t}{h}, \tag{4.71}$$

$$n^{(i)} \mapsto n^{(i)} \frac{h}{h + \Delta h}. \tag{4.72}$$

The latter is a good approximation if $\Delta h \ll h$, whence $\Delta t \ll h^2/(2D)$.

The above mappings are performed for all cells. The new carrier distribution is calculated via

$$n(z, t + \Delta t) = \sum_{i=1}^{N} n^{(i)}(t + \Delta t)\Theta_i(z, t + \Delta t). \tag{4.73}$$

For the next time step, new macroparticles are created from the former ones with the initial center $z^{(i)} = (i - \frac{1}{2})W/N$ and cell length $h = W/N$, but different weights, calculated through integration of (4.73):

$$n^{(i)} = \int_{z^{(i)}-h/2}^{z^{(i)}+h/2} n(z, t + \Delta t)\, dz. \tag{4.74}$$

From the point of view of the "particles", appropriate boundary conditions are given by vanishing particle flow at each boundary. So, if a macroparticle or a part of it attempts to cross the boundary of the sample, it is forced to remain in the cell C_1 or C_N, respectively.

From a macroscopic point of view, appropriate boundary conditions are fixed values of the transverse electric field at the boundaries $\mathcal{E}_z(0, t) = \mathcal{E}_z(W, t) = 0$. The above-mentioned conservation of particles implies global charge neutrality, i.e.

$$Q = \int_0^W \rho(z, t)\, dz = 0. \tag{4.75}$$

This together with (4.21) ensures that $\mathcal{E}_z(0, t) = \mathcal{E}_z(W, t)$. Since these values can be controlled via the integration constant $\mathcal{E}_z^{(0)}$, the "macroscopic" boundary conditions are satisfied.

For the intermediate-time-scale regime, the above algorithm is implemented as described. Additionally, the dynamics of the longitudinal field has to be taken into account.

For the lifetime regime, the difference between the time scales of the GR and transport processes does not allow a direct application of the algorithm. As described above, the dynamics of the electric field can be eliminated adiabatically. The remaining local dynamics (4.64) and (4.20) can be solved fast with a second-order Runge–Kutta method (Press *et al.* 1992). In addition to Gauss' law (4.21), eq. (4.63) has to be observed. This is done by a relaxation method with the aid of the above algorithm. The steady-state solution of the continuity equation

$$\dot{n} = \frac{\partial}{\partial z}\left(n\mu\mathcal{E}_z + D\frac{\partial n}{\partial z}\right) \tag{4.76}$$

is found with the particle-in-cell algorithm (the GR part being omitted). With the boundary condition $v_z(0) = v_z(W) = 0$ this solution satisfies (4.63).

Following analytic investigations (Schöll 1982b, 1983, 1987), we expect stable stationary filaments (see Section 4.4.2) for the case of *lifetime* semiconductors for infinite boundary conditions. If the dielectric relaxation of the electric field slows down, an oscillatory instability in terms of breathing filaments becomes possible, as discussed in Section 4.4.3. There we have found a Hopf bifurcation leading to periodic oscillations of the filament walls and hence of the current at low current densities J, while at higher current densities the filaments are stable.

Our numerical simulations corroborate these approximate analytic results. For higher current densities, in the lifetime regime, we do indeed find stable stationary filaments. Figure 4.11(a) shows a stable filament at the boundary of the sample. The time series of $\mathcal{E}_x$ and the transverse voltage $U_z = \int_0^W \mathcal{E}_z\, dz$ indicate that a stable structure is asymptotically approached after some transient oscillations. On lowering the current density, the filament walls become unstable which leads to small self-sustained oscillations of the macroscopic variables $\mathcal{E}_x$, $\langle n \rangle$, and U_z. This can be identified as the case of *breathing* filaments. A further reduction of the current-density results in a new spatio-temporal instability: The above-described breathing motion grows in time until the filament becomes detached from the boundary $z = W$ and travels to the other side ($z = 0$), where a similar breathing instability develops. As

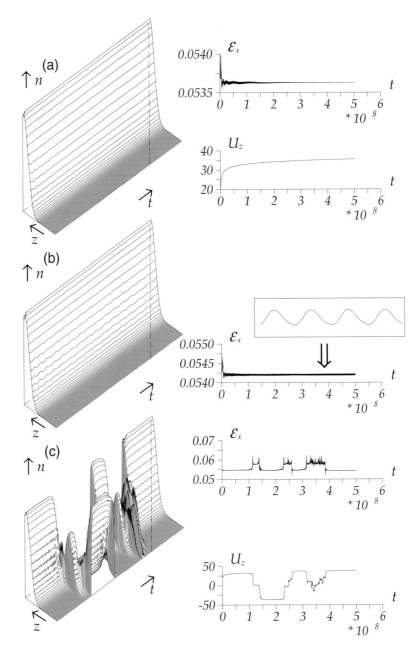

Figure 4.11. Complex dynamics of current filaments. In (a) the carrier density n is shown as a function of the lateral coordinate z and time t for the driving current density $j_0 = 0.000\,22$. The associated time series of the longitudinal electric field $\mathcal{E}_x$ and the transverse voltage U_z indicates that a stable filament is formed after a few transient oscillations. In (b) is shown the case of small periodic breathing oscillations for $j_0 = 0.000\,205$. The inset shows the self-sustained oscillations of $\mathcal{E}_x$ on an enlarged scale. The duration of the signal in the inset is $1.75 \times 10^6 \tau_{\mathrm{M}}$, the amplitude is $2 \times 10^{-5} \mathcal{E}_0$. In (c) is shown the case of an intermittent traveling filament for $j_0 = 0.0002$. The simulated time interval is $T = 5 \times 10^8 \tau_{\mathrm{M}} = 0.5$ ms; the width of the sample is $W = 1000 L_{\mathrm{D}} = 56$ μm. The units of $\mathcal{E}_x$ and U_z are $\mathcal{E}_0 = 60.8$ V cm^{-1} and $\mathcal{E}_0 L_{\mathrm{D}} = 0.34$ mV, respectively (after Hüpper *et al.* (1993a)).

a result, the whole filament travels intermittently from one side of the sample to the other and vice versa. The spatially averaged carrier density and the field $\mathcal{E}_x$ exhibit chaotic bursts with very low frequencies compared with those of the former regular breathing oscillations. During the laminar periods between the bursts the filament is pinned to either of the boundaries, which is indicated by an approximately constant positive or negative transverse voltage U_z, respectively.

The various possible spatio-temporal dynamic states of the filament which occur successively with decreasing current density for a wider sample are shown in Fig. 4.12: In (a) a stable filament is built up after transient oscillations; (b) shows a breathing filament, whose position is displaced erratically (intermittently) and; in (c) a filament arises and splits temporarily into two. Eventually, in (d) one of these vanishes due to a process of competition between these two filaments that is analogous to Ostwald ripening (Schimansky-Geier *et al.* 1991, Kunz and Schöll 1992). The

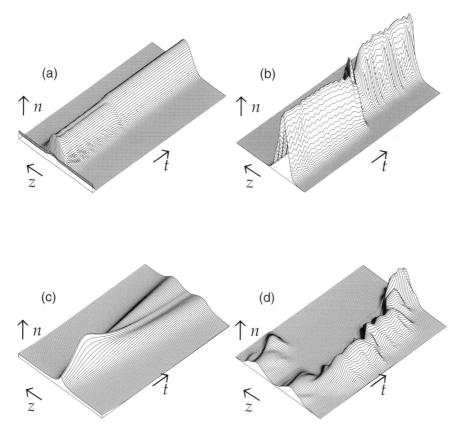

Figure 4.12. Various types of spatio-temporal dynamics of current filaments: (a) $j_0 = 0.00045$ (simulated time interval $T = 1.24 \times 10^8 \tau_M$), (b) $j_0 = 0.0001$ ($T = 1.6 \times 10^7 \tau_M$), (c) $j_0 = 0.00005$ ($T = 6 \times 10^6 \tau_M$), (d) the same as (c) but for the simulated time interval from $t = 10^7 \tau_M$ to $t = 2.3 \times 10^7 \tau_M$. The width of the sample is $W = 2000 L_D$ (after Hüpper *et al.* (1993a)).

motion of the remaining filament is accompanied by a change in its form and
position.

The stationary filament is symmetric, whereas the traveling filament is asymmetric.
Figure 4.13 shows the carrier density of a traveling filament (obtained with cyclic
boundary conditions in order to avoid pinning at the boundaries). Its leading edge is
steeper than its trailing edge. According to (4.23) $\mathcal{E}_z$ tends to change in time until

$$\mathcal{E}_z = -\frac{D}{\mu}\frac{\partial}{\partial z}\ln n \qquad (4.77)$$

holds. Thus the absolute value of the electric field is large at the leading edge of the
filament and small at its trailing edge. The higher electric field at the front end leads
to a higher impact-ionization rate and thus to an increase of the carrier density. At
the back end the carrier concentration decreases due to the lower electric field. As a
result both walls of the filament move in the same direction, i.e. the filament travels to
one side. This asymmetry of the concentration profile of traveling current filaments is
inverse to the asymmetry of the field profile in case of traveling field domains, e.g. in
the Gunn effect (Shaw *et al.* 1992).

In conclusion, the model exhibits a bifurcation scenario (with decreasing current)
from spatially homogeneous steady states to stable filaments, breathing filaments, and
traveling filaments, including complex spatio-temporal intermittent behavior in which
a breathing instability triggers spatio-temporal intermittency of filaments traveling
from one side of the sample to the other. It is suggested that this complex oscillatory
instability offers a detailed explanation of the chaotic spatio-temporal oscillations
observed in p-Ge at 4 K (Rau *et al.* 1991). The physical origin of the oscillatory
instability is based upon the autocatalytic impact-ionization of the acceptor's ground
and first excited states. A fluctuation of the carrier concentration in the filament wall
tends to increase due to impact ionization. In the lifetime regime, this excess carrier
density can be distributed effectively via drift and diffusion processes, leading to stable
filaments. For smaller current densities, the transverse electric field reacts too slowly

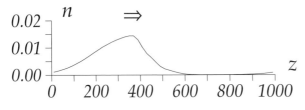

Figure 4.13. A laterally traveling filament. The carrier density n is shown as a function of the lateral coordinate z at fixed time for $j_0 = 0.0002$ with periodic boundary conditions. The arrow indicates the direction of the traveling filament. The width is $W = 1000L_D$ (after Hüpper *et al.* (1993a)).

upon such fluctuations, which allows the fluctuations to increase further and trigger instabilities.

It is worth noting that the time scales of the breathing filament oscillations and the traveling-filament oscillations are different. These time scales are connected with different dielectric relaxation times: The breathing-filament oscillation mode is localized at the filament wall, where the carrier density is orders of magnitude higher than it is outside the filament, leading to a short dielectric relaxation time and high-frequency oscillations. Outside the filament there are almost no free carriers, which results in very large GR and dielectric relaxation times. These very slow processes are responsible for the low frequencies observed. This explains the discrepancy between the frequencies obtained in experiments (Mayer *et al.* 1987) and those predicted by simple phenomenological theories (Schöll 1986*b*, Schöll and Drasdo 1990).

4.5   Traveling-wave instabilities

In this section we shall analyze pattern formation and nonlinear spatio-temporal dynamics in the longitudinal x-direction using (4.1)–(4.8). Although SNDC is commonly associated with transverse spatial instabilities leading to current filamentation (Ridley 1963), as demonstrated in the preceding section, we shall show here that longitudinal traveling-wave instabilities are also possible (Schöll 1987, Gaa and Schöll 1996).

4.5.1   Analysis of the linear stability

In the regime of low-temperature impurity breakdown, monotonic as well as S-shaped and N-shaped nonlinear current–voltage characteristics have been observed in various materials and under various conditions. Experiments have provided evidence for a menagery of transverse and also longitudinal instabilities, leading to current filamentation and field-domain formation, respectively. A variety of simple models for impurity-impact-ionization breakdown has been studied, including a single-level GR model with a monotonic $j(\mathcal{E})$ characteristic (Christen 1994*b*, 1995) and one with an NNDC characteristic (Cantalapiedra *et al.* 1993, Bergmann *et al.* 1996) explaining experimental findings of solitary longitudinal carrier-density waves and chaotic field domains in p-type Ge (Kahn *et al.* 1991). All these models have in common that they use simple phenomenological expressions for the GR rates, and that they are in principle of a single-level type. Thus they can not reproduce the SNDC characteristics observed, for example, in n-type GaAs (Spangler *et al.* 1992). Their behaviors in the nonlinear regime have been investigated either by simulation of the constitutive differential equations or by means of reduced amplitude equations, e.g. a complex Ginzburg–Landau equation (Christen 1995). Here, we shall investigate the two-level

GR model used throughout this chapter, incorporating elaborate Monte Carlo data for n-type GaAs as detailed in Section 4.2.3. We shall see that a longitudinal traveling-wave instability may also occur in the SNDC case, and that not even NDC is required.

First, the behavior of the dynamic system (4.1)–(4.8) in response to small perturbations is analyzed by a linear stability analysis in which we restrict ourselves to modes that propagate in the direction of the electric field. This analysis is aimed at samples whose transverse dimensions are much smaller than their longitudinal dimension. We follow the general scheme given by Schöll (1987), neglecting fluctuations of the magnetic field.

Small spatial and temporal fluctuations of the electric field and the electron concentrations around the homogeneous steady state $\{\mathcal{E}^0, n^0, n_1^0, n_2^0\}$ are described by

$$\delta\mathcal{E}(x, t) = \mathcal{E}(x, t) - \mathcal{E}^0, \tag{4.78}$$

$$\delta n(x, t) = n(x, t) - n^0, \tag{4.79}$$

$$\delta n_i(x, t) = n_i(x, t) - n_i^0, \qquad i = 1, 2, \tag{4.80}$$

where x is the longitudinal coordinate. Using the *Ansatz*

$$\delta\mathcal{E}, \delta n, \delta n_1, \delta n_2 \sim \exp(\lambda t) \tag{4.81}$$

with complex eigenvalue λ, we linearize the system (4.1)–(4.8). As in Section 4.4.1 dimensionless variables are used.

The linearization of eq. (4.8) with $\mu(\mathcal{E}) = \mu^0 + \delta\mu(\mathcal{E})$ and $\mu^0 \equiv \mu(\mathcal{E}^0)$ leads to

$$0 = \alpha\lambda \, \nabla \cdot \delta\mathcal{E} + \nabla \cdot \left(n^0\mu^0 \, \delta\mathcal{E} + n^0\mathcal{E}^0 \, \delta\mu + \mu^0\mathcal{E}^0 \, \delta n\right)$$
$$+ \nabla \cdot [(\nabla n^0) \, \delta\mu + \mu^0\nabla \, \delta n] \tag{4.82}$$

$$= \nabla \cdot \delta\mathbf{J}. \tag{4.83}$$

Here, in extension of (4.1)–(4.8), the parasitic wire and contact capacitances of the sample have been accounted for by adding a parallel external capacitance C_{ext} to the intrinsic capacitance $C_{\text{int}} = \epsilon_r\epsilon_0 A/L$ which introduces an additional time scale of dielectric relaxation $\alpha = 1 + C_{\text{ext}}/C_{\text{int}}$ (Wacker and Schöll 1995). A and L are the sample's cross-section and length, respectively.

From this equation we obtain the fluctuations of the total current density

$$\delta\mathbf{J} = (\alpha\lambda + n^0\mu^0) \, \delta\mathcal{E} + n^0\mathcal{E}^0 \, \delta\mu + \mu^0(\mathcal{E}^0 + \nabla) \, \delta n. \tag{4.84}$$

The next step is to substitute δn and $\delta\mu$ by the fluctuations of the electric field $\delta\mathcal{E}$. As before, we decompose the fluctuations of the electric field $\delta\mathcal{E} = (\delta\mathcal{E}_{\parallel}, \delta\mathcal{E}_{\perp})$ into components parallel and perpendicular to the applied static field $\mathcal{E}^0 \equiv (\mathcal{E}, 0)$. First we eliminate δn_1 and δn_2 using eq. (4.20):

$$\lambda \begin{pmatrix} \delta n_1 \\ \delta n_2 \end{pmatrix} = B \begin{pmatrix} \delta n_1 \\ \delta n_2 \end{pmatrix} + \boldsymbol{d} \, \delta n + \boldsymbol{f} \delta \mathcal{E}_{\parallel}, \tag{4.85}$$

where the matrix B and the vector $\boldsymbol{d}$ have been defined in eq. (4.25). The vector $\boldsymbol{f}$ is defined by $f_i = \partial \phi_i / \partial \mathcal{E}$. It follows that

$$\begin{pmatrix} \delta n_1 \\ \delta n_2 \end{pmatrix} = -\frac{\text{adj}(B - \lambda)}{G(\lambda)} (\boldsymbol{d} \, \delta n + \boldsymbol{f} \delta \mathcal{E}_{\parallel}), \tag{4.86}$$

where $G(\lambda) = \det(B - \lambda) = \lambda^2 - \lambda \, \text{tr} \, B + \det B$. With (4.86) we can rewrite the linearized version of Gauss' law (4.7)

$$\boldsymbol{\nabla} \cdot \delta \mathcal{E} = -(\delta n + \delta n_1 + \delta n_2) \tag{4.87}$$

as

$$\boldsymbol{\nabla} \cdot \delta \mathcal{E} = -\frac{H(\lambda)}{G(\lambda)} \delta n - \frac{F(\lambda)}{G(\lambda)} \delta \mathcal{E}_{\parallel}, \tag{4.88}$$

where $H(\lambda)$ has been defined in (4.28) and $F(\lambda) = -\sum_{i,j=1}^2 \text{adj}(B - \lambda)_{ij} \, f_j$. Solving (4.88) for δn and observing that $\delta \mu(\mathcal{E}) = (\partial \mu(\mathcal{E})/\partial \mathcal{E})\big|_{\mathcal{E}^0} \delta \mathcal{E}_{\parallel}$, we obtain

$$\delta \boldsymbol{J} = \left((\alpha \lambda + n^0 \mu^0) - \mu^0 (\mathcal{E}^0 + \boldsymbol{\nabla}) \otimes \frac{G(\lambda)}{H(\lambda)} \boldsymbol{\nabla} \cdot \right) \delta \mathcal{E}$$

$$+ \left(n^0 \frac{\partial \mu(\mathcal{E})}{\partial \mathcal{E}} \bigg|_{\mathcal{E}^0} \mathcal{E}^0 - \mu^0 (\mathcal{E}^0 + \boldsymbol{\nabla}) \frac{F(\lambda)}{H(\lambda)} \right) \delta \mathcal{E}_{\parallel}, \tag{4.89}$$

where $\otimes$ denotes the tensor product.

Since we are looking for traveling-wave instabilities, we introduce the Fourier transform of $\delta \mathcal{E}$ with regard to a wave vector $\boldsymbol{k}$ for which $\boldsymbol{k} \| \mathcal{E}^0$ must hold for the longitudinal fluctuations investigated here:

$$\delta \mathcal{E} = \int \delta \mathcal{E}(\boldsymbol{k}) \exp(i \boldsymbol{k} \cdot \boldsymbol{x}) \, d^3 k. \tag{4.90}$$

By combining this with Maxwell's law of induction, we can prove that the fluctuations $\delta \boldsymbol{J}$ are zero:

$$0 = \boldsymbol{\nabla} \times (\boldsymbol{\nabla} \times \delta \mathcal{E}) = -\frac{\partial}{\partial t} (\boldsymbol{\nabla} \times \delta \boldsymbol{B}) = -\mu_0 \lambda \, \delta \boldsymbol{J}. \tag{4.91}$$

On substituting (4.90) and (4.91) into (4.89) and observing that all perpendicular components vanish, we obtain

$$\left((\alpha \lambda + \tilde{n}) \frac{H(\lambda)}{G(\lambda)} + \mu k^2 - i v k - \frac{F(\lambda)}{G(\lambda)} (v + i k \mu) \right)_0 \delta \mathcal{E}_{\parallel} = 0, \tag{4.92}$$

where $v(\mathcal{E}) = \mu(\mathcal{E})\mathcal{E}$, $\tilde{n} = n(\partial v/\partial\mathcal{E})$, and the brackets are taken in the steady state. Equation (4.92) can be re-written as a complex polynomial in λ of order three, which determines the dispersion relation $\lambda(k)$:

$$(\alpha\lambda^3 + \lambda^2 t_2 + \lambda t_1 + t_0) + ik(-\lambda^2 v + \lambda u_1 + u_0) = 0, \tag{4.93}$$

where the coefficients t_0, t_1, t_2, u_0, and u_1 depend upon k and are given explicitly in Gaa and Schöll (1996).

From (4.93) we can determine the regime of n^0 and k values for which small longitudinal fluctuations grow, i.e. $\lambda = \lambda_0 + i\omega$ with positive λ_0. In this regime traveling-wave solutions ($k \neq 0$) or spatially homogeneous relaxation oscillations ($k = 0$) should exist. The numerical solution for n-type GaAs with the parameters of Table 4.2 reveals that parts of the negative-differential-conductivity branch, and also parts of the positive-differential high conductivity branch of the S-shaped $n^0(\mathcal{E}^0)$ characteristic, are unstable against the formation of traveling waves. With increasing α this unstable regime is enlarged.

For $\alpha = 1$ ($C_{\text{ext}} = 0$) the instability regime is shown in Fig. 4.14 as a dashed line. In real samples, however, α is much larger due to parasitic wire and contact capacitances. In the experimental setup of Margull (1996), for example, a minimum parasitic capacitance of 300 pF is found without an impedance converter. It is difficult to estimate the value of α since the intrinsic capacitance $C_{\text{int}} = \epsilon_r\epsilon_0 A/L$ is not known. A comparison of the calculated $j(\mathcal{E})$ characteristic (Fig. 4.14) with the measured current $I = jA$ versus voltage (Margull 1996) can give only a crude estimate of the effective current cross-section A. Using the experimental contact distance $L = 0.24$ cm and $A = 10^{-2}$ cm^2 yields an intrinsic capacitance on the order of 0.04 pF, and hence $\alpha \approx 10^4$. With this value of α, the instability regime is considerably enlarged, as shown by the dotted line in Fig. 4.14. In Fig. 4.15 the branch associated with an unstable eigenvalue is plotted for $n^0 \approx 1.05 \times 10^{-6}$ (corresponding

Table 4.2. *Material parameters for n-type GaAs*

Parameter	Symbol	Value
Donor concentration	N_D	7.0×10^{15} cm^{-3}
Acceptor concentration	N_A	2.0×10^{15} cm^{-3}
Lattice temperature	T_L	4.2 K
Low-field mobility	μ^*	1.54×10^4 cm^2 V^{-1} s^{-1}
Relative dielectric constant	ϵ_r	10.9
Dielectric relaxation time	τ_M	7.8×10^{-14} s
Debye length	L_D	6.6×10^{-7} cm
Ionization coefficient	X_1^S	9.1639×10^{-8} (in $1/\tau_M$)
Excitation coefficient	X^*	2.6317×10^{-10} (in $1/\tau_M$)
Capture coefficient	T^*	3.2113×10^{-6} (in $1/\tau_M$)

to point 'B' in Fig. 4.14) and $\alpha = 1$. It is undamped for finite wave vectors k in the range of $k_{c_1} < k < k_{c_2}$.

For larger α, e.g. $\alpha = 10^3$, we find a drastic change in the bifurcation scenario. For small n^0 in the vicinity of the threshold field, we still find the traveling-wave instability with one of the three branches of $\lambda(k)$ having a positive real part and thus being unstable between k_{c_1} and k_{c_2}. If n^0 is increased, k_{c_1} approaches zero. When k_{c_1} becomes zero, a second branch of the dispersion relation becomes undamped at $k = 0$. Beyond this point, for larger n^0, two branches are undamped in the wave vector intervals $0 < k < k_{c_1}$ (the first branch) and $0 < k < k_{c_2}$ (the second branch). In this regime spatially homogeneous relaxation oscillations (with $k = 0$) dominate the system. The existence of these homogeneous relaxation oscillations can

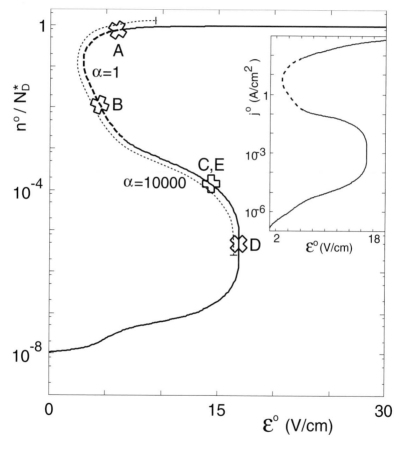

Figure 4.14. The stationary carrier density n^0 as a function of the electric field, calculated for n-type GaAs with the parameters given in Table 4.2. The dashed (dotted) line shows the region which is unstable for $\alpha = 1$ ($\alpha = 10\,000$). The letters A, B, C, D, and E denote the operating points for which time-dependent numerical solutions are shown in Fig. 4.16. (The inset shows the current density j versus the electric field $\mathcal{E}$.) (After Gaa and Schöll (1996).)

be physically understood as follows. These modes occur only in the NDC regime of the $n^0(\mathcal{E}^0)$ characteristic, which is always unstable under voltage control. From the current-conservation equation it follows that $\partial \mathcal{E}_x / \partial t = \alpha^{-1}(J_x - j_x)$, where the term on the right hand side of this equation is small for large α, even if the conduction current density j_x is not close to the total current density J_x. This causes a strong delay in the response of the electric field to changes in the drift–diffusion current density and therefore to changes in the electron concentration n. Because of the instability of the NDC branch and the delayed dielectric relaxation of $\mathcal{E}_x$, the electron concentration n approaches one of the stable branches with positive differential conductivity (PDC). However, with decreasing n the electric field must increase and vice versa since α is finite. If the electric field then leaves the bistability regime, e.g. by exceeding the threshold field, n increases strongly and the electric field drops below the holding field, whereupon the electron concentration relaxes to the low-conducting branch. In

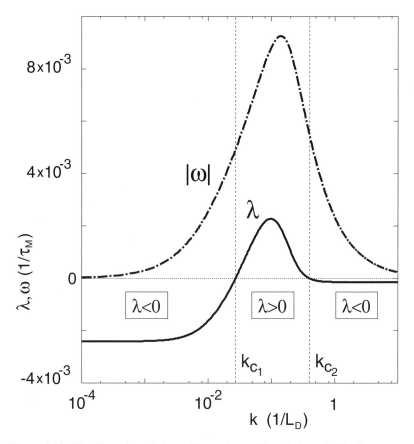

Figure 4.15. The dispersion relation of the undamped branch for point B of Fig. 4.14 ($n^0 \approx 1.05 \times 10^{-2}$ and $\alpha = 1$). The real part λ_0 (solid line) and the imaginary part ω (dashed line) of the complex eigenvalue λ are plotted as functions of the wave vector k (after Gaa and Schöll (1996)).

this way a cycle of spatially homogeneous field and electron-concentration oscillations is initiated around the stable operating point given by J_x. This mechanism of dielectric relaxation oscillations has previously been described in a variety of different models, see Section 4.3 (Schöll 1992, 1987).

4.5.2   Simulation of traveling carrier-density waves

Since the linear stability analysis has predicted the bifurcation of traveling carrier-density waves, we shall now investigate the behavior of the spatio-temporal solutions of the full *nonlinear* system. The complete nonlinear dynamic system consists of the equations (4.1)–(4.7) and the relation for the total current density $J = \alpha\epsilon\, \partial\mathcal{E}/\partial t + j$. From (4.8) it follows for a one-dimensional geometry that $J \equiv (J_x, 0, 0)$ is spatially constant. Since we are considering current control, J_x can be identified as the global control parameter. Since we are performing one-dimensional simulations, the notation simplifies as follows: $\nabla \to (\partial/\partial x, 0, 0), j \to (j_x, 0, 0)$, and $\mathcal{E} \to (\mathcal{E}_x, 0, 0)$. This simulation is not only a test of the results of the linear stability analysis, but also shows what kind of spatio-temporal scenarios are possible when the system's nonlinearities come into play.

We present simulation results for five operating points on the $n^0(\mathcal{E}^0)$ characteristic denoted by 'A', 'B','C', 'D', and 'E' in Fig. 4.14 (Fig. 4.16). Point A represents a state with positive differential conductivity at the onset of the high-conducting branch without an external capacitance, i.e. $\alpha = 1$. We choose a periodicity length of $L_x = 50L_D$, where L_D is the effective Debye length. The simulation results for the electron density for this case are shown in Fig. 4.16(a). We obtain traveling waves propagating at a constant phase velocity. However, at about $t = 200\tau_M$, where τ_M is the effective dielectric relaxation time, the initial spatial and temporal period is suddenly halved. This superharmonic frequency-doubling bifurcation appears to occur when the nonlinearities come into play. At the same time the donor levels are no longer populated homogeneously in space, since the initial perturbation of n induces spatial modulations of n_1 and n_2 due to their coupling by the GR processes.

Point B corresponds to a negative-differential-conductivity (NDC) state with $\alpha = 1$. The simulation results shown in Fig. 4.16(b) reveal a strong increase of the initial perturbation amplitude after $t \approx 2000\tau_M$, which is due to a decrease of the electric field below the holding field, where the negative-differential-conductivity branch no longer exists and the only stable state is the low-conducting state. Therefore the trapping of electrons is strongly enhanced. The conservation of the total current then again enforces an increase of the electric field, leading to strong spatio-temporal oscillations of n and $\mathcal{E}_x$.

For point C, which also lies on the NDC branch but at lower current density, an external capacitance is considered by using $\alpha = 100$. The amplitude of the initial perturbation of n, which is $1/1000$ of the steady-state electron concentration n^0, rapidly grows while the perturbation starts traveling through the sample (Fig. 4.16(c)).

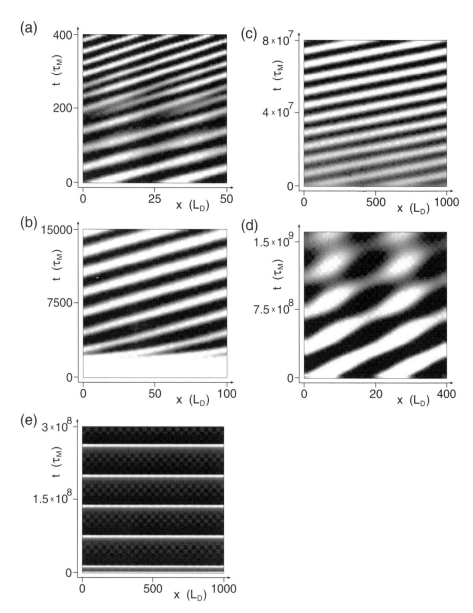

Figure 4.16. Simulation of traveling-wave instabilities in n-GaAs. The spatio-temporal dynamics of the carrier density n is shown for different points (labeled A, B, C, D, and E in Fig. 4.14 on the current–field characteristic. Darker areas indicate lower densities whereas brighter areas indicate higher densities. (a) $\mathcal{E}^0 = 6$ V cm^{-1}, $n^0 = 7.520 \times 10^{-1} N_D^*$; $\alpha = 1$ (b) $\mathcal{E}^0 = 4.5$ V cm^{-1}, $n^0 = 1.048 \times 10^{-2} N_D^*$; $\alpha = 1$ (c) $\mathcal{E}^0 = 14$ V cm^{-1}, $n^0 = 1.667 \times 10^{-4} N_D^*$; $\alpha = 100$ (d) $\mathcal{E}^0 = 17.029$ V cm^{-1}, $n^0 = 4.991 \times 10^{-6} N_D^*$; $\alpha = 100$ (e) $\mathcal{E}^0 = 14$ V cm^{-1}, $n^0 = 1.667 \times 10^{-4} N_D^*$, $\alpha = 10\,000$. Numerical parameters as in Table 4.2 (after Gaa and Schöll (1996)).

We find an oscillatory behavior with fixed period and a slightly increasing amplitude. After $t = 8 \times 10^7 \tau_M$ the amplitude is $6/100$ of n^0 and has thus grown by a factor of 60.

Point D lies on the NDC branch close to the threshold field beyond which impurity breakdown occurs. The electron concentration is as low as $n^0 \approx 5 \times 10^{-6} N_D^*$. The control parameter α is chosen to be $\alpha = 100$. Here we also find traveling waves with an even lower frequency (Fig. 4.16(d)). These waves exhibit additional modulations due to the coupling with the spatially inhomogeneous occupation of the donor ground state as a result of the spatially inhomogeneous initial perturbation. The reason is the very low electron density which makes the electrons rather susceptible to changes in n_1 that occur on a much longer time scale.

Point E corresponds to the same parameters as those for point C except for the external capacitance. Here α is set equal to $\alpha = 10\,000$, for which we expect the dominance of the $k = 0$ modes, leading to spatially homogeneous relaxation oscillations. These are indeed found in the simulations (Fig. 4.16(d)). Between the narrow white stripes and the broad dark areas in Fig. 4.16(e) the electron concentration changes by a factor of $\approx 10^5$, leading to sharp spikes in which the sample becomes highly conducting. This, together with the slow saw-tooth-like rise of the electric field, is typical of relaxation oscillations.

In conclusion, our simulations reveal a variety of longitudinal spatio-temporal instabilities ranging from the generation of sinusoidal harmonic and frequency-doubled traveling waves to spatially uniform large-amplitude relaxation oscillations. Furthermore, it is evident that the value of the contact and parasitic capacitance plays a crucial role in the bifurcation behavior and the system's stability. More-over, these simulations (Figs. 4.16(c) and (d)) offer an explanation of the sinu-soidal periodic oscillations with small amplitude ("precursor oscillations") which were observed experimentally in n-type GaAs just below the onset of breakdown (Margull 1996).

4.6 The dynamics of current-filament formation

4.6.1 Two-dimensional geometries with point contacts

In the previous sections only one spatial coordinate – either the transverse z-direction in case of current filamentation, or the longitudinal x-direction in case of traveling-wave instabilities – was taken into account. However, in particular in short samples with point contacts, the internal electric field is strongly distorted, so that the translational symmetry of current-filament profiles in the x-direction is broken. The dipolar electric field between two point contacts was included in a phenomenological model for current filaments in n-type GaAs in the regime of impurity breakdown, but the calculations were effectively reduced to a one-dimensional model, containing only the transverse z-direction, by several simplifying assumptions (Novák and Prettl 1995, Novák et al. 1995).

In this section we shall present two-dimensional simulations of the full nonlinear system considering the longitudinal x-direction as well as the transverse z-direction. In this way it is possible to model the dynamics of the breakdown process by which a current filament is formed in realistic two-dimensional sample geometries with point contacts (Gaa et al. 1996a, 1996b, 1996c).

We consider a square sample with side lengths $L_x = L_z = 0.02$ cm representing a thin n-GaAs film. We model point contacts by applying Dirichlet boundary conditions to two opposite regions of length $L_c = 8 \times 10^{-4}$ cm at the centers of the sample's edges parallel to the z-axis. At the contacts n is fixed to the large value $n_D = 5 \times 10^{15}$ cm^{-3} to model Ohmic contacts. All other boundaries are treated as insulating ones where the components of the current density j and the electric field $\mathcal{E}$ perpendicular to the boundaries vanish. Our numerical algorithm employs an implicit finite-element-scheme (Gajewski et al. 1991, Kunz et al. 1996). Owing to the strong nonlinearities and steep spatial gradients in the regime of impurity breakdown, an elaborate numerical algorithm with efficient time-step control and updating of the spatial grid is required. Our algorithm is not based on an explicit Euler scheme such as the particle-in-cell algorithms employed in Section 4.4 (Hüpper et al. 1993a, 1993b), but uses a semi-implicit scheme for the propagation of the solutions of (4.1)–(4.6), (4.8) in time. This guarantees a degree of numerical stability of the solutions of these highly complex nonlinear spatio-temporal partial differential equations, which an explicit Euler scheme could not provide without using unreasonably small iteration time steps and hence requiring amounts of CPU time that would make it infeasible to use simulations to study filament formation in two dimensions.

For simplicity, in the following a constant mobility of $\mu = 10^5$ cm^{2}V^{-1} s^{-1} will be used, but the impact ionization and capture coefficients are given by the elaborate microscopic Monte Carlo data derived in Section 4.2.3. In contrast to the previous sections we assume voltage control.

4.6.2   The nascence of current filaments

We study the nascence of current filaments when the applied voltage is switched rapidly to a value above the breakdown threshold so that the semiconductor is forced from the nearly insulating state to a highly conducting state. Within 1 ps the voltage is linearly increased from $U = 0$ V to $U = 0.48$ V, corresponding to an average field of $\mathcal{E} = 24$ V cm^{-1}. Owing to the Ohmic nature of the contacts, we find small regions in the vicinity of the contacts in which the electron concentration n in the conduction band is greatly enhanced compared with that in the bulk, which is very low (Fig. 4.17(a)). Practically all carriers are bound in the donor ground state. The electron temperature in the whole sample is equal to the lattice temperature T_L (Fig. 4.18(a)). Owing to the assumption of voltage control, the electric field

$\mathcal{E}$ reacts quasi-instantaneously forming a dipole-like electric-field distribution (Fig. 4.19(a)) and inducing the formation of enlarged areas of high electron density both at the cathode and at the anode. The current density $j \equiv |\mathbf{j}|$ (Fig. 4.20(a)) is very low and remains so during the increase in voltage. Subsequently impact-ionization multiplies the electron concentration at the cathode (the top contact at $x = 0.02$ cm) and establishes a front that moves toward the anode (the bottom contact; Figs. 4.17(b) and 4.20(b)). The propagation of the front is accompanied by a high-field domain

Electron density in the conduction band

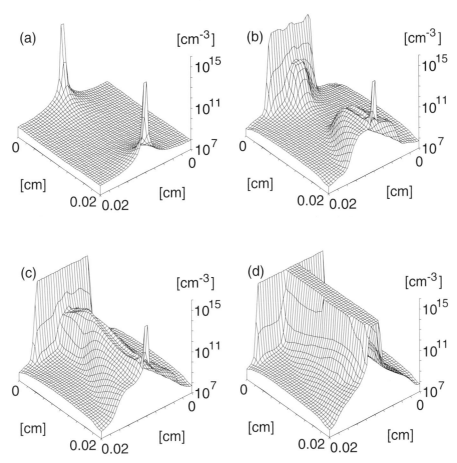

Figure 4.17. A simulation of the formation of a current filament in a thin film of n-GaAs. The temporal evolution of the electron density $n(x, z)$ is shown as a function of the spatial coordinates x and z for a square sample with two point contacts: (a) $t = 1$ ps, (b) $t = 0.5$ ns, (c) $t = 1.0$ ns, and (d) $t = 2.5$ ns. The applied voltage $U = 0.48$ V corresponds to an average field $\mathcal{E}_{av} = 24$ V cm^{-1}, i.e. above the threshold field $\mathcal{E}_{th} = 17$ V cm^{-1} of the homogeneous $n(\mathcal{E})$ characteristic (calculated with the parameters of Table 4.2, but with $\mu = 10^5$ cm^2 V^{-1} s^{-1}) (after Gaa et al. (1996b)).

associated with a slightly increased electron temperature T_e (Fig. 4.18(b)). Although
the electric field behind the front is smaller than that in front of it, for reasons of current
conservation the increased electron density in regions passed by the front is almost
conserved because recombination is a much slower process than is generation. Hence
impact ionization downstream is encouraged, whereas further generation upstream
is inhibited. When the front meets the region of increased carrier density around
the anode, a rudimentary filament is formed, albeit with a carrier density several
orders of magnitude lower than that corresponding to completely ionized donor states
of density $N_D^* = 5 \times 10^{15}$ cm^{-3} (Fig. 4.17(c)). Now donor-impact ionization
is becoming enhanced in the rudimentary filament because the excited level n_2 is
increasingly populated. The current density (Fig. 4.20(c)) and the electron temperature
in the rudimentary filament are growing significantly (Fig. 4.18(c)). The potential
distribution Φ is being deformed (Fig. 4.19(c)) by the nascent filament. Finally

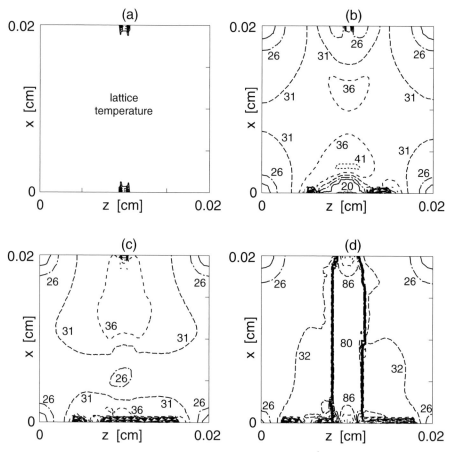

Figure 4.18. The temporal evolution of the electron temperature $T_e(x, z)$. The
timesteps and voltage are the same as those in Fig. 4.17. Contour plots are shown, on
which the numbers denote T_e in kelvins (after Gaa et al. (1996a)).

impact ionization leads to a uniform increase of electron density until the filament attains its mature state (Fig. 4.17(d)) in which almost all donors are ionized, and the carrier density corresponds to the upper branch of the homogeneous steady-state characteristic $n(\mathcal{E})$. Within the filament the excited donor level is much more highly populated than it is outside; nevertheless, still only about 2% of the band carriers are trapped in the excited level, while the ground level is *completely* depleted inside the filament. Thus the population ratio between ground and excited levels is inverted in the filament. Also the current density j and the electron temperature T_e are much larger inside the filament than they are outside (Figs. 4.20(d) and 4.18(d)). The deformation of the potential distribution is completed (Fig. 4.19(d)).

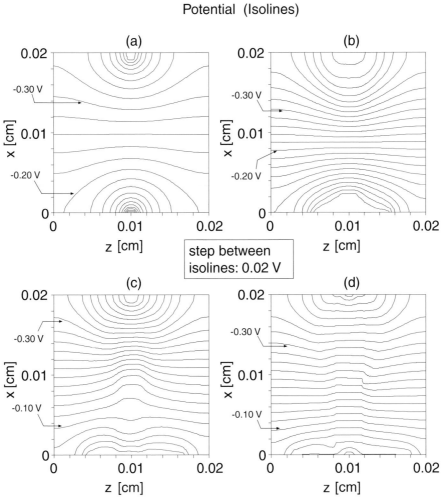

Potential (Isolines)

Figure 4.19. The temporal evolution of the potential $\Phi(x, z)$. The timesteps and voltage are the same as those in Fig. 4.17. Potential isolines are shown (after Gaa *et al.* (1996*a*)).

Hence, from our simulation, we find three stages of impact-ionization breakdown: A stage of front creation and propagation from the injecting contact, i.e. the cathode (stage I), followed by a stage of stagnation after the front has reached the anode (stage II), and a final stage during which the rudimentary filament grows into a mature filament (stage III). The first and the last of those predicted stages of the nucleation of a current filament have indeed been observed experimentally in n-GaAs at 4.2 K by quenched photoluminescence using a pulsed electric field, and the front velocity has been estimated to be on the order of several times 10^5 cm s^{-1} (Aoki and Fukui 1998).

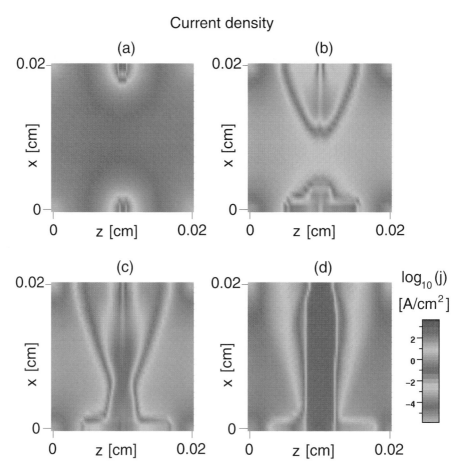

Figure 4.20. The temporal evolution of the current density $j(x, z)$. The timesteps and voltage are the same as those in Fig. 4.17. The grayscale of the density plots corresponds to a logarithmic scale of j. Note that the injecting cathode is at the top (after Gaa *et al.* (1996a)).

4.6.3 Fully developed filaments

Our simulations in the previous section have shown that the final high-current steady state reached after breakdown is characterized by an inhomogeneous, filamentary current-density distribution, as was indeed found experimentally by spatially resolved measurements at liquid-helium temperature. In the past several methods have been applied to reconstruct the spatial distribution of inhomogeneous current flow under nonequilibrium conditions. High-resolution images of impurity-breakdown-induced current filaments were obtained by using a scanning electron microscope (Mayer *et al.* 1987, 1988, Aoki and Kondo 1991). The effect of an external magnetic field on current filaments in n-GaAs epitaxial layers at low temperatures has been investigated by using a scanning laser microscope (Brandl *et al.* 1989, Spangler *et al.* 1994). In both cases a focussed electron or laser beam, respectively, is scanned line by line across the surface of a thin sample and the response of the current is synchronously recorded as a function of the focal position. Irradiation destabilizes the current flow at the filament borders, yielding structures in the spatially resolved response that indicate the extent of the current flow in the sample. Owing to the sequential nature of the procedure, only stationary and sufficiently stable filamentary patterns may be reconstructed. Furthermore, focussed irradiation may change the spatial distribution of the current flow itself, leading to artifacts (Aoki *et al.* 1990*b*) due to the interaction of the focussed beam with the current filament.

A less invasive method for visualizing impurity-breakdown-induced current filaments has been developed on the basis of quenched photoluminescence by Eberle *et al.* (1996). The presence of free electrons in the current filament reduces the intensity of excitonic and impurity luminescence due to a depletion of excitons and donor and acceptor levels by enhancement of impact ionization (Bludau and Wagner 1976, Ryabushkin and Bader 1991, Karel *et al.* 1992). This effect has been used to spatially resolve the distribution of free carriers constituting a filament in homogeneously illuminated samples. Such sophisticated techniques for mapping two-dimensional carrier-density distributions, including quenched photoluminescence (Eberle *et al.* 1996, Aoki 1999*b*) and near-field scanning photoluminescence (Bar-Joseph 1999), have not only provided the possibility of directly observing current filaments in n-GaAs thin films due to impact ionization of carriers from shallow donors, but might also lead to new insights into nonuniform carrier-density distributions (Kaya *et al.* 1998) and current filamentation (Tsemekhman *et al.* 1997, Eaves 1998) in quantum-Hall-effect breakdown.

We are now in a position to compute the stable parts of the S-shaped stationary current–voltage characteristic from the simulations of the two-dimensional sample described in the previous section. The procedure used to calculate these points is to increase the voltage in a stepwise manner and to wait after each step until the system returns to its steady state. We start at zero voltage and increase the voltage until switching from the low-conductive state to the highly conductive state occurs, as described in Section 4.6.2. Then, at a fixed voltage on the upper branch of the characteristic, the filament develops until it becomes stationary. Afterwards the voltage is decreased in a stepwise manner and again after each step we allow the system to relax to a steady state. This is done until the voltage drops below the holding voltage, whereupon the upper branch becomes unstable and returns to the lower branch.

For this simulation we use the same initial conditions as those described above. The points on the lower branch are calculated in voltage steps of $\Delta U = 0.025$ V, starting from $U = 0$ V. The voltage increase of ΔU is done linearly within 1 ps. The time for relaxation to the steady state turns out to be less than 1 ns. For the calculation of the points on the upper branch we switch to the highly conductive state

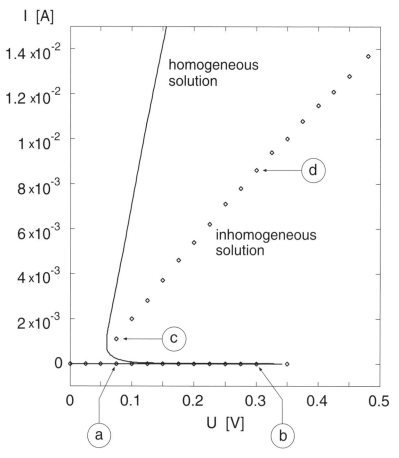

Figure 4.21. The stationary current–voltage characteristic calculated with the parameters of Table 4.2 and Fig. 4.5 from the spatially inhomogeneous solution for the two-dimensional sample (diamonds), in comparison with the homogeneous solution (full lines). The letters correspond to the plots shown in Figs. 4.22–4.24 below. (The thickness of the film is 1.4×10^{-3} cm.) (After Gaa *et al.* (1996c).)

at $U = 0.5$ V. Voltage is decreased in steps of $\Delta U = 0.025$ V in the same manner as before.

The resulting points are plotted in Fig. 4.21 in comparison with the spatially homogeneous steady-state solution. Obviously, the slope of the homogeneous solution is larger by a factor of about five than that of the inhomogeneous solution on the highly conducting branch, whereas on the low-conducting branch the current in both cases is almost zero. The reason for the reduction in slope is that the current is carried only by the filamentary channel, which has a smaller cross-section. Both solutions have almost constant slopes on the highly conducting branch, which is obvious since the carrier density does not change significantly in this regime.

For four different points on the current–voltage characteristic (labeled a–d in Fig. 4.21) the calculated spatial distributions of the electron density in the conduction band $n(x, z)$ (Fig. 4.22) and in the excited donor level $n_2(x, z)$ (Fig. 4.23), the potential

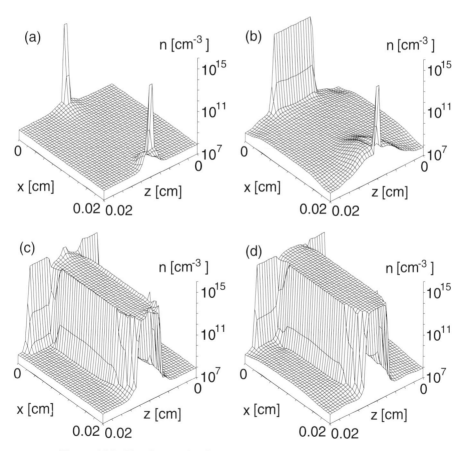

Figure 4.22. The electron density $n(x, z)$ in the steady state corresponding to the points labeled a–d in the current–voltage characteristic of Fig. 4.21: (a) $U = 0.075$ V (lower branch), (b) $U = 0.3$ V (lower branch), (c) $U = 0.075$ V (upper branch), and (d) $U = 0.3$ V (upper branch) (after Gaa et al. (1996c)).

distribution $\Phi(x, z)$ (Fig. 4.24), and the current-density distribution (Fig. 4.25) are displayed.

Regarding the electron densities n on the lower branch (Figs. 4.22(a) and (b)), there is just a slight increase in electron concentration in the vicinity of the contacts. On the upper branch (Figs. 4.22(c) and (d)) we have a fully developed filament that exhibits a slight decrease in electron concentration as the voltage is reduced on going from (d) to (c). Regarding the electron concentration in the excited donor level n_2, Figs. 4.23(a) and (b)) show that practically all electrons are in the donor ground state on the lower branch, whereas on the upper branch at low voltage (c) the excited level is considerably populated since the GR cycle there runs predominantly via the excited level. With increasing voltage (d) the concentration n_2 in the interior of the filament decreases due to the decrease in recombination with growing electron temperature, and the excited state is fully impact ionized. Note that n_2 assumes its peak values at the filament boundaries and is somewhat smaller in the interior of the filament. This corresponds to population inversion between the ground and excited levels in the filament boundary. Outside the filament, practically all carriers are trapped in

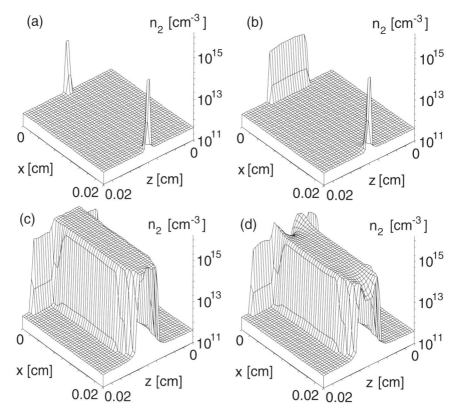

Figure 4.23. The same as Fig. 4.22 but for the electron density in the excited donor level $n_2(x, z)$ (after Gaa *et al.* (1996c)).

the donor ground level, which is, in turn, strongly depleted inside the filament. The simulations thus give detailed microscopic information about the density profiles of carriers in band states and in ground and excited donor states in a cross-section of the filament. This corrects simpler phenomenological models in which one assumed either a peak of n_2 in the filament wall but no population inversion inside the filament (Brandl and Prettl 1991), or monotonic wall profiles of n_2 and n_1, and values of n much lower than n_1 and n_2 inside the filament (Novák *et al.* 1995).

The potential $\Phi(x, z)$ (Fig. 4.24) exhibits a dipole-like distribution for the lower branch and a deformation caused by the filament on the upper branch. There the electric field is of the same order inside and outside the filament, but distinctly higher at the filament boundaries. The current density (Fig. 4.25) exhibits a slight increase in a broad smeared-out central region for the points on the lower branch, and a strong

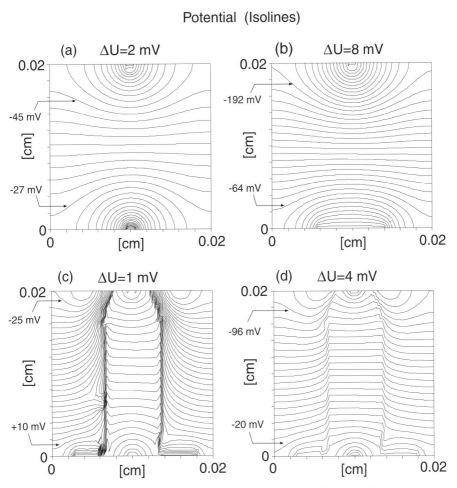

Figure 4.24. The same as Fig. 4.22 but for the potential $\Phi(x, z)$. ΔU denotes the steps between the potential isolines shown. (After Gaa *et al.* (1996c).)

increase in a narrow filamentary channel for the points on the upper branch. All values correspond to the regime of bistability between the holding field and the threshold field of the $n(\mathcal{E})$ characteristic, thus allowing spatial coexistence of the low-conducting and the highly conducting state.

Experimental results for a homogeneously doped n-GaAs layer prepared by vapor-phase epitaxy on a semi-insulating substrate have been published (Spangler *et al.* 1994, Eberle *et al.* 1996, Gaa *et al.* 1996c). Two Ohmic point contacts were alloyed onto opposite corners of the sample with a separation of 4.0 mm. The samples were immersed in superfluid helium in an optical cryostat and biased in series with a load resistor of 2.3 kΩ. Photoluminescence was excited by the diffuse radiation of four red-light-emitting diodes fixed close to the sample inside the cryostat. The luminescent surface of the samples was imaged onto a linear array of photodiodes through a long-wave-pass filter to suppress the visible excitation light. A full image of the sample was taken by scanning the array line by line across the image of the sample.

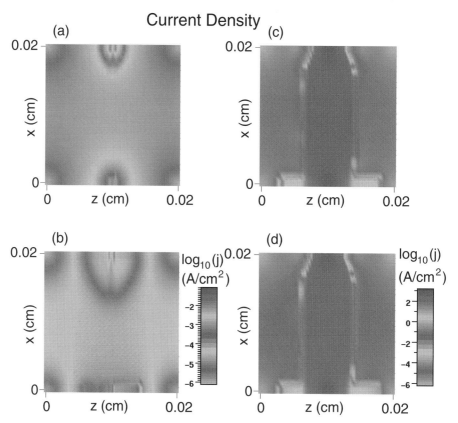

Figure 4.25. The same as Fig. 4.22 but for the current density $j(x, z)$ (after Gaa *et al.* (1996c)).

Figure 4.26 shows the measured current–voltage characteristic of a sample display-ing the typical behavior of low-temperature impurity breakdown observed in n-GaAs (Brandl and Prettl 1991): A low-conducting lower branch up to the threshold voltage V_{th}, a range of negative differential resistance between V_{th} and the holding voltage V_h, and a high-conducting upper branch. The bistable region of inhomogeneous current flow is between V_h and V_{th}. In the lower and upper branches the current flow is stable whereas in the region of negative differential resistance the current through the sample oscillates. These oscillations have been time averaged yielding the curve shown here. The extent of the oscillations in the current–voltage plane is shaded in Fig. 4.26. In comparison with the calculated current–voltage characteristic, Fig. 4.21, it must be emphasized that the distance between the contacts of the sample and the sample's width are 20 times larger than those assumed in the model calculation. This leads to higher values both of the threshold voltage V_{th} and of the sustaining voltage V_h, as well as of the current.

In Fig. 4.27 six different images of the current distribution for various biasing conditions are shown. The labeling of the images corresponds to points indicated by (a)–(f) in the current–voltage characteristic of Fig. 4.26. The dark pixels indicate regions of strong luminescence quenching. The measurements show that some concentration of electrons is already present well below V_{th} and that the electron

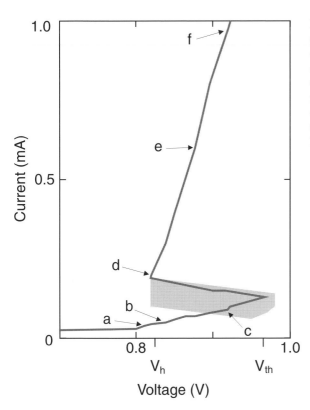

Figure 4.26. The experimental current–voltage characteristic of a square sample with two point contacts (n-GaAs epitaxial film). The characters a–f correspond to the plots shown in Fig. 4.27 (after Brandl and Prettl (1991)).

distribution is inhomogeneous (Figs. 4.27(a) and (b)). A high free-electron density is found in the vicinity of the point contacts where the electric-field strength is large. The free-carrier distribution is determined by the inhomogeneous electric field due to the point contacts. This experimental result agrees quite well with the calculations shown in Figs. 4.22–4.25(a) and (b). In the range of negative differential resistance the image of the current flow is very wide and diffuse (Fig. 4.27(c)). The voltage across the sample oscillates with frequencies in the range of several hundred kilohertz. Since the exposure time of the recording (1–4 s) is much longer than the oscillation period, the image is an average over many periods. In this bias regime the broad and diffuse distribution of quenched luminescence indicates lateral oscillations of the current flow. Such spatial oscillation modes in the form of breathing filaments had been predicted previously (Schöll and Drasdo 1990, Hüpper *et al.* 1993*a*), as discussed in Sections 4.4.3 and 4.4.5.

On the upper branch of the characteristic the lateral distribution of electrons exhibits sharp borders, indicating self-organized current filamentation (Figs. 4.27(d)–

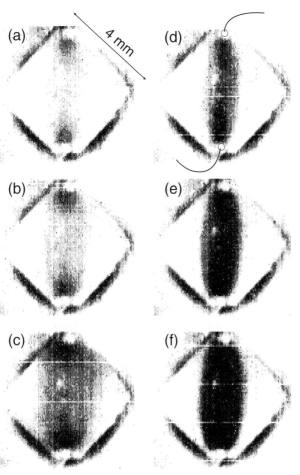

Figure 4.27. Spatially resolved photoluminescence intensity plots for the operating points a–f on the I–V characteristic in Fig. 4.26 (after Gaa *et al.* (1996*c*)).

(f)). With rising current the width of the filament increases. These reconstructions of the fully developed filament agree well with the simulations in Figs. 4.22–4.25(c) and (d).

4.6.4   Samples with inner point contacts

In this section we shall present two-dimensional simulations for epitaxial samples with a different point contact geometry (Schöll 1999, Schwarz *et al.* 2000*b*). Two small circular Ohmic contacts of diameter 0.5 mm, spaced by a distance of 3.5 mm in the *x*-direction, are placed in the interior of a rectangular sample with side lengths $L_x = 5$ mm and $L_z = 7.5$ mm. As before, at the Ohmic contacts *n* is fixed to $n_D = 5 \times 10^{15}$ cm^{-3}. At all other boundaries the components of the current density *j* and the electric field $\mathcal{E}$ perpendicular to the boundaries are set equal to zero.

Let us study the formation of current filaments when a bias voltage U_0 is applied to the sample in series with a load resistor of resistance *R*. The calculated current–voltage characteristic is shown in Fig. 4.28 on a semilogarithmic scale for various values of U_0 (full circles) and *R* (open circles). For each value of U_0 and *R* the operating point (I, U) is determined by the intersection of the load line $U_0 = U + RI$ and the characteristic. The current *I* has been calculated self-consistently by integrating over the current-density distributions corresponding to the respective values of the voltage *U* dropping across the sample. The upper branch of the S-shaped current–voltage characteristic corresponds to filamentary current-density distributions, whereas the lower branch represents the homogeneous low-conductivity state. In those two-dimensional simulations the sample's thickness L_y and the electron mobility μ act merely as scaling parameters for the value of the total current *I*; the physics is unaffected if the load resistance *R* is simultaneously rescaled by $1/(L_y\mu)$.

Figure 4.29 shows the current-density distributions corresponding to five different operating points (a)–(e) on the characteristic in Fig. 4.28. They represent stationary current filaments for fixed $U_0 = 50$ V and five values of the load resistance *R* increasing from (a) to (e). At $R = 100\ \Omega$ (a) the filament is very wide and plum-shaped (convex). With increasing *R* the operating point on the characteristic is shifted on the upper branch toward lower sample voltage and current, and the width of the filament decreases. Finally, at $R = 10$ MΩ (e) the operating point is located on the intermediate branch with negative differential conductance, and the filament becomes very unstable and thin, and its current density drops by several orders of magnitude. Upon a further increase of *R* (or likewise a decrease of U_0 at fixed *R*) the filament eventually fades away. The changes of the shape and the width of the filament describe a transition from a pattern reflecting the dipolar electric field between the contacts (a) to a self-organized straight filament with parallel borders and constant width (d). The curving of the filament boundaries around the circular contacts with increasing current can be understood as a geometric effect within a simple reduced model (Novák *et al.* 1998*b*).

The corresponding calculated spatial distributions of the electron densities in the conduction band, and donor ground and excited levels, of the potential and of the modulus of the electric field are shown in Fig. 4.30 for the case (c) of Fig. 4.29. The shape of the filament is visualized by the sharply increased free electron concentration (Fig. 4.30(a)), the depleted donor ground state (Fig. 4.30(b)) and the increased population of the donor excited state (Fig. 4.30(c)). Within the filament still only about 2% of the band carriers are trapped in the excited level, while the ground level is *completely* depleted inside the filament. Note that n_2 assumes its peak value at the filament's borders. Thus the population ratio of the ground and excited levels

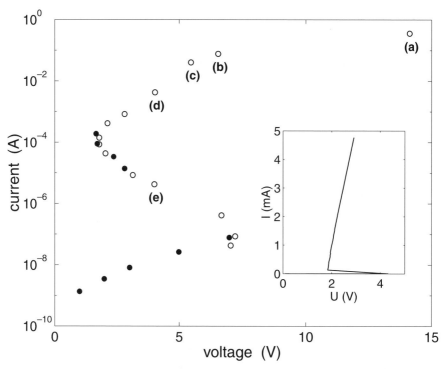

Figure 4.28. The simulated filamentary current–voltage characteristic for a rectangular sample (5 mm × 7.5 mm, layer thickness 14 μm) with two inner point contacts (diameter 0.5 mm, spaced 3.5 mm apart) operated in a resistive load circuit. The points on the characteristic are calculated with fixed bias $U_0 = 50$ V and varying load $R = 100\ \Omega$ (a), 500 Ω (b), 1 kΩ (c), 10 kΩ (d), 50 kΩ, 100 kΩ, 300 kΩ, 500 kΩ, 1 MΩ, 5 MΩ, and 10 MΩ (e), 100 MΩ, 500 MΩ, and 1 GΩ (open circles), and with fixed load $R = 470$ kΩ and varying bias $U_0 = 1, 2, 3, 5, 7, 10, 20,$ 50, and 100 V (full circles). The letters (a)–(e) refer to the current-density distributions depicted in Fig. 4.29 (calculated for n-GaAs at $T_L = 4.2$ K with the parameters of Table 4.2, except for $\mu = 10^5$ cm^2 V^{-1} s^{-1}, and generation–recombination coefficients taken from the Monte Carlo simulation in Fig. 4.5). In the inset the measured $I(U)$ characteristic of sample PC06 (Table 4.3) is shown (on a linear scale). (After Schwarz *et al.* (2000*b*).)

is inverted in the filament and, in particular, at the filament's border. The potential distribution (Fig. 4.30(d)) represents that of a dipole modified inside the filament, where the equipotential lines are parallel and equidistant, and at the boundaries of the filament, where local neutrality is violated and a transverse electric field is induced by the space charge in the filament boundary. This is also reflected by the increased modulus of the electric field at the boundaries (Fig. 4.30(e)); the space-charge field is superimposed upon the normal dipolar field distribution due to the contact geometry, which leads to enhancement of the field near the contacts. It should be noted that the nascence of the filaments is similar to Section 4.6.2 except that here *two* fronts moving toward each other originate from *both* contacts as a result of strong impact ionization in the regions of enhanced field near both contacts for the geometry considered in this section (Schöll 1999).

In Fig. 4.31 experimental luminescence images comparable to these simulations in the regime of filamentary current flow for various currents are shown. The shapes of the filaments and their changes with rising current are in good agreement with the calculations. In spite of the inhomogeneous field distribution imposed by the point contacts, the filaments are formed as stripe-like structures with clearly defined borders. The borders remain parallel at lower currents and are continuously transformed to a convex shape at higher current levels.

The transition from a parallel-bordered to a convex-shaped filament can also be demonstrated, both in experiments and in simulations, by considering the dependency of the filament's width on the current, Fig. 4.32. As long as the filament's width is small in comparison with the contact distance, the filament's borders are parallel and the width increases proportionally with the current. At higher currents, when

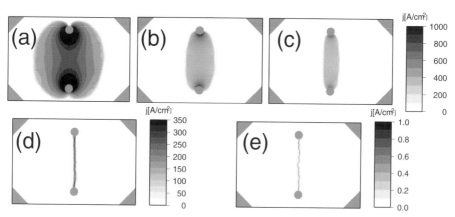

Figure 4.29. Simulated current filaments in a rectangular sample (5 mm × 7.5 mm) with two inner point contacts for an applied bias voltage $U_0 = 50$ V and various load resistances R. The stationary current density $|j(x, z)|$ is shown as a density plot. (a) $R = 100 \; \Omega$, (b) $R = 500 \; \Omega$, (c) $R = 1 \; k\Omega$, (d) $R = 10 \; k\Omega$, and (e) $R = 10 \; M\Omega$ (calculated with the same parameters as those in Fig. 4.28). (After Schwarz *et al.* (2000*b*).)

the filament's width becomes comparable to the contact distance, the borders of the filament become more and more convex and the width increases sublinearly with the current. The linear width-versus-current dependency indicates the existence of

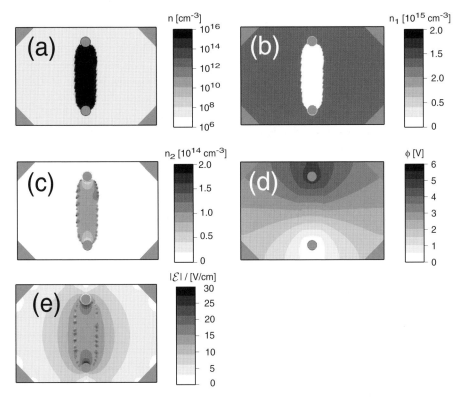

Figure 4.30. Simulated density plots of the distributions (a) of electrons in the conduction band, (b) of electrons in the donor ground level, (c) of electrons in the donor excited level, (d) of the potential, and (e) of the modulus of the electric field, for the current filament shown in Fig. 4.29(c) (after Schwarz *et al.* (2000*b*)).

Table 4.3. *Sample dimensions and material parameters of different samples used in the experiments by Hirschinger et al. (1997a, 1997b), and Schwarz et al. (2000b)*

Sample	PC06	C5	C6
Contact geometry	Points	Corbino	Ring
Contact distance (mm)	3.05	1.0	0.5
Contact diameter (mm)	0.3	0.08 / 2.1	2.0 / 3.0
Layer thickness (μm)	4.3	3	3
n at 77 K (10^{15} cm^{-3})	1.9	5.0	5.0
μ at 77 K (10^3 cm^2 V^{-1} s^{-1})	43	28	28

a constant critical current density inside the filament in the parallel-bordered regime. Indeed, in Figs. 4.29(b)–(d), a uniform constant current density of about 350 A cm^{-2} can be seen inside the filaments. On re-scaling the electron mobility used in the simulations to the experimental value of Table 4.3, one obtains 150 A cm^{-2}. This value is still too high if it is compared with the 60 A cm^{-2} obtained experimentally from the slope of the measured width-versus-current dependency in Fig. 4.32. However, perfect agreement both for the value of the critical current density and for the shape of the width-versus-current dependency can be achieved by assuming a somewhat lower electron mobility, 17×10^3 cm^2 V^{-1} s^{-1}, and a somewhat lesser sample thickness, 2.9 μm. Both assumptions are physically substantiated: The former reflects the expected decrease in mobility with decreasing temperature (Novák *et al.* 1998*b*); and the latter corresponds to a reduction of the thickness of the electrically active layer due to the presence of the depleted regions on the surface and the interface with the substrate.

4.6.5 Spontaneous symmetry breaking in Corbino disks

Next, we present simulations of a sample with circular contact symmetry, i.e. a Corbino disk (Schwarz and Schöll 1997, Schöll 1999, Schwarz *et al.* 2000*a*, 2000*b*). Such geometries offer the possibility of studying pattern formation in samples without lateral boundaries, and are therefore of considerable interest. Spontaneous symmetry breaking of the current-density distribution and various self-organized patterns of

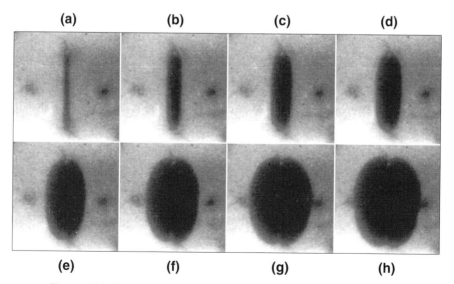

Figure 4.31. Experimental luminescence images of current filaments in point contact sample PC06 (Table 4.3) at 1.8 K with current increasing from (a) to (h): $I = 0.40$, 1.11, 1.57, 2.05, 3.42, 5.29, 6.70, and 8.40 mA (after Schwarz *et al.* (2000*b*)).

single and multiple current filaments have been observed experimentally in such samples (Hirschinger *et al.* 1997*a*, Aoki and Fukui 1999).

First we present simulations for a sample that is confined by an outer ring electrode of diameter 2.1 mm (the anode) and has a concentric inner point contact of 80 μm diameter (the cathode) in the center of the disk, and a thickness of 3 μm. A constant bias voltage $U_0 = 1.95$ V is applied to the Corbino disk via a load resistor of resistance $R = 10$ kΩ. Initially the sample is in the insulating state, and the equipotential lines are concentric circles corresponding to a radial electric field decreasing with distance. Subsequently impact ionization multiplies the electron concentration at the injecting inner point contact, the cathode, where the electric field is highest due to the circular sample geometry, and establishes a radially symmetric front that moves toward the circular anode, expanding in diameter (Fig. 4.33(a)). The electric field behind the front is decreased due to the higher carrier density, i.e. lower resistance, such that almost all the voltage drop occurs in the nonconducting region outside the circular front. The equipotential lines indicate that the field is particularly high just ahead of

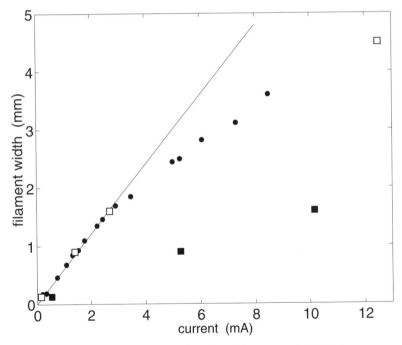

Figure 4.32. The filament width as a function of the current. Solid circles correspond to the measured dependency for sample PC06 (Table 4.3), the solid line is a guide to the eye for the linear dependency in the regime of a parallel bordered filament. Solid squares depict the simulated dependency, rescaled to sample parameters according to Table 4.3. Open squares depict the same dependency rescaled in order to account for the reduced mobility and sample thickness. (After Schwarz *et al.* (2000*b*).)

the front, encouraging further downstream impact ionization and acceleration of the front. At a certain radius the circular front breaks up into several streamers (b), each of which, upon reaching the anode, forms a rudimentary radial current filament (c). This spontaneous symmetry-breaking instability can be understood by inspecting the equipotential line which approximately follows the contour of the front, corresponding to the interface between the conducting interior and the nonconducting exterior region. A small local perturbation of the front contour will result in a squeezing of the equipotential lines such that the field becomes highest at the point of maximum curvature. Therefore impact ionization will be enhanced there, and the bulge of the front will grow further, forming a streamer. Note that the streamer expansion is driven by the collective impact-ionization process, not by individual drifting carriers; it is about an order of magnitude faster than the carrier drift velocity. Similar self-generated streamer instabilities of *planar* fronts have been known also in plasma physics (Ebert *et al.* 1996, 1997). The spontaneous breaking of the rotational symmetry of the carrier density front is reminiscent of the Sekerka–Mullins instability (Pelcé 1988).

The symmetry of the pre-filamentary state, i.e. the number of rudimentary filaments, depends on parameters of the sample as well as on the applied voltage. The pre-filaments then start growing in current while retaining approximately their original width (d). This leads to a rising total current I and thus, via the external load

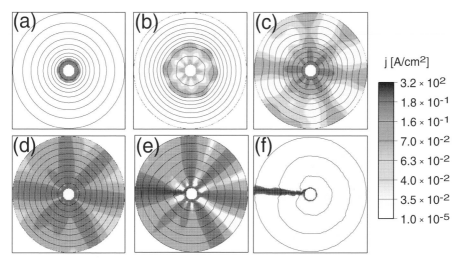

Figure 4.33. Simulation of the formation of a current filament in a Corbino disk. The temporal evolution of the current density $|j(x, y)|$ for a circular sample of diameter 2.1 mm and an inner contact diameter of 80 µm is shown. (a) $t = 0.2$ ns, (b) $t = 0.8$ ns, (c) $t = 2.3$ ns, (d) $t = 4.0$ ns, (e) $t = 8.0$ ns, and (f) $t = 20.0$ ns. Equipotential lines are plotted, spaced by 0.2 V. The bias voltage $U_0 = 2$ V and the load resistance $R = 10$ kΩ correspond to an operating point with an average field $\mathcal{E}_{av} = 5.2$ V cm^{-1}, i.e. between the holding field $\mathcal{E}_h = 4$ V cm^{-1} and the threshold field $\mathcal{E}_{th} = 17$ V cm^{-1} of the homogeneous $j(\mathcal{E})$ characteristic (calculated with the same material parameters as those in Fig. 4.28). (After Schwarz *et al.* (2000*b*).)

resistance, which acts as a global constraint due to Kirchhoff's law $U_0 = U + RI$, to a reduction in the sample voltage U. The pre-filaments therefore enter into a competition as a result of which only the filament which was the first to reach the outer contact survives (e). The other filaments slowly decay into the low-conducting state and vanish (f). This winner-takes-all dynamics is characteristic of a global coupling imposed by the load resistance, and is in line with the general discussion of global constraints in Sections 3.1 and 4.4.4 (Schimansky-Geier *et al.* 1991).

Note that the initial stage of an expanding radial front is always initiated, regardless of the bias polarity, at the central contact, due to the higher electric field there, even if this is the noninjecting anode. In that case, the front propagation is even considerably faster due to the larger fields and hence strongly increased impact ionization (Schwarz *et al.* 2000*a*).

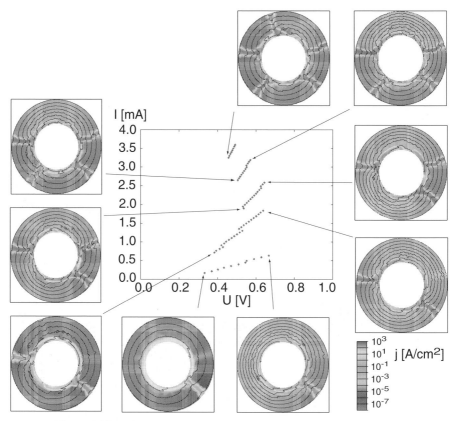

Figure 4.34. The simulated current–voltage characteristic of a circular sample with two concentric contacts of diameters 1 and 2.1 mm. The insets show the current density $|j(x, y)|$ for various operating points indicated by arrows. The bias voltage U_0 is swept from 2.0 to 36.5 V and the load resistance is $R = 10 \,\mathrm{k\Omega}$ (calculated with the same material parameters as those in Fig. 4.28). (After Schwarz *et al.* (2000*b*).)

If the operating point is shifted to larger currents, more than one filament can survive. This is shown in Fig. 4.34 for a sample with a larger central contact diameter. The bias voltage U_0 is increased from 2.0 to 36.5 V. The insets show the stationary current-density distributions at various operating points for a sample with two concentric contacts with diameters 1 and 2.1 mm. As the current increases, the current density in the filament grows up to a threshold at which a second filament is formed. This decreases the sample's resistance, resulting in a switch-back of the sample voltage according to Kirchhoff's law. Thus a second branch of the current–voltage characteristic arises, corresponding to two filaments. Similarly, successive branches with three, four, five, etc. filaments occur. Each jump in the current–voltage characteristic is associated with the generation of an additional filament. Hysteresis with respect to bias sweep-up and sweep-down is also found. With increasing bias the transition from one to two filaments, for example, occurs at a threshold sample voltage of about 0.7 V, whereas upon decreasing the bias the two-filament state is sustained down to sample voltages of about 0.17 V because the sustaining voltage is substantially lower than the threshold voltage for impact ionization. Similar hysteresis

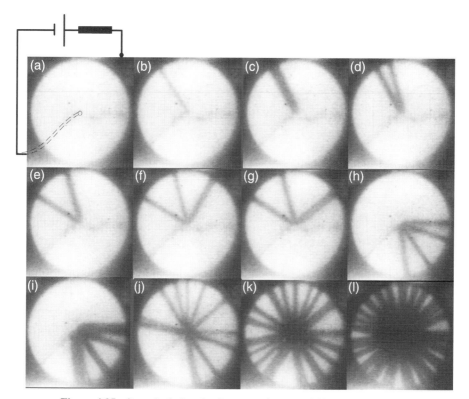

Figure 4.35. Quenched photoluminescence images of filaments in a Corbino sample (sample C5, see Table 4.3) with current increasing from (a) to (l): $I = 0, 0.10, 0.43,$ 0.45, 0.47, 0.58, 0.77, 0.80, 1.54, 1.71, 4.08, and 7.08 mA (after Schwarz *et al.* (2000*b*)).

cycles are found for higher numbers of filaments. Different branches with different numbers of filaments can thus be reached at the same bias voltage by appropriate sweeps.

Thus the Corbino sample is a highly multistable system, and various multiple-filament states can be reached at the same bias voltage by appropriate sweeps. This multistability has an advantage over the multistable filamentary current–voltage characteristics which have been realized by having arrays of point contacts in rectangular n-GaAs samples (Kostial *et al.* 1995*b*), since here the multiple filaments arise in a self-organized way and do not need complicated structuring of multiple-point-contact geometries. Switching between different multifilamentary states in the structures investigated by Kostial *et al.* (1995*b*) has been discussed in the context of data storage and synergetic computers. Other semiconductor systems in which multistable spatial patterns have attracted interest because of their potential prospects for use as multi-bit memory elements include field domains in semiconductor superlattices, see Chapter 6 (Kastrup *et al.* 1994).

Such behavior is in good agreement with experiments (Hirschinger *et al.* 1997*a*). Stationary quenched photoluminescence images of a Corbino sample are shown in Fig. 4.35. The figure shows the formation of multifilamentary patterns upon increasing

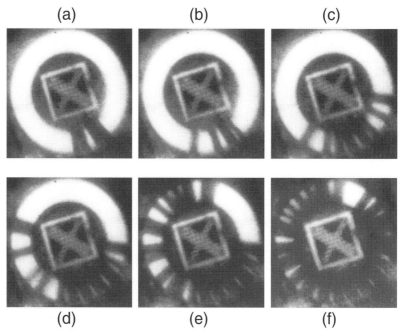

Figure 4.36. Quenched photoluminescence images of filaments in the circular sample corresponding to Fig. 4.34 (sample C6, see Table 4.3). The current increases from (a) to (f): $I = 1.54$, 1.58, 3.73, 5.73, 9.54, and 13.0 mA. (After Schwarz *et al.* (2000*b*).)

the current. At a critical value of the dc bias the originally uniform state bifurcates into a single current filament being formed at a random position (b). As in the case of two point contacts, self-organization yields a parallel-bordered filament even if the undisturbed electric fields are strongly inhomogeneous. To enforce the formation of a filament with a practically constant width along the radial direction, large redistributions of space charge must occur. Upon a further increase of the current, the filament approximately doubles its width (c) and a second bifurcation from one filament to two filaments occurs (d). Typically a new filament is separated from the original one by an angular distance in the range 25°–90°. The subsequent generation of additional filaments upon increasing the dc bias occurs in the same way, (e–l). With increasing current all filaments grow in width and in most cases suddenly one of the filaments splits, resulting in two filaments of smaller width. The angular distribution of the filaments becomes more and more symmetric with increasing number of filaments. Experimental images of a sample geometry similar to that in Fig. 4.34 are shown in Fig. 4.36. In contrast to the simulation shown in Fig. 4.34, each newly generated filament emerges in the vicinity of an existing one. The difference might result from the different dynamic conditions of the experiment and the simulations due to parasitic reactive circuit components. Hysteresis of multiple filamentary states upon bias sweep-up and sweep-down is observed, as in the simulations.

Chapter 5

Nonlinear carrier dynamics in crossed electric and magnetic fields

It has been found experimentally that an external magnetic field applied perpendicular to the electric field can sensitively affect the spatio-temporal instabilities in the regime of impurity-impact ionization. For instance, it has been observed that even a relatively weak magnetic field can induce complex chaotic current oscillations. In this chapter we study the nonlinear and chaotic dynamics of carriers in crossed electric and magnetic fields. We present a general framework for the description of a dynamic Hall instability, and apply it in particular to the conditions of low-temperature impurity breakdown. We show that chaos control by time-delayed feedback can stabilize those chaotic oscillations. Furthermore, we discuss the complex spatio-temporal dynamics of current filaments in a Hall configuration, resulting either in lateral motion or in a deformation of the filamentary patterns.

5.1   Introduction

Ever since E. H. Hall discovered in 1879 the effect which bears his name, galvano-magnetic phenomena in solids have received significant attention. The Hall effect has been used as an important probe of material properties in many branches of solid-state physics, and more than 200 million components of devices successfully utilize the Hall effect (Chien and Westgate 1980).

Along with the development of practical uses of the Hall effect, the theoretical foundation of galvanomagnetic phenomena has been established (Madelung 1957). Interest has recently been revived strongly by the extension of the classical Hall effect into the regime of the integral (von Klitzing 1990) or fractional (Eisenstein and

Störmer 1990) quantum Hall effect, and by the discovery of the negative Hall effect and chaotic dynamics in lateral superlattices (Fleischmann *et al.* 1994, Schöll 1998*b*).

In this chapter we focus on a different extension of the classical Hall effect, viz. into the *nonlinear dynamic* regime (Hüpper and Schöll 1991, Hüpper *et al.* 1992). This is connected with a class of nonlinear dynamic phenomena that have been studied extensively both theoretically and experimentally (Glicksman 1971, Požela 1981, Abe 1989, Schöll 1992, Shaw *et al.* 1992, Peinke *et al.* 1992, Aoki 2000): Self-generated chaotic current or voltage oscillations in high-purity semiconductors under dc bias. In many experiments one crucial requirement for such behavior is the simultaneous application of a parallel or perpendicular static magnetic field (Held *et al.* 1984, Seiler *et al.* 1985, Song *et al.* 1989, Aoki *et al.* 1990*a*); in others they are sensitively modified thereby (Peinke *et al.* 1989, Brandl *et al.* 1990). Even a relatively weak perpendicular magnetic field can induce chaotic current oscillations (Peinke *et al.* 1985, Seiler *et al.* 1985, Brandl *et al.* 1987, Yamada *et al.* 1988, Fujii *et al.* 1989, Spinnewyn *et al.* 1989, Brandl *et al.* 1990, Aoki and Kondo 1991, Rau *et al.* 1991). Stationary current filaments can either be destabilized, resulting in nonlinear oscillations due to transversally traveling filaments (Clauss *et al.* 1991, Hirsch *et al.* 1994, Spangler *et al.* 1994, Hirschinger *et al.* 1998, Niedernostheide *et al.* 1998, Aoki and Fukui 1998) or the filament patterns are deformed by the Lorentz force preserving their stability (Novák *et al.* 1995, 1998*a*, Schöll *et al.* 1998*a*). Theoretical work has been carried out in order to investigate the effects of a magnetic field in the breakdown regime. The dynamic Hall effect has been proposed as a mechanism by means of which to induce chaotic temporal oscillations (Hüpper and Schöll 1991). The transverse periodic or chaotic motion of a current filament under a crossed electric and magnetic field has been analyzed numerically for p-type Ge by one-dimensional simulations based on the semiclassical semiconductor-transport equations neglecting the longitudinal spatial dependency (Hüpper *et al.* 1993*b*). The spatial deformation of current filaments in the presence of a transverse magnetic field has been explained by an effectively one-dimensional drift–diffusion model assuming a dipole-like electric field between two point contacts (Novák *et al.* 1995), and by microscopic simulations on two-dimensional spatial domains with various contact geometries (Kunihiro *et al.* 1997, Schwarz *et al.* 2000*b*).

Our aim here is to extend the classical *static* theoretical treatment of the Hall effect into the fully dynamic regime. This yields an explanation of the influence of a transverse magnetic field in terms of a classical magneto-transport model, which is elaborated here for the case of low-temperature impurity breakdown although it may be applied to more general situations. The basic new idea is to consider both the applied drift field and the induced Hall field as dynamic variables whose temporal dependencies are governed by dielectric relaxation, while the magnetic field will be considered as an external control parameter. Further dynamic degrees of freedom are provided by the carrier densities, which are coupled to the electric fields by strongly nonlinear interactions as a result of impact ionization of shallow impurity levels.

In our treatment (Hüpper *et al.* 1992) we neglect quantum effects leading to the Zeeman splitting of impurity levels, and the splitting of the conduction band into Landau subbands of spacing $\Delta E = \hbar\omega_c$, where $\omega_c = eB/m^*$, with magnetic induction B and elementary charge e, in an effective-mass approximation with an effective mass m^*. In the impurity-impact-ionization regime the mean energy E per carrier is comparable to the ionization energy E_{th} of the impurities. A classical treatment will thus be applicable in the magnetic-field regime in which

$$\Delta E \ll E_{th} \tag{5.1}$$

holds.

5.2   The dynamic Hall effect

5.2.1   Constitutive model equations

Our approach to magnetotransport in semiconductors is based upon the Boltzmann equation for the semiclassical carrier-distribution function of one band $f(r, k, t)$, where r is the spatial coordinate and k is the wave vector, which was introduced in Section 2.3.2, eq. (2.18):

$$\frac{\partial f}{\partial t} + v_g \nabla_r f + \frac{q}{\hbar}(\mathcal{E} + v_g \times B)\nabla_k f = \left(\frac{\partial f}{\partial t}\right)_{coll}. \tag{5.2}$$

Here v_g is the group velocity, and $\mathcal{E}$ and B are the electric field and the magnetic induction, respectively. The collision integral $(\partial f/\partial t)_{coll}$ includes in principle all dissipative processes such as phonon and impurity scattering, and generation and recombination (GR) processes involving localized energy levels, in particular impact ionization of impurities. The carrier charge is $q = \pm e$ for holes and electrons, respectively.

Attention is now confined to a nondegenerate, isotropic parabolic band structure characterized by an effective mass m^*: $E(k) = \hbar^2 k^2/(2m^*)$. Hydrodynamic balance equations for the slow macroscopic observables such as the carrier density $n(r, t) = \int f(r, k, t)z\, d^3k$, the mean momentum per carrier $p(r, t) = n^{-1}\int \hbar k\, f(r, k, t)z\, d^3k$, and the mean energy per carrier $w_n(r, t) = n^{-1}\int E(k)\, f(r, k, t)z\, d^3k$ are obtained by multiplying (5.2) by appropriate functions of k and integrating over the first Brillouin zone (z is the density of states in k-space):

$$\frac{\partial n}{\partial t} + \nabla_r(nv) = \int \left(\frac{\partial f}{\partial t}\right)_{coll} z\, d^3k, \tag{5.3}$$

$$\frac{\partial(np)}{\partial t} + [\nabla_r(nv)]p + (nv\,\nabla_r)p + \nabla_r(nk_B T_e) - qn\,(\mathcal{E} + v \times B)$$

$$= \int \hbar k\left(\frac{\partial f}{\partial t}\right)_{coll} z\, d^3k, \tag{5.4}$$

$$\frac{\partial(nw_n)}{\partial t} + \boldsymbol{\nabla}_r(n\boldsymbol{v})w_n + \boldsymbol{\nabla}_r(nk_{\mathrm{B}}T\boldsymbol{v}) - \boldsymbol{\nabla}_r\boldsymbol{j}_Q - qn\boldsymbol{v}\mathcal{E}$$

$$= \int E(\boldsymbol{k})\left(\frac{\partial f}{\partial t}\right)_{\mathrm{coll}} z\,d^3k \tag{5.5}$$

with heat flux $\boldsymbol{j}_Q = \frac{1}{2}nm^*\langle(\boldsymbol{v}_{\mathrm{g}} - \langle\boldsymbol{v}_{\mathrm{g}}\rangle)^2(\boldsymbol{v}_{\mathrm{g}} - \langle\boldsymbol{v}_{\mathrm{g}}\rangle)\rangle$ where the brackets denote the ensemble average $\langle A\rangle = n^{-1}\int A(\boldsymbol{k})\,f(\boldsymbol{r},\boldsymbol{k},t)z\,d^3k$. The mean group velocity $\boldsymbol{v}(\boldsymbol{r},t) = \langle\boldsymbol{v}_{\mathrm{g}}\rangle$ is related to the carrier temperature $T_{\mathrm{e}} = [m^*/(3k_{\mathrm{B}})]\langle(\boldsymbol{v}_{\mathrm{g}} - \langle\boldsymbol{v}_{\mathrm{g}}\rangle)^2\rangle$ and the mean momentum and energy by

$$\boldsymbol{p} = m^*\boldsymbol{v}, \qquad w_n = \frac{m^*}{2}v^2 + \frac{3}{2}k_{\mathrm{B}}T_{\mathrm{e}}. \tag{5.6}$$

In order to obtain a closed set of balance equations for n, $\boldsymbol{p}$, and w_n we approximate the heat flux by $\boldsymbol{j}_Q = -\kappa\,\boldsymbol{\nabla}_r T_{\mathrm{e}}$ with thermal conductivity κ and express the collision integrals, i.e. the right-hand sides of (5.3) and (5.5), in terms of the common generation–recombination (GR) rate and energy-relaxation rate via

$$\phi(n,\boldsymbol{n}_t,w_n) = \int\left(\frac{\partial f}{\partial t}\right)_{\mathrm{coll}} z\,d^3k \tag{5.7}$$

$$-n\frac{w_n - w_{n0}}{\tau_e(w_n)} = \int E(\boldsymbol{k})\left(\frac{\partial f}{\partial t}\right)_{\mathrm{coll}} z\,d^3k. \tag{5.8}$$

Here τ_e is the mean-energy-dependent energy-relaxation time (Nougier *et al.* 1981), $w_{n0} = \frac{3}{2}k_{\mathrm{B}}T_{\mathrm{L}}$, where T_{L} is the lattice temperature, and $\boldsymbol{n}_t = (n_1,\ldots,n_M)$ are the densities of carriers bound at the M trap states (the ground state and excited states) of an impurity, for which similar GR rate equations (with the rates ϕ_t) are obtained, cf. (2.40) and (2.42):

$$\frac{\partial n}{\partial t} + \boldsymbol{\nabla}_r(n\boldsymbol{v}) = \phi(n,w_n), \tag{5.9}$$

$$\frac{\partial\boldsymbol{n}_t}{\partial t} = \boldsymbol{\phi}_t(n,\boldsymbol{n}_t,w_n), \tag{5.10}$$

$$\frac{\partial w_n}{\partial t} + (\boldsymbol{v}\,\boldsymbol{\nabla}_r)w_n + \frac{1}{n}\boldsymbol{\nabla}_r(nk_{\mathrm{B}}T_{\mathrm{e}}\boldsymbol{v}) - \frac{\kappa}{n}\Delta T_{\mathrm{e}} - q\boldsymbol{v}\mathcal{E} = -\frac{(w_n - w_{n0})}{\tau_e(w_n)}. \tag{5.11}$$

Note that the introduction of a macroscopic momentum-relaxation time proposed by Nougier *et al.* (1981) is not adequate in magnetotransport problems. This will be discussed later. Contrary to this approach, we make the following assumptions for the momentum-balance equation:

(i) Momentum relaxation occurs faster than do all other processes, so that $\boldsymbol{p}$ can be eliminated adiabatically from (5.4) by setting $d\boldsymbol{p}/dt \equiv \partial\boldsymbol{p}/\partial t + (\boldsymbol{v}\,\boldsymbol{\nabla}_r)\boldsymbol{p} = 0$.

(ii) The integral

$$\iint W(k', k) f(r, k', t) kz \, d^3k' \, z \, d^3k$$

is approximated by 0. This holds exactly, for example, if the relaxation-time approximation

$$\left(\frac{\partial f}{\partial t}\right)_{\text{coll}} = -\frac{(f - f_0)}{\tau(k)}$$

is used in the Boltzmann equation.

Thus (5.4) leads to a current-density field relation

$$j = qnv = nK^{(1)}F - nK^{(2)}(B \times F) + nK^{(3)}B(BF) \tag{5.12}$$

with the kinetic coefficients (Madelung 1957)

$$K^{(i)} = \left\langle \frac{(q\tau/m^*)^i}{1 + q^2\tau^2 B^2/m^{*2}} \right\rangle \tag{5.13}$$

and $F = q\mathcal{E} - (1/n)\nabla_r(nk_B T_e)$, $1/\tau(k) = \int W(k, k') f(r, k', t) z \, d^3k'$. Note that this does not require any further assumption about f, so the model is not restricted to weak electric or magnetic fields.

Equation (5.12) can be cast into a more convenient form with the following definitions of the drift mobility, μ_d, and the Hall mobility, μ_H, respectively:

$$\mu_H = \frac{q}{e}\frac{K^{(2)}}{K^{(1)}}, \qquad \mu_d = \frac{q}{e}K^{(1)}\left[1 + (\mu_H B)^2\right], \tag{5.14}$$

which yields

$$j = en\mu_B\frac{F}{q} - qn\mu_H\mu_B\left(B \times \frac{F}{q}\right) + nK^{(3)}B(BF) \tag{5.15}$$

with $\mu_B = \mu_d/[1 + (\mu_H B)^2]$. The nomenclature μ_d and μ_H is suited to the conventional Hall configuration, i.e. an externally applied magnetic field perpendicular to the applied electric field. This configuration will be assumed in the remainder of this chapter. Note that the introduction of a macroscopic mean-energy-dependent momentum-relaxation time, similar to the case without a magnetic field (Nougier et al. 1981), will lead to $\mu_H = \mu_d$, which neglects the "statistics factor" $r_H = \mu_H/\mu_d$.

The dynamics of $\mathcal{E}$ and B are determined by Maxwell's equations:

$$\nabla_r \times \mathcal{E} = -\frac{\partial B}{\partial t}, \qquad\qquad \nabla_r B = 0, \tag{5.16}$$

$$\frac{1}{\mu_0}\nabla_r \times B = \epsilon\frac{\partial \mathcal{E}}{\partial t} + j \equiv j_0, \qquad \epsilon\nabla_r\mathcal{E} = \rho, \tag{5.17}$$

with the local charge density $\rho = e(N_D - N_A) + q(n + \sum_{i=1}^{M} n_i)$, where N_D and N_A are the donor and acceptor concentrations, respectively, and $\epsilon = \epsilon_0\epsilon_r$ and μ_0 are the permittivity and permeability, respectively. If the semiconductor is connected to a constant voltage source U_0 via a load resistor of resistance R and a capacitor of capacitance C, Kirchhoff's law leads to

$$U_0 = RI(t) + U(t), \qquad (5.18)$$

$$I(t) = I^{\text{tot}}(t) + C\frac{\partial U}{\partial t}, \qquad (5.19)$$

with current $I(t) = \int j(\mathbf{r}, t)\,df$, $U(t) = \int \mathcal{E}\,dx$. However, in practical cases, $\mathbf{B}$ does not depend on time (Schöll 1987), so we assume always a curl-free electric field: $\nabla_r \times \mathcal{E} = 0$. Equations (5.9)–(5.11) and (5.16)–(5.19) together with appropriate boundary conditions form a closed set of model equations.

The time scales on which the variables n, w_n, and $\mathcal{E}$ vary are determined by the GR lifetime, the energy-relaxation time, and Maxwell's dielectric relaxation time, respectively. For the purpose of this treatment we assume an energy-relaxation time much smaller than the other time scales. Neglecting temporal and spatial derivatives in (5.11) leads to a local mean-energy–field relation:

$$w_n = w_{n0} + \frac{\eta}{2}m^*\mu_B\mu_d\mathcal{E}^2 \qquad (5.20)$$

with $\eta = 2e\tau_e/(\mu_d m^*)$. One of the impurity levels, for instance n_M, can be eliminated via Gauss' Law (5.17),

$$n_M = -\frac{e}{q}(N_D - N_A) - n - \sum_{i=1}^{M-1} n_i + \frac{\epsilon}{q}\nabla_r\mathcal{E}. \qquad (5.21)$$

A magnetic field applied perpendicular to the current density gives rise to a Hall field, which leads to a spatially inhomogeneous transverse carrier-density profile. In the simplest model, for moderate magnetic fields, we neglect these nonuniformities. For $M \geq 2$ the static spatially homogeneous $j(\mathcal{E})$ characteristic can become S-shaped, and self-organized spatial structures (current filaments) are possible, see Chapter 4. The dynamics of these filaments will be discussed in Section 5.4 below.

Choosing the coordinate system such that the drift current and the drift field ($\mathcal{E}_x$) are in the x-direction, the magnetic field is in the y-direction, and the Hall field ($\mathcal{E}_z$) is in the z-direction (Fig. 5.1), Eqs. (5.9), (5.10), and (5.16)–(5.19) reduce to

$$\frac{\partial n}{\partial t} = \phi(n, n_1, \ldots, n_{M-1}, \mathcal{E}; B), \qquad (5.22)$$

$$\frac{\partial n_i}{\partial t} = \phi_i(n, n_1, \ldots, n_{M-1}, \mathcal{E}; B), \qquad i = 1, \ldots, M-1, \qquad (5.23)$$

$$c\epsilon\frac{\partial \mathcal{E}_x}{\partial t} = J_0 - (en\mu_B + \sigma_L)\mathcal{E}_x + qn\mu_B\mu_H B\mathcal{E}_z, \qquad (5.24)$$

$$\epsilon \frac{\partial \mathcal{E}_z}{\partial t} = -en\mu_B\mathcal{E}_z - qn\mu_B\mu_H B\mathcal{E}_x, \tag{5.25}$$

where $J_0 = U_0/(RA)$, $\sigma_L = L/(RA)$, $c = 1 + CL/(\epsilon A)$, L is the sample length and A is the cross-section. Note that the GR rates depend upon $\mathcal{E} \equiv (\mathcal{E}_x^2 + \mathcal{E}_z^2)^{1/2}$ via the GR coefficients, in particular impact ionization.

5.2.2   Linear stability analysis

Analytic conditions for the onset of oscillatory instabilities via a Hopf bifurcation can be derived in a straightforward but lengthy way by a linear stability analysis (Schöll 1987) of (5.22)–(5.25). Equations (5.22)–(5.25) can be put into the generic form

$$\frac{\partial q}{\partial t} = \Phi(q, \mathcal{E}_x) \tag{5.26}$$

$$c\epsilon \frac{\partial \mathcal{E}_x}{\partial t} = J - (\sigma_L + en\mu_B)\mathcal{E}_x, \tag{5.27}$$

with $q = (n, n_1, \ldots, n_{M-1}, \mathcal{E}_z)$ and $J = J_0 + qn\mu_B\mu_H B\mathcal{E}_z$. The fixed point of (5.22)–(5.25) (denoted by a superscript $*$) reproduces the classical static Hall effect

$$\mathcal{E}_z^* = -\frac{q}{e}\mu_H B\mathcal{E}_x^*, \tag{5.28}$$

$$J_0 = (en^*\mu_d + \sigma_L)\mathcal{E}_x^*. \tag{5.29}$$

The eigenvalues λ_i of the Jacobian matrix $\mathcal{A}$ obtained by linearizing (5.26) and (5.27) around this fixed point determine the stability of the steady state. The product of these eigenvalues can be expressed in terms of the differential conductivity $\sigma_{\mathrm{diff}} \equiv dj/d\mathcal{E}_x$, see Chapter 3 (Schöll 1989),

$$\det \mathcal{A} = \prod_{i=1}^{M+2} \lambda_i = -\det \tilde{A}\,\frac{\sigma_L + \sigma_{\mathrm{diff}}}{c\epsilon}, \tag{5.30}$$

where $\tilde{A}$ is the Jacobian of the subsystem describing the dynamics of q with fixed $\mathcal{E}_x$.

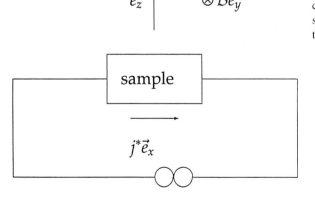

$\vec{e}_z$ $\otimes B\vec{e}_y$

Figure 5.1. The Hall configuration of the sample; showing the orientation of the axes.

sample

$j^*\vec{e}_x$

For $M = 1$ (i.e. a single impurity level) $\det \tilde{A} = (\tau_{GR}\tau_M^*)^{-1}$ holds with the effective GR lifetime $\tau_{GR} \equiv -(\partial\phi/\partial n)^{*-1}$ and the effective dielectric relaxation time $\tau_M^* = \epsilon(1 + \mu_H^2 B^2)/(en^*\mu_{diff})$. Since here we are not interested in mobility-induced instabilities, we assume always a positive differential mobility $\mu_{diff} \equiv dv/d\mathcal{E}_x$. The GR lifetime is always positive, hence $\det \tilde{A} > 0$. A Hopf bifurcation ($\lambda_1 = \lambda_2^*$ purely imaginary, $\lambda_3 < 0$) requires $\det \mathcal{A} < 0$, thus $\sigma_{diff} > -\sigma_L$. For current control ($\sigma_L = 0$) this means positive differential conductivity (PDC). A detailed analysis for the general case ($\sigma_L \neq 0$) has been carried out (Hüpper and Schöll 1991). For the special case $\sigma_L = 0$ (current-controlled conditions) with $c = 1$ (no external capacitor) this yields the necessary but not sufficient conditions

$$\frac{1}{\tau_{GR}} < \frac{(\mu_H B)^2 - 1}{\tau_M^*}, \qquad \mu_H B > 1, \tag{5.31}$$

which requires a large GR lifetime, and a minimum value of B. The oscillation frequency at the onset of the instability is given by

$$\omega_{cr} = \sqrt{-\frac{\partial\phi}{\partial n}\left(\frac{2}{\tau_M^*} - n^*\frac{\partial\mu_d}{\partial\mathcal{E}}\right) + \frac{1}{\tau_M^*}\left(\frac{1}{\tau_{M_0}^*} - n^*\frac{\partial\mu_d}{\partial\mathcal{E}}\right) + n^*\frac{\partial\phi}{\partial\mathcal{E}}\mu_d\mathcal{E}^*}, \tag{5.32}$$

where $\tau_{M_0}^* = \tau_M^*(B = 0)$. For $M \geq 2$ (i.e. more than one impurity level) an S-shaped current–voltage characteristic with a negative-differential-conductivity branch (NDC) is possible even with constant mobility. Thus, in order to describe SNDC and bistability of the static current–field characteristic associated with low-temperature impurity breakdown, at least one excited impurity level has to be taken into account. For $M = 2$ we have $\det \tilde{A} = -(\det A_{GR})\tau_M^{*-1}$ with A_{GR} the Jacobian describing fluctuations in the carrier densities only: $\det A_{GR} = (\partial\phi/\partial n)(\partial\phi_1/\partial n_1) - (\partial\phi_1/\partial n)(\partial\phi/\partial n_1)$, which can be positive or negative. Thus a magnetic field-induced Hopf bifurcation can be obtained in the PDC and also in the NDC region. Self-generated and chaotic oscillations even in the absence of a magnetic field are possible on the NDC branch if

$$\tau_M^{*-1} < \lambda_{GR}, \tag{5.33}$$

where $\lambda_{GR} > 0$ is the largest eigenvalue of A_{GR}. These are GR-induced oscillations (Schöll 1987), which require a small (differential) mobility, in contrast to the magnetic-field-induced oscillations, characterized by (5.31).

Some insight into the qualitative behavior of the *nonlinear* system can be gained from a discussion of the flow in phase space (Hüpper and Schöll 1991). For the sake of simplicity we restrict ourselves to a constant mobility and $\sigma_L = 0$ with $c = 1$. In an $n = $ constant hyperplane in phase space, the $\dot{n} = 0$ isocline is a circle and the $\dot{\mathcal{E}}_x = 0$ isocline is a tangential straight line, whereas the $\dot{\mathcal{E}}_z = 0$ isocline forms a perpendicular, radial line. Without the time dependency of n, the evolution of the system would be determined by the complex eigenvalues $\lambda_{1,2} = -(1 \pm i\mu_H B)/\tau_M^*$, namely by (5.24) and (5.25). This oscillatory behavior persists in the case of a time-dependent n, but inside the circle the carrier density decreases, whereas it increases outside the circle. This leads to a variation of the curvature of the orbit. Because of the different curvatures in the different regimes of phase space, phase points that are initially close may become separated later. This indicates the possibility of chaos, as a result of the Rössler-type structure of the dynamic system: A two-variable oscillator ($\mathcal{E}_x$ and $\mathcal{E}_z$) combined with a switching-type submanifold (n) (Rössler 1976).

The numerical simulation of the nonlinear system is in agreement with this qualitative discussion. We shall demonstrate this in the following two sections.

5.2.3 Chaotic Hall instability with positive differential conductivity

In this section we present numerical simulations for the single-level model ($M = 1$):

$$\frac{\partial n}{\partial t} = \phi(n, \mathcal{E}_x, \mathcal{E}_z; B), \tag{5.34}$$

$$\epsilon \frac{\partial \mathcal{E}_x}{\partial t} = J_0 - (en\mu_B)\mathcal{E}_x + qn\mu_B\mu_H B\mathcal{E}_z, \tag{5.35}$$

$$\epsilon \frac{\partial \mathcal{E}_z}{\partial t} = -en\mu_B\mathcal{E}_z - qn\mu_B\mu_H B\mathcal{E}_x, \tag{5.36}$$

where we have neglected external capacitances and assumed current-controlled conditions. An example of a model for the GR kinetics is

$$\phi(n, \mathcal{E}_x, \mathcal{E}_z; B) = X^S(N_A^* - n) - T^S(\mathcal{E})n(n + N_t) + X(\mathcal{E})n(N_A^* - n), \tag{5.37}$$

where $N_A^* \equiv |N_A - N_D|$ is the effective doping concentration, N_t is the concentration of the compensating minority carrier dopant, X^S is the emission coefficient,

$$T^S(\mathcal{E}) = T_0^S\left(\frac{E_0}{E_0 + (\eta + 1)E'}\right)^{3/2} \tag{5.38}$$

is the capture coefficient,

$$X(\mathcal{E}) = X_0\Sigma^{1/2}\left/\left\{(1 + \Sigma)\left[1 + \exp\left(\frac{3(E_{th} - E')}{2(E_0 + \eta E')}\right)\right]\right\}\right. \tag{5.39}$$

is the impact-ionization coefficient (Westervelt and Teitsworth 1985), with $E = E_0 + \eta E'$ with $E' = \frac{1}{2}m^*\mu_B\mu_d\mathcal{E}^2$, and $\Sigma = E'/E_{th}$. For material parameters corresponding to various materials and physical situations (see Table 5.1), Figs. 5.2(a) and (b) show the regime of oscillatory instabilities. The insets show the static $j(\mathcal{E}_x^*)$ characteristics.

Crossing the boundary of the instability regime, a stable limit cycle is generated by a supercritical Hopf bifurcation. This simple periodic oscillation undergoes a period-doubling route to chaos upon further increase of $\mathcal{E}_x^*$, followed by further bifurcations that depend upon the specific material. Figures 5.3(a) and (b) show the resulting bifurcation diagrams: Figs. 5.4(a) and (b) the Lyapunov exponents and Kaplan–Yorke dimension as functions of $\mathcal{E}_x^*$. Regimes with positive Lyapunov exponents and nonzero fractal dimensions refer to chaotic states. Note that one Lyapunov exponent is always zero, and one always remains negative. If the bifurcation diagram is plotted as a function of B it displays a period-doubling scenario with *decreasing B* (Schöll *et al.* 1992). The period-doubling route is also visualized in the frequency regime. Figures 5.5(a) and (b) show the power spectra for a periodic and a chaotic oscillation, respectively. The peaks in the power spectra for different $\mathcal{E}_x^*$ are sampled in a frequency bifurcation diagram in Fig. 5.5(c), displaying period-doublings and chaotic bands. In Fig. 5.6 typical chaotic time series and phase portraits are shown.

5.2.4 The intermittency route to chaos

In several of the periodic windows of the control parameter space Pomeau–Manneville routes to chaos via intermittency are found. It is convenient to use the magnetic field as a control parameter. For instance, for fixed $\mathcal{E}_x^*$ and increasing B, just before the onset of the period-three window an intermittency route occurs. Three intermittent time series for different magnetic fields are shown in Fig. 5.7. The return map shows

Table 5.1. *Material parameters for the one-level model (ϵ_0 is the permittivity of the vacuum and m_0 is the free-electron mass)*

	p-Ge	n-InSb
T_0^s	3×10^{-6} cm^3 s^{-1}	1.5×10^{-6} cm^3 s^{-1}
X^s	10^{-4} s^{-1}	10 s^{-1}
X_0	6×10^{-6} cm^3 s^{-1}	6×10^{-5} cm^3 s^{-1}
μ_d	$\mu_0 [1 + 13.2(\mathcal{E}/\mathcal{E}_0)^{0.7}]^{-1}$	μ_0
μ_0	10^6 cm^2 V^{-1} s^{-1}	3×10^6 cm^2 V^{-1} s^{-1}
$\mathcal{E}_0$	66 V cm^{-1}	
μ_H	μ_d	μ_d
N_A	10^{14} cm^{-3}	2.9×10^{13} cm^{-3}
N_D	10^{11} cm^{-3}	3.2×10^{14} cm^{-3}
ϵ	$16\epsilon_0$	$18\epsilon_0$
T_L	4 K	4 K
E_{th}	10 meV	0.7 meV
m^*	$0.35m_0$	$0.013m_0$
η	1	1

tangent bifurcations indicating type-I intermittency. The probability distribution $P(l)$ of various lengths of the laminar regions and the universal scaling of the mean laminar length $\langle l \rangle \propto (B - B_c)^{-1/2}$ as a function of the control parameter B clearly confirm type-I intermittency (Figs. 5.8(a) and (b)), cf. Section 1.3.1. Additional critical scaling laws proposed for higher cumulants (Rein *et al.* 1993) of l are also found. Figure 5.8(c) shows the scaling law of the variance of the laminar length $\langle (\Delta l)^2 \rangle \propto (B - B_c)^{-1}$, where B_c is the critical control parameter for which the motion becomes periodic.

These simulations are in good agreement with experiments performed in p-Ge (Clauss *et al.* 1991, Rau *et al.* 1991) and n-InSb (Seiler *et al.* 1985, Song *et al.* 1989), in which the magnetic-field or current dependency of the chaotic scenarios, the B threshold for the onset of oscillatory instabilities, and type-I intermittency (Richter *et al.* 1991, 1992) have indeed been found.

(a)

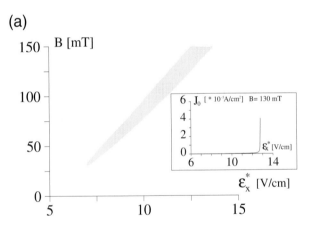

(b)

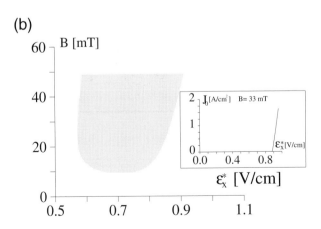

Figure 5.2. The dynamic Hall effect; showing the regime of oscillatory instability in the $(\mathcal{E}_x^*, B)$ control parameter plane for (a) p-Ge and (b) n-InSb. The insets show the corresponding static current-density versus electric-field characteristics. (After Hüpper *et al.* (1992).)

5.2.5 Chaotic Hall instability with negative differential conductivity

Next we consider a model with two impurity levels. The GR kinetics is given by

$$\frac{\partial n}{\partial t} = X_1^S n_2 - T_1^S n(n + N_t) + X_1 n n_1 + X_1^* n n_2,$$ (5.40)

$$\frac{\partial n_1}{\partial t} = T^* n_2 - X^* n_1 - X_1 n n_1,$$ (5.41)

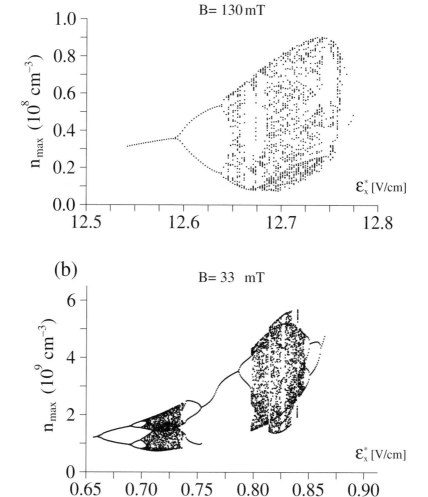

Figure 5.3. The route to chaos: the bifurcation diagram of the maxima of the carrier density n_{max} versus the control parameter $\mathcal{E}_x^*$ for (a) p-Ge and (b) n-InSb (after Hüpper *et al.* (1992)).

with $n_2 = N_A^* - n - n_1$, where X_1^S, T_1^S, X_1, X_1^*, X^*, and T^* denote the appropriate GR coefficients (see Fig. 4.1) and $X_1 = X_1^0 \exp(-0.016\alpha)$, $X_1^* = X_1^{*0} \exp(-4 \times 10^{-3}\alpha)$, and $\alpha = (\mu_B/\mu_d)^{1/2}(\mathcal{E}_1/\mathcal{E})$ are used for the impact-ionization coefficients. Typical material parameters are listed in Table 5.2. The two different proposed oscillation mechanisms discussed in Section 5.2.2, eqs. (5.33) and (5.31), are realized in different physical situations: (i) oscillations arising for $B = 0$ on the NDC branch (Schöll 1986b), and (ii) oscillations due to the dynamic Hall effect, arising for B larger than a threshold value on the low-conductivity PDC branch, and extending also to the middle SNDC branch and the high-conductivity PDC branch with increasing B. The regime of oscillatory instability for case (ii) is shown in Fig. 5.9. The inset shows the static SNDC characteristic which is independent of B in the simple approximation $\mu_d = \mu_H$.

A different oscillation mechanism is clearly seen in the time series and phase portraits for case (i) (Fig. 5.10). In contrast to Fig. 5.6 the Hall field $\mathcal{E}_z$ does not play an active part in the oscillation cycle, but follows with some inertia $\mathcal{E}_x$. The low mobility needed for mechanism (i) inhibits the oscillations of case (ii).

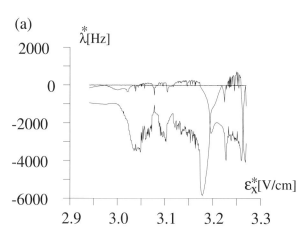

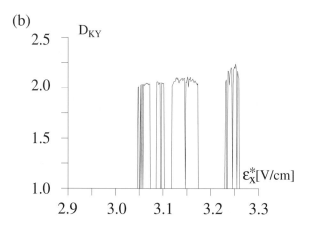

Figure 5.4. (a) Lyapunov exponents λ^* and (b) the Kaplan–Yorke dimension D_{KY} as functions of the control parameter $\mathcal{E}_x^*$ for the dynamic Hall instability in p-Ge with $B = 100$ mT and mobility $\mu_d = \mu_0$ (after Hüpper et al. (1992)).

In the SNDC regime current filaments are likely to form. While these have been shown to be associated with breathing oscillations of type (i) for $B = 0$, the coupling with the dynamic Hall effect for $B \neq 0$ can give rise to a much richer turbulent spatio-temporal dynamics. In Section 5.4 the dynamic Hall effect will be extended to filamentary conduction.

5.3   Chaos control

In this section we shall demonstrate that, by using the method of time-delayed feedback control (time-delay autosynchronization, see Section 1.3.2), a stable tunable semiconductor oscillator can be designed on the basis of chaos control of the dynamic Hall effect (Schöll and Pyragas 1993).

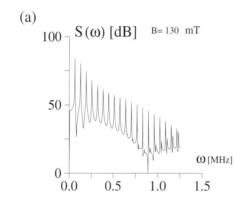

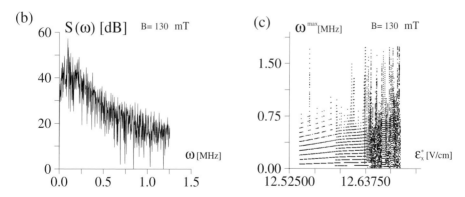

Figure 5.5. Power spectra $S(\omega)$ as a function of the angular frequency of the voltage oscillations in the dynamic Hall effect. (a) $\mathcal{E}_x^* = 12.52$ V cm^{-1} (a limit cycle), (b) $\mathcal{E}_x^* = 12.65$ V cm^{-1} (chaos), and (c) shows the maxima in the power spectra as a function of $\mathcal{E}_x^*$. All data refer to p-Ge. (After Hüpper *et al.* (1992).)

(a)

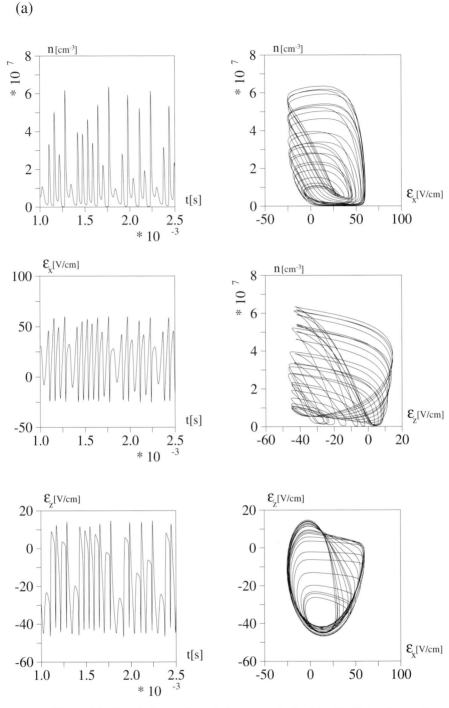

Figure 5.6. Chaotic time series and phase portraits for (a) p-Ge (first and second columns) with $\mathcal{E}_x^* = 12.52$ V cm^{-1} and $B = 130$ mT and for (b) n-InSb (third and fourth columns) with $\mathcal{E}_x^* = 0.72$ V cm^{-1} and $B = 33$ mT. The temporal evolution of the three variables (first and third columns) and various projections of the phase portraits (second and fourth columns) are shown. (After Hüpper *et al.* (1992).)

5.3.1   Stabilizing unstable orbits by delayed feedback control

As discussed in Section 1.3.2, any chaotic attractor of a nonlinear dynamic system contains an infinite dense set of unstable periodic orbits (UPOs). In the pioneering work by Ott, Grebogi and Yorke (Ott *et al.* 1990) (OGY) it was shown that any

(b)

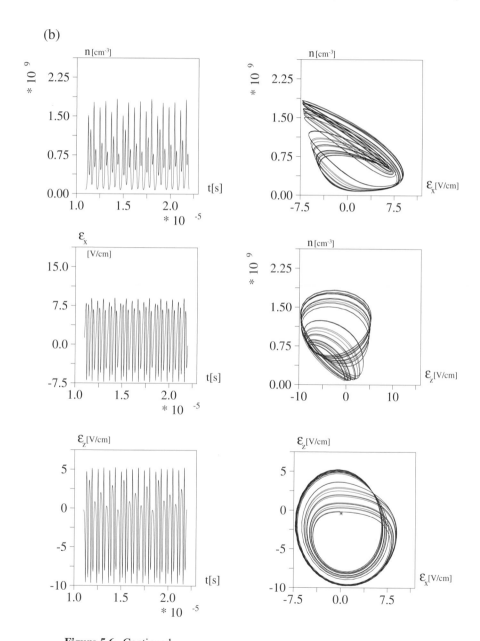

Figure 5.6. Continued.

of these UPOs may be stabilized by applying a small time-dependent perturbation to the control parameter of the system such that the trajectories are thrown onto the stable manifold of the particular UPO. However, a disadvantage of the OGY method is that the feedback control is applied only at discrete times given by the return times of the Poincaré map of the dynamic flow, which has to be computed for this purpose. The method can thus stabilize only those UPOs whose largest Lyapunov exponent is small compared with the reciprocal time interval between the parameter changes. Moreover, it requires a large computational effort, which makes it difficult to control fast processes in real time. Therefore we shall apply here the time-continuous

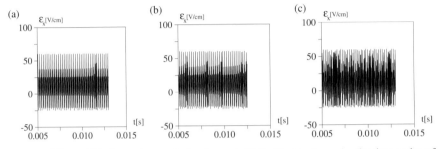

Figure 5.7. Intermittency in the dynamic Hall effect is shown by the time series of the drift field $\mathcal{E}_x$ for (a) $B = 130.046$ mT, (b) $B = 130.050$ mT, and (c) $B = 130.065$ mT, and $\mathcal{E}_x^* = 12.664$ V cm^{-1} (p-Ge) (after Rein et al. (1993)).

Table 5.2. *Material parameters for the two-level model for n-GaAs; for case (ii) only the differences with respect to case (i) are shown*

	Case (i)	Case (ii)
T_1^s	10^{-6} cm^3 s^{-1}	
T^*	10^6 s^{-1}	
X_1^s	10^4 s^{-1}	5×10^5 s^{-1}
X^*	10^4 s^{-1}	4.7×10^5 s^{-1}
X_1^0	5×10^{-8} cm^3 s^{-1}	
X_1^{*0}	10^{-6} cm^3 s^{-1}	
$\mathcal{E}_1$	256 V cm^{-1}	
μ_d	$\mu_0 \frac{\arctan(3000\mathcal{E}/\mathcal{E}_1)}{3000\mathcal{E}/\mathcal{E}_1}$	μ_0
μ_0	500 cm^2 V^{-1} s^{-1}	10^5 cm^2 V^{-1} s^{-1}
μ_H	μ_d	
N_D	1.3×10^{15} cm^{-3}	
N_A	3×10^{14} cm^{-3}	
T_L	4 K	

autosynchronization method proposed by Pyragas (1992). The stabilization of UPOs is achieved by adding a delayed self-controlled feedback to one of the dynamic equations, as described by (1.36).

5.3.2 The tunable semiconductor oscillator

We will now apply time-continuous chaos control to the dynamic Hall instability for the single-impurity-level model (5.34)–(5.36):

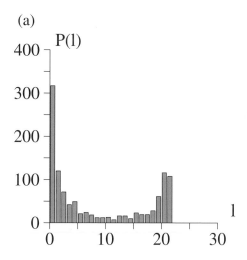

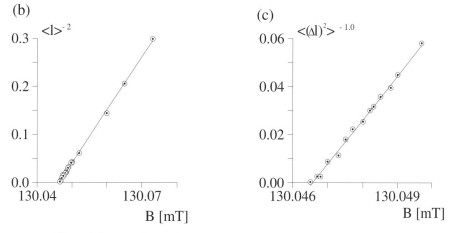

Figure 5.8. Intermittency in the dynamic Hall effect. (a) A histogram of the distribution of the laminar lengths $\langle l \rangle$ for p-Ge at $B = 130.0467$ mT (200 000 periods were used in the simulations). (b) The mean laminar length $\langle l \rangle^{-2}$ versus the control parameter B. (c) The variance $\langle (\Delta l)^2 \rangle^{-1}$ versus B. The numerical parameters are the same as those in Fig. 5.7. (After Rein *et al.* (1993).)

$$\frac{\partial n}{\partial t} = \phi(n, \mathcal{E}_x, \mathcal{E}_z; B), \tag{5.42}$$

$$\epsilon \frac{\partial \mathcal{E}_x}{\partial t} = J - (en\mu_B)\mathcal{E}_x + en\mu_B\mu_H B\mathcal{E}_z, \tag{5.43}$$

$$\epsilon \frac{\partial \mathcal{E}_z}{\partial t} = -en\mu_B\mathcal{E}_z - en\mu_B\mu_H B\mathcal{E}_x. \tag{5.44}$$

The voltage across the sample $U(t) = \mathcal{E}_x(t)L$, where L is the sample's length, is a dynamic output variable that is readily accessible in experiment. Control can be achieved by coupling this output signal back onto the external current J which is chosen as an input signal. The current is thus modulated according to

$$J = J_0 + \delta J(t), \tag{5.45}$$

$$\delta J(t) \equiv K[\mathcal{E}_x(t - \tau) - \mathcal{E}_x(t)]$$

with control amplitude K and delay time τ. If $\mathcal{E}_x(t)$ is a UPO with period τ, $\delta J \equiv 0$ and stabilization can be achieved for an appropriate choice of K. In Fig. 5.11 we show the simulation of the system (5.42)–(5.44) for material parameters corresponding to p-Ge at 4 K (see Table 5.1) in the chaotic regime. The effect of the delayed feedback control switched on at $t = 4.5$ ms for various values of τ is shown. In (a), after a very short transient, the system is forced to a period-one UPO of the unperturbed system. As expected, the relative amplitude of the control signal $\delta J/J_0$ decays rapidly. In (b) a period-two UPO is stabilized. The characteristic values of K and τ are on the order

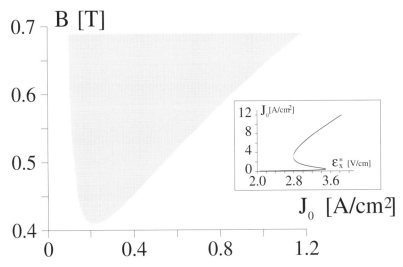

Figure 5.9. The dynamic Hall effect; showing the regime of oscillatory instability in the (J_0, B) control parameter plane for n-GaAs. The numerical parameters correspond to case (ii) of Table 5.2. The inset shows the corresponding static current density versus electric-field characteristics. (After Hüpper *et al.* (1992).)

of the sample's conductivity and the inverse maximum frequency of the output signal, respectively. For optimization of the control parameters, the variance of the control signal

$$D^2 = \left\langle \left(\frac{\delta J}{J_0} \right)^2 \right\rangle_t \equiv \frac{K^2}{J_0^2} \frac{1}{T} \int_0^T [\mathcal{E}_x(t) - \mathcal{E}_x(t-\tau)]^2 \, dt \qquad (5.46)$$

is calculated. D^2 exhibits a typical resonance structure as a function of τ (Fig. 5.12) with distinct mimina corresponding to the periods of the UPOs. Furthermore, D^2 is small only if the coupling constant K lies within a finite range. If K is too small, the feedback is too weak to stabilize the UPO. If, on the other hand, K is too large, the changes of the variable $\mathcal{E}_x$ are very fast so that the other variables $\mathcal{E}_z$ and n have no time to follow them. This is illustrated in Fig. 5.13 by plotting the largest nonzero

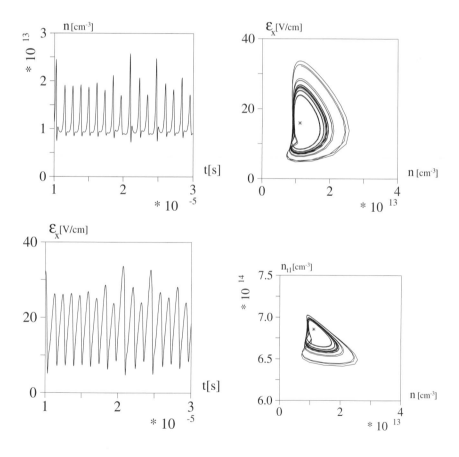

Figure 5.10. The dynamic Hall effect: chaotic time series (left-hand column) and phase portraits (right-hand column) for n-GaAs with $J_0 = 0.1198$ mA cm^{-2} and $B = 2$ T for case (ii) of Table 5.2 are shown (after Hüpper *et al.* (1992)) (continued overleaf).

Lyapunov exponent λ as a function of K. Note that the range of stabilization, i.e. negative λ, is smallest for the period-two UPO.

Different modes of self-generated periodic voltage oscillations can be selected by choosing an appropriate delay time of the control signal, corresponding to an integer multiple of the driving period. Thus the oscillator can be conveniently tuned between a set of modes. Such flexibility can not be achieved with a *linear* electronic oscillator. The occurrence of chaos thus offers useful applications in the field of semiconductor devices. Experimentally, only a simple delay line and a difference amplifier are necessary. The method can also be extended to control *transient chaos* (Reznik and Schöll 1993), whereby the UPOs of a chaotic *repeller* can be stabilized by a delayed feedback, or to chaos induced by a periodic driving force, as has been demonstrated for a real-space-transfer semiconductor oscillator (Cooper and Schöll 1995).

Finally we note that the stabilization of UPOs can also give insight into the dynamic structure of the uncontrolled chaotic system, since the UPOs constitute the "building bricks" of the chaotic attractor. The method of delayed feedback control can also be used to stabilize a UPO in a nonchaotic regime, and thus visualize, for example, an unstable orbit that has appeared in a saddle-node bifurcation of two limit cycles.

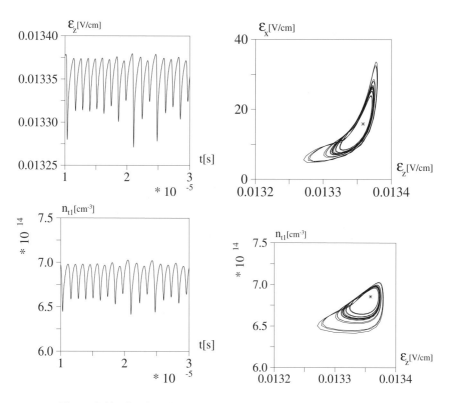

Figure 5.10. Continued.

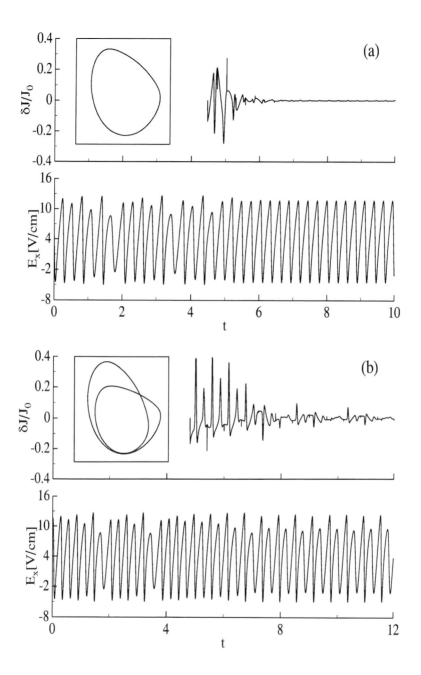

Figure 5.11. Chaos control by time-delayed feedback in the dynamic Hall instability for $K = 7 \times 10^{-8} \ \Omega^{-1} \ cm^{-1}$ and delay times (a) $\tau = 0.28$ ms and (b) $\tau = 0.6$ ms. The figure shows the temporal evolution of the electric field $\mathcal{E}_x$ and the current feedback control $\delta J/J_0$ which is switched on at $t = 4.5$ ms. The insets depict phase portraits of the unstable periodic orbits which are stabilized by the control, projected onto the Hall field $\mathcal{E}_z$ versus the drift field $\mathcal{E}_x$. The material parameters are given in Table 5.1 (p-Ge) with $\mu_d = \mu_0$, $\mathcal{E}_x^* = 3.27$ V cm^{-1}, and $B = 30$ mT. (After Schöll and Pyragas (1993).)

5.4   Transverse motion of current filaments

As we have seen in Section 5.2, in crossed electric and magnetic fields a dynamic Hall instability occurs (Hüpper and Schöll 1991) due to the nonlinear interaction of dielectric relaxation both of the applied electric field and of the induced Hall field combined with the generation–recombination kinetics, in particular impact ionization of impurities. In case of SNDC (Section 5.2.5) current filamentation is expected. In this section we analyze the complex nonlinear spatio-temporal dynamics of current filaments under crossed electric and magnetic fields within a simple model that takes into account only the transverse coordinate perpendicular to the current flow (Hüpper *et al.* 1993*b*). It thus neglects the influence of the contacts. It will be shown that the current filaments may travel laterally in a regular or intermittent way.

5.4.1   Laterally traveling filaments

As in the previous sections of this chapter, we consider doped bulk semiconductors at liquid-helium temperatures. As discussed earlier, high-purity semiconductors under the simultaneous application of parallel or crossed electric and magnetic fields have been observed to exhibit self-generated chaotic current or voltage oscillations under dc bias in the regime of impurity-impact-ionization breakdown. Such oscillatory

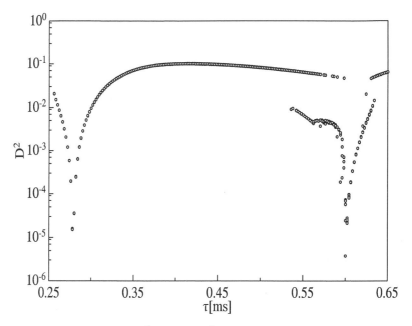

Figure 5.12. The variance $D^2 = \langle (\delta J / J_0)^2 \rangle_t$ of the control signal as a function of the delay time τ for $K = 7 \times 10^{-8} \ \Omega^{-1} \ \text{cm}^{-1}$ and the parameters of Fig. 5.11. For each value of τ, ten different initial conditions were chosen. (After Schöll and Pyragas (1993).)

instabilities are often connected with SNDC, which gives rise to current filamentation in homogeneously as well as in inhomogeneously (Kostial *et al.* 1993*a*, Kehrer *et al.* 1995*b*) doped material. For perpendicular electric and magnetic fields, the filaments either exhibit asymmetric breathing oscillations (Brandl *et al.* 1990), or travel transversally in the direction of the Lorentz force across the sample (Clauss *et al.* 1991, Hirsch *et al.* 1994, Spangler *et al.* 1994, Hirschinger *et al.* 1998, Nieder-nostheide *et al.* 1998, Aoki and Fukui 1998).

Whereas for positive differential conductivity, a mechanism for spatially homo-geneous magnetic-field-induced oscillations in terms of the dynamic Hall effect can describe the main features of the observed behavior, as we have seen in Section 5.2, an extension of this model to filamentary conduction is necessary in order to understand these complex spatio-temporal instabilities. Such a model, which is able to account for the spatio-temporal degrees of freedom of unipolar nonlinear filamentary conduction and instabilities in crossed electric and magnetic fields, will be presented here.

In the regime in which we are interested, a constant momentum-relaxation time τ_m may be assumed. In the presence of a magnetic field the current density j is then given by (5.15)

$$j = en\mu_B \frac{F}{q} - qn\mu\mu_B \left(B \times \frac{F}{q} \right) + n\mu^2 \mu_B B(BF), \qquad (5.47)$$

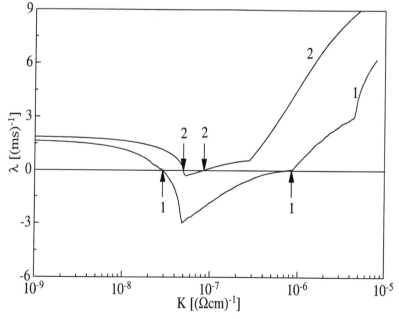

Figure 5.13. The Lyapunov exponent λ for the period-one (1) and period-two (2) unstable periodic orbits as a function of the control amplitude K for the parameters of Fig. 5.11 and the optimum value of τ. The arrows mark the boundaries of the control domain. (After Schöll and Pyragas (1993).)

where n is the density of carriers, $q = \pm e$ is their charge, $\mu = \tau_m e / m^*$ is the mobility for $B = 0$, $\mu_B = \mu / (1 + \mu^2 B^2)$, $F = q\mathcal{E} - eDn^{-1}\mu_B^{-1}\nabla_r n$, and D is the diffusion constant.

As the relevant dynamic variables we choose, besides the electric field $\mathcal{E}$ and n, the densities of carriers bound at two shallow impurity levels n_1 and n_2 corresponding to ground and excited states, respectively, of donors or acceptors. Transitions between these levels and the band states are possible due to generation–recombination (GR) processes including impact ionization. It is important that both the applied drift field $\mathcal{E}_x$ and the induced Hall field $\mathcal{E}_z$ are treated dynamically reflecting their finite dielectric relaxation time, while B (applied in the y-direction) is considered as a control parameter. Since the dominant spatial inhomogeneity in the SNDC regime occurs perpendicular to the drift field in the form of current filaments, we assume spatial inhomogeneity only in the z-direction. The dynamic equations for a sufficiently long sample with ideal planar contacts connected in series with a voltage source U_0 and a load resistor of resistance R_L and in parallel with a capacitance C are the continuity equations

$$\frac{\partial n}{\partial t} + \frac{\partial (n v_z)}{\partial z} = \phi(n, \boldsymbol{n}_t, \mathcal{E}), \qquad \frac{\partial \boldsymbol{n}_t}{\partial t} = \boldsymbol{\phi}_t(n, \boldsymbol{n}_t, \mathcal{E}), \tag{5.48}$$

and the dielectric-relaxation equations

$$c_r \frac{\partial \mathcal{E}_x}{\partial t} = j_0 - (\mu_B \langle n \rangle + \sigma_L)\mathcal{E}_x - \mu_B \mu B \left(\frac{\Delta n}{W} - \langle n \mathcal{E}_z \rangle \right), \tag{5.49}$$

$$\frac{\partial \mathcal{E}_z}{\partial t} = -n v_z, \tag{5.50}$$

supplemented by Maxwell's equations specialized for the configuration considered:

$$\frac{\partial \mathcal{E}_x}{\partial z} = 0, \qquad \frac{\partial \mathcal{E}_z}{\partial z} = \rho \equiv n + \sum_i^M n_i - 1, \tag{5.51}$$

where $c_r = 1 + LC/(\epsilon A)$, $A = bW$ is the lateral cross-section of the sample, L is the length in the x-direction, W is the width in the z-direction and b is the thickness in the y-direction, ϵ is the permittivity, $j_0 = U_0/(R_L A e \mu_1 N_A^* \mathcal{E}_0)$, $\sigma_L = \tau_M L/(R_L A \epsilon)$, ρ is the local charge density, ϕ and ϕ_t are the GR rates, and

$$v_z = \mu_B \left(\mathcal{E}_z - \frac{1}{n} \frac{\partial n}{\partial z} + \mu B \mathcal{E}_x \right) \tag{5.52}$$

is the mean velocity in the z-direction. The brackets $\langle \; \rangle$ denote the spatial mean value $\int_0^W dz/W$, and $\Delta n \equiv n(W) - n(0)$. Note that these equations hold for holes ($q = e$), but can easily be adapted to electrons by inverting the sign of $\mathcal{E}_z$. All quantities are given in dimensionless units, i.e. $\mu, t, z, \mathcal{E}$, and n and $\boldsymbol{n}_t$ are scaled by the low-field mobility μ_1, the dielectric relaxation time $\tau_M = \epsilon/(e\mu_1 N_A^*)$, the Debye length

$L_D = (k_B T_L \epsilon / (e^2 N_A^*))^{1/2}$, the thermal field $\mathcal{E}_0 = k_B T_L / (e L_D)$ and the effective acceptor concentration $N_A^* = N_A - N_D$, respectively. The classical static Hall effect is reproduced by the homogeneous steady states (denoted by an asterisk *): $\mathcal{E}_z^* = -\mu B \mathcal{E}_x^*$ and $j_0 = n^* \mu \mathcal{E}_x^*$.

We now consider the case that the static $j_0(\mathcal{E}_x^*)$ characteristic exhibits SNDC. The stability of the homogeneous stationary states $\Phi^* = (n^*, n_t^*, \mathcal{E}_x^*, \mathcal{E}_z^*)$ with respect to fluctuations $\delta\Phi(z,t) \propto e^{ikz} e^{\lambda t}$ can be investigated by linearization of eqs. (5.48)–(5.51). Owing to eq. (5.51) only homogeneous modes ($k = 0$) allow $\delta\mathcal{E}_x \neq 0$. This motivates a separate treatment for homogeneous and inhomogeneous modes.

The homogeneous modes yield oscillatory instabilities previously found for the dynamic Hall effect, see Section 5.2. For the inhomogeneous case we obtain a new spatio-temporal instability: $\lambda(k) = ivk - \tilde{D}k^2$ with

$$v = (\mu\mu_B)^{1/2} B \frac{(\partial\rho/\partial\mathcal{E})[(\partial\phi_1/\partial n_1)(\partial\phi_2/\partial n_2) - (\partial\phi_2/\partial n_1)(\partial\phi_1/\partial n_2)]}{n^*[-(\partial\phi_1/\partial n_1) - (\partial\phi_2/\partial n_2) - (\partial\phi/\partial n)]} \qquad (5.53)$$

and an effective diffusion coefficient $\tilde{D}$. It can be shown that, under general conditions, v has the opposite sign to that of B. Thus the fluctuation moves *transversally in the direction of the Lorentz force*. The velocity is proportional to B for small B in agreement with experiment (Hirsch et al. 1994), until μ_B differs significantly from μ.

To get a physical idea of this remarkable transverse motion of the fluctuations, we consider a carrier-density fluctuation around the unstable homogeneous state with and without a magnetic field. This fluctuation will grow in both cases due to the GR instability. Thus a fluctuation of the transverse electric field will be built up according to eq. (5.50), so that the drift and diffusion currents will tend to cancel each other out: $n \delta\mathcal{E}_z = \partial\delta n/\partial z$. In the case $B = 0$ this will lead to a *symmetric* change of the absolute value of the electric field, since $\mathcal{E}_z^* = 0$. This local increase of the electric field results in a symmetric increase of the fluctuation due to the GR processes. In the case $B \neq 0$ the fluctuation $\delta\mathcal{E}_z$ will *reduce* the absolute value of the electric field on one side, while it will *increase* it on the opposite site. Owing to the GR processes the side of the fluctuation with lower electric field will decrease, and the opposite side will increase. As a result, the fluctuation will move in the direction of the Lorentz force. Note that, in this picture, only the shape of the fluctuation moves transversally, while the carriers themselves remain at their transverse positions, performing free-to-bound transitions and vice versa. This corrects the former view of traveling filaments being due to individual carriers moving in the direction of the Lorentz force (Clauss et al. 1991). Rather, the situation is analogous to the motion of water molecules in a water wave.

For the numerical simulation we choose material parameters corresponding to p-Ge at 4 K with $c_r = 1$ and $\sigma_L = 0$ (current control) in the impurity-breakdown regime with GR rates $\phi = X_1^S n_2 - T_1^S n(1 + c - n_1 - n_2) + X_1 n n_1 + X_1^* n n_2$, $\phi_1 = T^* n_2 - X^* n_1 - X_1 n n_1$, and $\phi_2 = -\phi - \phi_1$, where $c = N_D/N_A^*$ is the compensation. The GR coefficients have been obtained from a spatially homogeneous Monte Carlo simulation with $B = 0$, see Section 4.2.3 (Quade et al. 1994a), simply by scaling the electric field

$\mathcal{E}$ by a factor of $(1 + \mu^2 B^2)^{-1/2}$. This is possible because we have assumed constant mobility. The resulting $j_0(\mathcal{E}_x^*)$ characteristic exhibits SNDC (Fig. 5.14). The Monte Carlo data have been fitted by smooth functions and substituted into the differential equations (5.48)–(5.51) which have been solved with the aid of the method of particles (Hockney and Eastwood 1981) using periodic transverse boundary conditions. Our nonlinear spatio-temporal analysis predicts the formation of current filaments on the NDC branch of the current-density–field characteristic. These filaments move in the direction of the Lorentz force (Fig. 5.15). After some transients the filament travels with a constant velocity v. Figure 5.16 shows v as a function of the applied magnetic field. Additionally, the velocity obtained by the linear stability analysis is plotted. The latter is larger because it describes only *small* fluctuations from the *NDC state*. Note that the velocity ($\approx 10^2$–10^3 cm s^{-1}) is much smaller than the drift velocity of the carriers ($\approx 10^5$–10^6 cm s^{-1}), in agreement with experiment (Hirsch *et al.* 1994). An analytic expression for the lateral velocity of the filaments, associated with the nonuniformity of the Hall angle, has also been derived within this model (Christen 1994a).

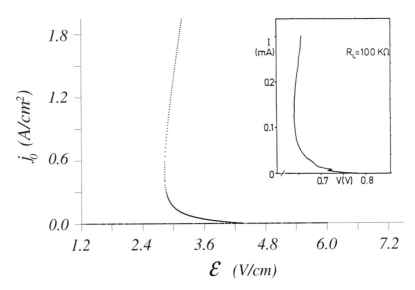

Figure 5.14. The static current-density–field characteristic calculated for p-Ge at 4 K with $N_A = 10^{14}$ cm^{-3}, $N_D = 5 \times 10^{12}$ cm^{-3}, $\epsilon = 16\epsilon_0$, $\mu = \mu_1 = 10^5$ cm^2 V^{-1} s^{-1}, $\tau_M = 10^{-12}$ s, $L_D = 5.6 \times 10^{-6}$ cm, $\mathcal{E}_0 = 60.8$ V cm^{-1}, $X^* = 10^{-15}$, $T^* = 7.21 \times 10^{-5}$, $X_1^S = 1.4 \times 10^{-6}$, $\alpha = 60.8(1 + \mu^2 B^2)^{-1/2}$, $X_1(\mathcal{E}) = 7.85 \times 10^{-4} \exp[-11.3(\alpha\mathcal{E})^{-0.745}]$, $X_1^*(\mathcal{E}) = 4.18 \times^{-2} \exp[-3.72(\alpha\mathcal{E})^{-0.66}]$, and $T_1^S(\mathcal{E}) = -1.2 \times 10^{-3} \exp[-0.2(-0.254 + \alpha\mathcal{E})^2] + 1.73 \times 10^{-3}(0.421 + \alpha\mathcal{E})^{-0.887}$ (after Quade *et al.* (1994a)). The inset shows a current–voltage characteristic measured in p-Ge at 4.18 K with a 100 kΩ load resistance (Rau *et al.* 1991).

5.4.2   Chaotic transverse motion

For a finite sample with Dirichlet boundary conditions (Hüpper *et al.* 1993*a*) $\mathcal{E}_z(0, t) = \mathcal{E}_z(W, t) = 0$ the filament travels to the boundary and is pinned there.

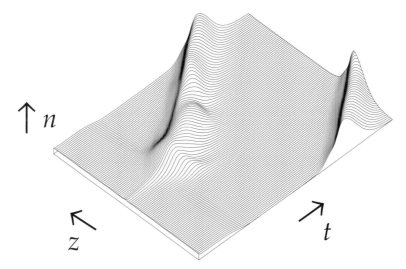

Figure 5.15. The spatio-temporal dynamics of the carrier density $n(z, t)$ in the presence of an externally applied magnetic field under cyclic boundary conditions for $j_0 = 34$ mA cm^{-2} and $B = 10$ mT with planar contacts. The total time is 8.1×10^{-6} s; the sample's width is $W = 90$ μm. (After Hüpper *et al.* (1993*b*).)

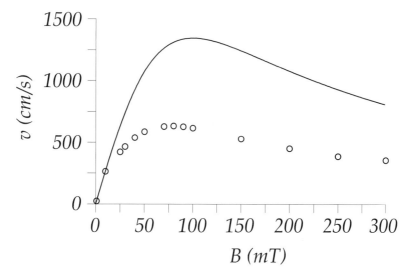

Figure 5.16. The transverse velocity of the traveling filaments as a function of the magnetic field for $j_0 = 93.76$ mA cm^{-2} and $W = 120$ μm from the nonlinear simulation (circles). The full line shows the velocity obtained by the linear stability analysis for the same control parameters. (After Hüpper *et al.* (1993*b*).)

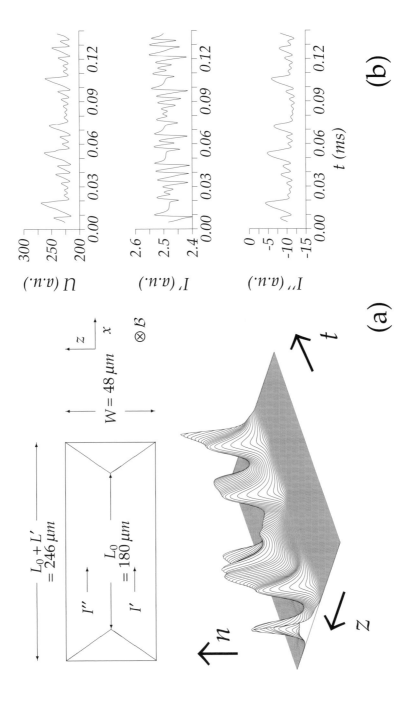

Figure 5.17. Traveling filaments for $j_0 = 0.85$ mA cm^{-2} and $B = 20$ mT with triangular contacts. The inset shows the sample geometry. (a) The spatio-temporal dynamics of $n(z, t)$. (b) Time series of the sample voltage $U(t)$ and currents I' and I'' (see the inset), all in arbitrary units. (After Hüpper *et al.* (1993*b*).)

This behavior changes if other contact geometries are used. We demonstrate this for triangular contacts (Fig. 5.17), which may also serve as a model for *nonideal* planar Ohmic contacts (Hirsch *et al.* 1994).

In this case the length of the sample $L = L_0 + 2|z - W/2|(L'/W)$ is a function of the transverse coordinate. Owing to our assumption of homogeneity in the x-direction we restrict ourselves to small contact angles, i.e. L'/W should be small relative to logarithmic derivatives like $(\partial \ln n/\partial z)$. Qualitatively, these contacts yield the following scenario: The free carriers inside the moving filament will be spread over an increasing effective sample length $L(z)$, which reduces the maximum carrier density in the filament. Additionally, the drift field and therefore the generation rate decrease. Both effects tend to destroy the filament. As a result, the current through the sample decreases, which increases the voltage of the sample and thus enhances the filament due to an increase in generation rate. The time scales of these processes are given by the time the filament needs to travel across the sample, the GR time, and the dielectric relaxation time of the electric field.

Depending upon the ratios of these time scales and the geometry of the contacts, various states of the filament are possible. For small magnetic fields and comparable time scales the filament rests in the middle of the sample and exhibits asymmetrically breathing filament boundaries such as have been observed in n-GaAs (Brandl *et al.* 1989, 1990).

A more complex dynamic behavior is found if the dielectric relaxation and the destruction of the filament through the GR processes are slower than the transverse motion of the filament. Such a situation is shown in Fig. 5.17(a). The filament travels in the direction of the Lorentz force. Owing to the inertia of the dielectric relaxation the drift field increases slowly. This supports destruction of the filament, so that, with increasing voltage, a new filament can be generated in the middle of the sample. This successive nucleation, traveling, and destruction of filaments leads to slow chaotic voltage and current oscillations (Fig. 5.17(b)) and has indeed been found in experiments on p-Ge (Hirsch *et al.* 1994) and n-GaAs (Spangler *et al.* 1994, Hirschinger *et al.* 1998, Aoki 1999*a*).

As a function of the magnetic field, above a minimum threshold B, we find successively asymmetric breathing oscillations (Fig. 5.18(a)); small regular oscillations due to periodic nascence and destruction of traveling filaments (Fig. 5.18(b)); chaotic oscillations due to intermittent nascence and destruction (Fig. 5.18(c)); and pinning of the filaments at the transverse boundary after some transient traveling sequences (Fig. 5.18(d)).

5.5   Deformation of current filaments in point–contact samples

In order to fully understand the effects of a transverse magnetic field on the filament properties, two-dimensional analyses including the spatial dependencies both on the longitudinal x- and on the transverse z-direction are indispensable since the Lorentz force acts on charge carriers moving between the contacts, and points in the direction perpendicular to their motion. Therefore, in this section we shall present two-dimensional simulations of the formation of current filaments in n-type GaAs under a transverse magnetic field (Kunihiro *et al.* 1997). They are appropriate to thin epitaxial layers, for which the third dimension is irrelevant (Hirschinger *et al.* 1997*a*, 1997*b*).

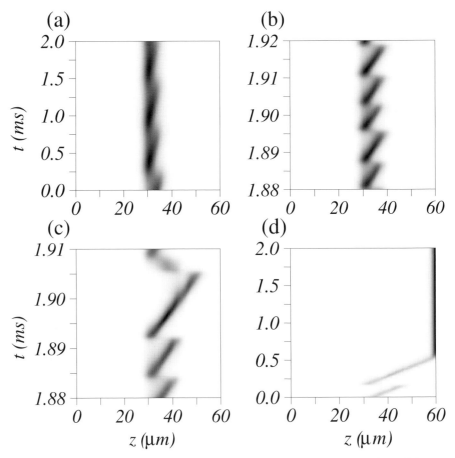

Figure 5.18. Density plots of the carrier density $n(z, t)$ for various values of the magnetic field: (a) $B = 11$ mT, (b) $B = 15$ mT, (c) $B = 20$ mT, and (d) $B = 25$ mT. The numerical parameters are the same as those in Fig. 5.17. (After Quade *et al.* (1994*a*).)

The current density $j(B)$ in the presence of a magnetic field B normal to the current flow is given for n-type material by

$$j(B) = \frac{j(0) + \mu B \times j(0)}{1 + |\mu B|^2},$$

(5.54)

where $j(0)$ is the current density for $B = 0$ given by eq. (5.47).

5.5.1 The nascence of bent filaments

We investigate the formation of a current filament upon application of a voltage ramp as in Chapter 4, Section 4.6. Now a magnetic field is applied in the direction perpendicular to the sample plane pointing downwards with a strength of $B = 5$ mT.

The dynamic processes of current-filament formation under a transverse magnetic field are shown in Fig. 5.19. After the voltage is applied at the anode, a dipole-like electric field is formed quasi-instantaneously between the two point contacts, inducing electron injection from the cathode. Then a charge-carrier front is formed near the cathode and propagates toward the anode (Fig. 5.19(a)). Though the carrier front is driven by impact ionization, it is not strong enough to affect the carrier population in the shallow impurities. In other words, almost all the carriers in the bulk are still bound at the donor ground level. The carrier density in the vicinity of the noninjecting anode contact is also increased by weak impact ionization, since a relatively large fraction of the applied voltage drops at the junction between the n-type anode and the insulating bulk in a manner similar to that in a reversely biased n–p junction, leading to a higher electric field here. Note that the growth of the carrier front on the anode side, which is mainly driven by diffusion, has already been affected by the Lorentz force (Fig. 5.19(a)).

In order to conserve the total current density J longitudinally following from eq. (4.8), which is approximately given by $J = en\mu\mathcal{E}$, the electric field $\mathcal{E}$ ahead of the carrier front, where the electron density n is still low, becomes higher than that behind it. Hence impact ionization downstream is encouraged. When the injecting carrier front finally reaches the other front near the anode contact (Fig. 5.19(b)), most of the applied voltage is sustained here and the electric-field strength becomes high enough to trigger strong impact ionization from the donor ground level.

The increased free-carrier concentration relaxes the electric field near the anode, then the high-field domain starts moving back from the anode to the cathode, again for the reasons of current conservation, in a see-saw-like mode (Kunz and Schöll 1996). Hence impact ionization propagates longitudinally, following the motion of the high-field domain, and multiplies the number of free carriers uniformly by several orders of magnitude, finally forming a thin homogeneous straight filament (Fig. 5.19(c)). Inside the filament, the excited donor levels are also populated uniformly whereas the ground level is almost depleted.

In the next stage, the straight filament is gradually widened and bent in the direction of the Lorentz force (Fig. 5.19(d)). The density of carriers trapped in the excited donor level is also shifted transversally with slightly increasing population toward the filament border. On the other hand, the electron temperature in the expanded area of the bent filament decreases slightly. Direct experimental evidence of this nucleation process has been obtained by quenched photoluminescence mapping (Aoki and Fukui 1998).

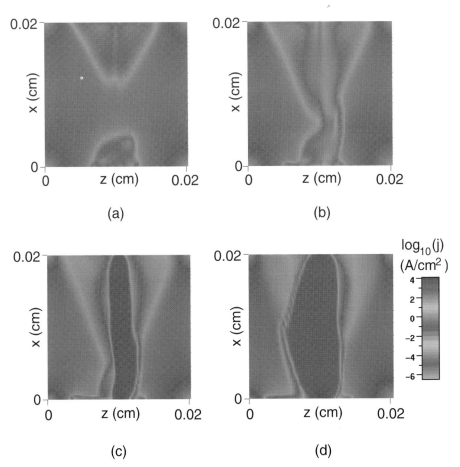

Figure 5.19. A simulation of the formation of a current filament in n-GaAs under crossed electric and magnetic fields. The temporal evolution of the current density $|j(x, z)|$ for a square sample with two point contacts for $B = 5$ mT and $\mathcal{E}_{av} = 24$ V cm^{-1} is shown. (a) $t = 0.5$ ns, (b) $t = 1.0$ ns, (c) $t = 1.6$ ns, and (d) $t = 2.5$ ns (after Kunihiro *et al.* (1997)).

5.5.2 Lorentz-force effects

In order to confirm that the lateral growth of current filaments can be attributed to the Lorentz force, the orientation of an applied magnetic field has been inverted in the simulations. We find that inverting the magnetic field generates a symmetric pattern of current flow with respect to the $z = 0.01$ cm plane, as expected from the reversed Lorentz force. Figure 5.20 shows the spatial distribution of the electric-field strength at $t = 2.5$ ns, i.e. when a steady state has practically been reached, for $B = 0$ and $B = \pm 5$ mT, wherein a magnetic field pointing downward is defined as positive and one pointing upward as negative, and the cathode is always at the top. Compared with the symmetric distribution of the electric field for $B = 0$, the fields exhibit strong

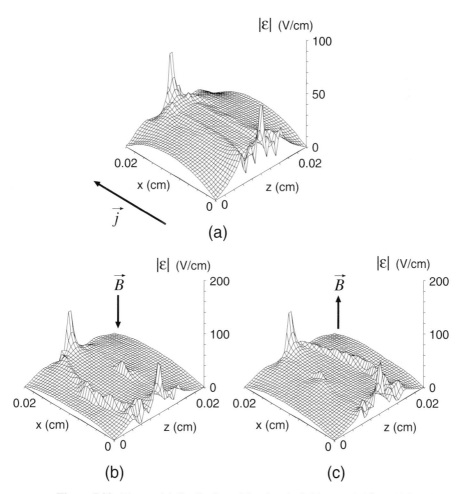

Figure 5.20. The spatial distribution of the electric-field strength $|\mathcal{E}(x, z)|$ for (a) $B = 0$, (b) $B = 5$ mT, and (c) $B = -5$ mT in the quasisteady state ($t \approx 2.5$ ns), $V = 0.48$ V. A magnetic field pointing downward is defined as positive and one pointing upward as negative. (After Kunihiro *et al.* (1997).)

asymmetry at opposite filament boundaries for $B = \pm 5$ mT. The electric fields at the filament borders pointing in the direction of the Lorentz force are much stronger than those at the opposite borders. These asymmetric fields are in good agreement with asymmetric ridge-like structures observed by scanning laser microscopy under a transverse magnetic field (Brandl et al. 1989, Spangler et al. 1994, Novák et al. 1995). Also, the simulations and the experiments show that the width and curvature of the filaments become larger with increasing magnetic-field strength.

The total current during the formation of current filaments is shown in Fig. 5.21 for various magnetic fields. The current-time characteristics consist of four stages corresponding to the formation processes of a current filament shown in Figs. 5.19(a)–(d): (a) a low-current state corresponding to a stage of carrier-front creation and propagation from the cathode, (b) a slight increase and stagnation in the current corresponding to the formation of a rudimentary filament followed by the backward motion of a high-field domain, (c) a strong increase in the total current accompanied by the establishment of a homogeneous straight filament, and (d) an almost saturated current with a filament bent by the Lorentz force. It is remarkable in Fig. 5.21 that higher magnetic fields accelerate the transition from almost insulating to highly conducting states. This is because a transverse magnetic field induces a finite lateral

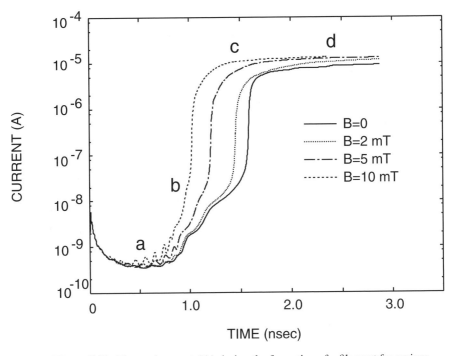

Figure 5.21. The total current $I(t)$ during the formation of a filament for various magnetic fields. The letters a–d correspond to the stages of filament formation shown in Figs. 5.19 (a)–(d). (After Kunihiro et al. (1997).)

Hall field inside the filament, thus increasing the modulus of the electric field. As a result transitions from almost insulating to highly conducting states are accelerated, since the probability of impact ionization rapidly increases with the local electric-field strength. However, experimentally the opposite effect has been found (Aoki and Fukui 1998), which is probably due to an increase in magnetoresistance.

The physical processes involved in filament formation under a transverse magnetic field can be understood as follows. Figure 5.22 shows the cross-sections of the modulus of the electric-field strength with and without a magnetic field at $x = 0.01$ cm, corresponding to each stage of filament formation shown in Figs. 5.19(a)–(d). Cross-sections of the lateral electric field $\mathcal{E}_z$, and the corresponding space charge $\rho = \epsilon \nabla \cdot \mathcal{E} \approx \epsilon \, \partial \mathcal{E}_z / \partial z$, are also shown in Figs. 5.23 and 5.24, respectively, in the quasisteady state, wherein the shaded areas indicate the cross-sections of the electron concentration.

For $B = 0$, the electric-field profile is completely symmetric with respect to the center of the sample ($z = 0.01$ cm) during all stages of filament formation (Fig. 5.22(a)). In the first stage the electric field exhibits a gentle peak at the middle of the sample due to the dipole-like electric field induced between the point contacts. After a charge-carrier front passes by, the field profile then develops two maxima which arise from the lateral fields localized in the transition zone between high- and low-electron-density regions (Fig. 5.23(a)) and its formation is explained as follows. In the steady state, charge neutrality holds far from the filament borders since the negative charge of free electrons compensates for positive ionized donors inside, and almost all donors are frozen out, i.e. neutral, outside the filament. In the transition zone, electrons tend to diffuse from high- to low-density regions, leading to the weak carrier accumulation at the outside edge and the slight depletion at the inside edge of the filament walls. Hence charge neutrality is broken here because immobile ionized donors are left inside the filament borders (Fig. 5.24(a)). The lateral electric field induced by this local space-charge dipole prevents the electrons from diffusing outward, thus stabilizing the filament.

For $B = 5$ mT, the electric field also retains an almost symmetric profile until the stage of a rudimentary filament ($t \approx 1.0$ ns). However, once impurity breakdown sets in and a straight filament has started to grow, the electric field exhibits strong asymmetry between the opposite filament boundaries (Fig. 5.22(b)). This arises from the asymmetric lateral field (Fig. 5.23(b)) and its origin can be explained by invoking similar processes to those for $B = 0$. On the filament border pointing in the direction of the Lorentz force, the diffusion of the electrons out of the filament is enhanced. As a result more negative and positive space charges are generated by mobile electrons and by immobile ionized donors, respectively, on this border than are generated for $B = 0$ (Fig. 5.24(b)), giving rise to a strong local lateral field. The high-field border is shifted by the Lorentz force, inducing impact ionization in the region behind it. Though the average electric-field strength inside the filament becomes weaker as the filament becomes wider due to the point-contact geometry, the free-electron density is still high

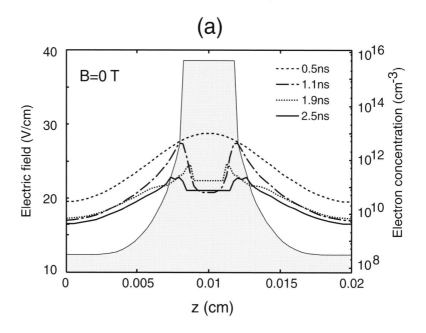

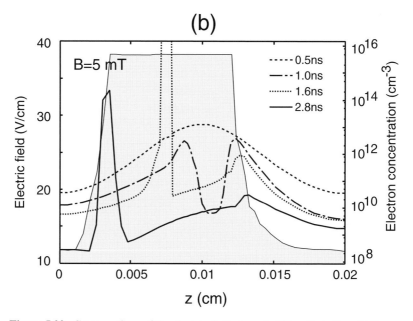

Figure 5.22. Cross-sections of the electric-field strength $|\mathcal{E}(x, z)|$, in the middle between the two point contacts ($x = 0.01$ cm) for (a) $B = 0$ and (b) $B = 5$ mT. The various times correspond to the stages of filament formation shown in Figs. 5.19(a)–(d). Shaded areas indicate the electron concentration in the quasisteady state. ($V = 0.48$ V, after Kunihiro *et al.* (1997).)

as long as the electric field is higher than the threshold field, $\mathcal{E}_h$, for impact ionization from the excited donor level, which is much smaller than that for the ground level.

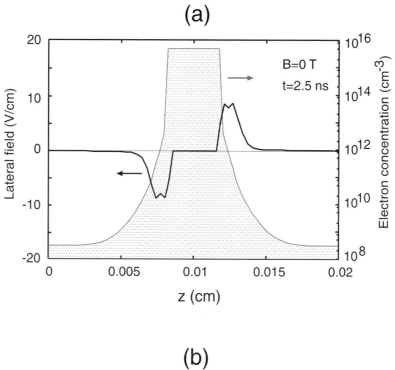

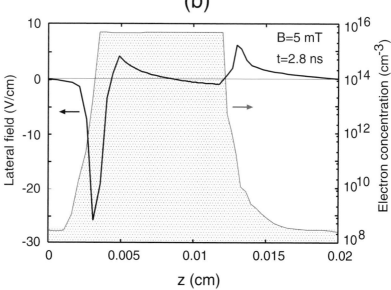

Figure 5.23. The same as Fig. 5.22 but for the lateral electric field $\mathcal{E}_z(x, z)$ (after Kunihiro *et al.* (1997)).

The lateral electric field serves as an attractive force compensating for the outward Lorentz and diffusion forces, and if balance between them holds, the filament border is stabilized under a transverse magnetic field.

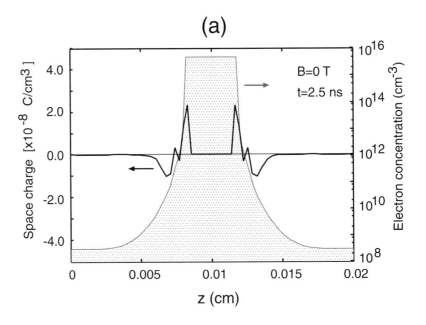

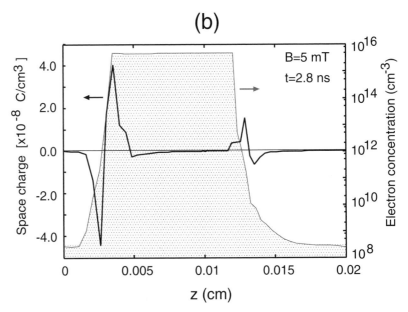

Figure 5.24. The same as Fig. 5.22 but for the space-charge density (after Kunihiro *et al.* (1997)).

On the other filament border, electrons diffusing outward are impeded by the Lorentz force pointing in the opposite direction. Therefore less space charge and a weaker lateral field are created on this border than occur for $B = 0$ (Figs. 5.23(b) and 5.24(b)). This complementary action of lateral fields at the opposite boundaries can explain the asymmetric signals observed by laser scanning microscopy, in which the signal of one filament border grows whereas the signal of the other border decays with increasing magnetic field (Brandl *et al.* 1989). It is suggestive that this asymmetric lateral field is also the source of the finite Hall voltage across the sample found in experiments (Brandl *et al.* 1990) and it may play a crucial role in inducing chaotic oscillations (Brandl *et al.* 1987).

As explained above, spatial patterns of current filaments under a transverse magnetic field are determined by a local balance between the lateral diffusion current and the lateral drift current resulting from the lateral electric field, which may be conceptualized as a superposition of the dipole field due to the space charge already present in the filament walls at $B = 0$ and the Hall field. This physical picture is consistent with the reasoning developed previously for the one-dimensional model in Section 5.4 (Hüpper *et al.* 1993*b*) leading to transversally traveling filaments in the absence of a point-contact-induced dipolar electric field. In the simulations presented here, the filament border pointing in the direction of the Lorentz force is displaced transversally and the other border is almost fixed.

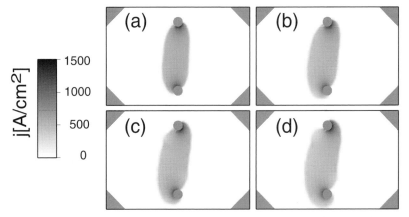

Figure 5.25. Simulated current filaments in a rectangular sample (5 mm × 7.5 mm) with two inner point contacts for an applied bias voltage $U_0 = 20$ V, $R = 100$ Ω, and various perpendicular magnetic fields B: (a) $B = 30$ mT, (b) $B = 60$ mT, (c) $B = 90$ mT, and (d) $B = 120$ mT. Note that the injecting cathode is at the top. (Calculated for n-GaAs at $T_L = 4.2$ K with the same parameters as those in Fig. 4.29.) (After Schwarz *et al.* (2000*b*).)

5.5.3 Magnetic-field-induced distortion of filaments

Now let us consider the influence of a magnetic field in sample geometries with two small circular contacts in the interior of the sample, as studied in Section 4.6.4

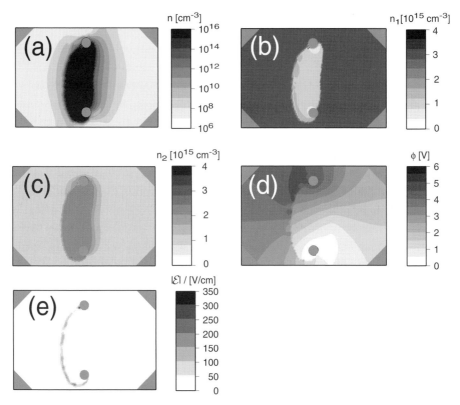

Figure 5.26. Simulated density plots of the distributions (a) of electrons in the conduction band, (b) of electrons in the donor ground level, (c) of electrons in the donor excited level, (d) of the potential, and (e) of the modulus of the electric field, for a current filament with $U_0 = 50$ V, $R = 1$ kΩ, $B = 60$ mT (after Schwarz *et al.* (2000*b*)).

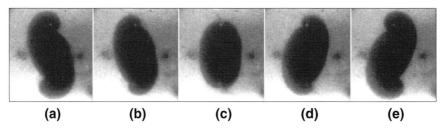

(a)   (b)   (c)   (d)   (e)

Figure 5.27. Quenched photoluminescence images of current filaments at fixed current $I = 5.1$ mA and various magnetic fields: (a) $B = -600$ mT, (b) $B = -300$ mT, (c) $B = 0$ mT, (d) $B = 300$ mT, and (e) $B = 600$ mT. (Sample PC06, see Table 4.3). (After Schwarz *et al.* (2000*b*).)

(Schwarz *et al.* 2000*b*). Figure 5.25 shows the stationary current filaments under the action of a magnetic field for the same sample geometry as in Figs. 4.29 and 4.30. With increasing B the filaments become more and more asymmetrically distorted by the Lorentz force, and, at large B and large current, they assume an inverted S shape. The corresponding spatial distributions of the electron densities in the conduction band and the donor ground and excited levels (Figs. 5.26(a)–(c)) reflect the shape of the filament, indicating donor depletion, and population inversion between the ground and excited states. The dipolar potential distribution (Fig. 5.26(d)) is strongly distorted by the action of the magnetic field which induces a Hall field in the filament boundaries. This Hall field points in the same direction at both boundaries, while the lateral electric field induced by the space charges in the filament boundaries already at $B = 0$ always points out of the filament, i.e. to the left at the left-hand boundary, and to the right at the right-hand boundary. The superposition of the two fields leads to strongly asymmetric net electric fields in the two filament boundaries. Therefore, at the left-hand filament boundary, which is the side to which the Lorentz force points for the polarity of the magnetic field chosen, the magnitude of the resulting field is much larger than that at the right-hand boundary. This electric field localized at the left-hand filament border can be clearly seen in Fig. 5.26(e). Since the carriers drifting from the top cathode to the bottom anode contact are deflected toward the left-hand border by the Lorentz force, it is plausible that this filament border is elongated, winding around the circular bottom contact. This explains the asymmetric shape of the filaments.

In Fig. 5.27 the experimental quenched photoluminescence images of the sample in a normal magnetic field for magnetic fields pointing up and down are shown. There is good agreement with the simulations of Fig. 5.25. It can be seen that the inversion of the magnetic field tilts the body of the filament in the opposite direction, yielding a pair of mirror-inverted structures at the same field strengths. It should be noted that this behavior is observed only with wide, convex filaments. For small currents, narrow self-organized filaments are laterally elongated in the direction of the Lorentz force (Hirschinger *et al.* 1997*a*, 1997*c*).

Chapter 6

Stationary and oscillating domains in superlattices

In this chapter we study vertical high-field transport in semiconductor superlattices. Depending upon the circuit conditions and the material parameters, e.g. the mean doping density N_D, either stable stationary domains (for high N_D), or self-sustained oscillations of the domains (for intermediate N_D) are found. We shall see that this behavior is strongly affected by growth-related imperfections such as small fluctuations of the doping density, or the barrier and quantum-well widths, and that weak disorder on microscopic scales can be quantitatively detected in the global macroscopic current–voltage characteristics. The bifurcations which occur and the roles of the various realizations of the microscopic disorder are discussed, as is the dynamics of domain formation.

6.1   Introduction

In Section 2.2.1 (Fig. 2.5) it was mentioned that vertical high-field transport in GaAs/AlAs superlattices is associated with NNDC and field-domain formation induced by resonant tunneling between adjacent quantum wells. This was observed experimentally by many groups (Esaki and Chang 1974, Kawamura *et al.* 1986, Choi *et al.* 1987, Helm *et al.* 1989, Helgesen and Finstad 1990, Grahn *et al.* 1991, Zhang *et al.* 1994, Merlin *et al.* 1995, Kwok *et al.* 1995, Mityagin *et al.* 1997). Those domains are the subject of the present chapter. The field domains may be either stationary, leading to characteristic sawtooth current–voltage characteristics (Esaki and Chang 1974), or traveling, associated with self-sustained current oscillations (Kastrup *et al.* 1995, Hofbeck *et al.* 1996). In strongly coupled superlattices, i.e. superlattices with small barrier widths, oscillations above 100 GHz at room temperature (Schomburg *et al.* 1998, 1999) have been realized experimentally, whereas in weakly coupled superlattices the frequencies are many orders of magnitude lower (Kastrup *et al.* 1997). In the former case the impact of electron heating may cause SNDC in addition to the NNDC occurring at lower fields, and the combination of S- and N-type instabilities then leads to a modified structure of the high-field domains associated with self-generated gigahertz oscillations (Steuer *et al.* 1999, 2000). The existence of domain states and of multistable, sawtooth-like current–voltage characteristics consisting of many branches has been explained for structurally perfect, weakly coupled samples by phenomenological (Korotkov *et al.* 1993, Miller and Laikhtman 1994, Prengel *et al.* 1994, Bonilla *et al.* 1994*b*) and microscopic (Wacker 1998, Cao and Lei 1999) models. The standard approaches of miniband transport, Wannier–Stark hopping, and sequential tunneling have been derived from a full quantum-transport theory using the framework of nonequilibrium Green functions as limit cases depending on the mutual ratios of three energy scales: (i) the miniband width as a measure for the coupling between the wells, (ii) the potential drop across one superlattice period, and (iii) the scattering width (Wacker and Jauho 1998*b*).

It has also been found in computer simulations that small amounts of imperfections and disorder, such as fluctuations of doping and of well or barrier widths, sensitively influence the domain states (Wacker *et al.* 1995*c*, Schwarz *et al.* 1996*a*, 1996*b*, 1998, Schöll *et al.* 1998*b*, Steuer *et al.* 1999). Experimentally, the global properties have been related directly to the extent of disorder by X-ray analysis of superlattices (Grenzer *et al.* 1998). The role of interface roughness has been studied within a quantum-transport approach (Wacker and Jauho 1998*a*, Rauch *et al.* 1998, Wacker and Hu 1999). Chaotic dynamics has also been found both theoretically (Bulashenko and Bonilla 1995, Bulashenko *et al.* 1996, Bonilla *et al.* 1996) and experimentally (Zhang *et al.* 1996, Luo *et al.* 1998*b*, Bulashenko *et al.* 1999), if an ac driving bias is applied.

In this chapter we use a simple phenomenological model (Prengel *et al.* 1994, Schwarz and Schöll 1996, Patra *et al.* 1998) to demonstrate the various types of stationary and dynamic patterns that may arise in weakly coupled superlattices under high electric fields. We shall investigate the bifurcations of stationary and traveling field domains and their dependencies upon doping fluctuations as well as upon various boundary conditions. It should be noted that similar nonlinear behaviors are found in other models that have been studied widely on various levels of sophistication, and hence our approach is characteristic of nonlinear vertical high-field transport in superlattices.

6.2   High-field domains

6.2.1   A sequential resonant-tunneling model

First, we introduce a model for vertical transport and electric-field-domain formation in doped superlattices that is the basis of the following simulations. The superlattice considered here consists in a sequence of layers of two different periodically alternating semiconductor materials A and B, e.g. GaAs and AlAs, with layer thicknesses l and b, respectively. They are typically grown by molecular beam epitaxy (MBE), which makes it possible to obtain layer widths on the order of a few monolayers. Since the two materials have different fundamental band gaps, a discontinuity occurs both in the valence and in the conduction band at each heterojunction. For n-doped superlattices at low temperatures (a few kelvins) we can limit our description to the conduction band, neglecting hole transport.

Assuming the effective-mass approximation, we obtain a Hamiltonian with a periodic potential in the growth direction consisting of quantum wells (GaAs layers) separated by barriers (AlAs layers). We use the usual (approximate!) *Ansatz* to separate the eigenfunctions into a parallel and a vertical part. The first part is solved by a two-dimensional quasi-free electron gas, whereas for the latter the energy spectrum is split into a miniband structure.

In case of a strong applied vertical electric field, however, the minibands break up in favor of electronic states that are localized in the wells. This effect is especially important for weakly coupled quantum wells due to their small miniband widths of typically only a few milli-electronvolts. Our model, therefore, as a first approximation, assumes isolated quantum wells with bound energy levels E_j, $j = 1, 2, \ldots$. The interaction with the neighboring wells is then treated as a perturbation (Prengel *et al.* 1994).

Using the two-dimensional electron concentrations (per unit area) $n_j^{(i)}$ in the jth bound level of the ith quantum well as the dynamic variables of our system, we can describe the charge transport by writing a set of rate equations. For simplicity we will limit our simulations to the two lowest energy levels in each well, $j = 1$ and 2. The temporal evolution of the formal vector $\boldsymbol{n}^{(i)} \equiv \left(n_1^{(i)}, n_2^{(i)}\right)$ of electron concentrations in the ith well can be expressed as

$$\dot{\boldsymbol{n}}^{(i)} = \mathbf{A}^i(F^{(i)})\boldsymbol{n}^{(i-1)} + \left(-\mathbf{B}^i(F^{(i)}) - \mathbf{C}^i(F^{(i+1)}) + \mathbf{T}\right)\boldsymbol{n}^{(i)} + \mathbf{D}^i(F^{(i+1)})\boldsymbol{n}^{(i+1)} \quad (6.1)$$

with matrices

$$\mathbf{A}^i := \left(R_{r,kj}^{(i)}\right)_{jk}, \tag{6.2}$$

$$\mathbf{B}^i := \left(\sum_m R_{l,mj}^{(i)} \delta_{jk}\right)_{jk}, \tag{6.3}$$

$$\mathbf{C}^i := \left(\sum_m R_{r,jm}^{(i+1)} \delta_{jk}\right)_{jk}, \tag{6.4}$$

$$\mathbf{D}^i := \left(R_{l,jk}^{(i+1)}\right)_{jk}. \tag{6.5}$$

$R_{r,jk}^{(i)}$ stands for the transition probability per unit time (i.e. the reciprocal average tunneling time) for tunneling from the jth level of the $(i-1)$th quantum well into the kth energy level of the ith well across the ith barrier separating them, i.e. to the right. $R_{l,jk}^{(i)}$ denotes the reverse process. The transport coefficients are functions of the local electric field $F^{(i)}$ across the barrier. For transitions between equivalent levels of neighboring wells the miniband-conduction formalism (Ignatov *et al.* 1991) is used to calculate them, whereas transport between different energy levels is derived from quantum-mechanical perturbation theory (Prengel *et al.* 1994). The connection with the notation used elsewhere (Wacker *et al.* 1995c) is $R_{r,jj}^{(i)} = R_j^{(i)}$, $R_{r,12}^{(i)} = X_r^{(i)}$, $R_{r,21}^{(i)} = Y_r^{(i)}$, $R_{l,12}^{(i)} = X_l^{(i)}$, and $R_{l,21}^{(i)} = Y_l^{(i)}$. For the intersubband relaxation from the first excited energy level $E_2^{(i)}$ into the ground state of the same well $E_1^{(i)}$, we assume a constant relaxation time of $\tau_{21} = 1$ ps (Rühle *et al.* 1991), which in eq. (6.1) is expressed by the constant matrix

$$\mathbf{T} := \begin{pmatrix} 0 & 1/\tau_{21} \\ 0 & -1/\tau_{21} \end{pmatrix}. \tag{6.6}$$

In order to deal with structurally imperfect, weakly disordered superlattices, we allow for frozen-in fluctuations of the individual well and barrier widths, l_i and b_i, respectively, and introduce a "local" superlattice constant $d_i := (l_{i-1} + l_i)/2 + b_i$ to be employed in the transport coefficients for tunneling across the ith barrier (cf. Fig. 6.1). Thus, $R_{r,jk}^{(i)}$ and $R_{l,jk}^{(i)}$ explicitly depend on the local layer widths. The

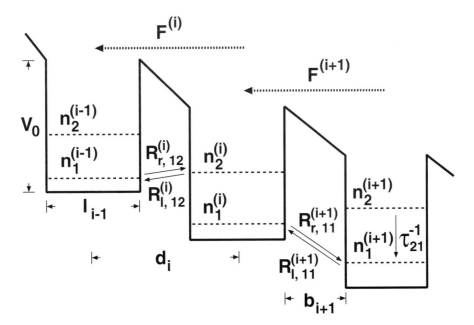

Figure 6.1. The model for charge transport in a superlattice with different widths of the individual quantum wells and barriers. Various transitions are indicated.

(three-dimensional) doping concentration $N_\text{D}^{(i)}$ in the ith well is also allowed to vary from well to well.

The electric fields $F^{(i)}$ have to be calculated self-consistently from the electron concentrations using the discrete version of Gauss' law

$$\epsilon\left(F^{(i+1)} - F^{(i)}\right) = e\left(n_1^{(i)} + n_2^{(i)} - l_i N_\text{D}^{(i)}\right) \equiv \rho_i^{(2\text{D})} \tag{6.7}$$

where e is the elementary charge, ϵ is the permittivity, and $\rho_i^{(2\text{D})}$ is the (two-dimensional) charge density. Since the electric fields $F^{(i)}$ must satisfy the condition

$$U = \sum_{i=1}^{N+1} F^{(i)} d_i = \text{constant} \tag{6.8}$$

for the total applied voltage U in a superlattice with $N + 1$ barriers, we find

$$F^{(i)} = \frac{1}{\sum_{n=1}^{N+1} d_n}\left\{U + \frac{1}{\epsilon}\left[\sum_{m=1}^{i-1} \rho_m^{(2\text{D})}\left(\sum_{n=1}^{m} d_n\right) - \sum_{m=i}^{N} \rho_m^{(2\text{D})}\left(\sum_{n=m+1}^{N+1} d_n\right)\right]\right\}. \tag{6.9}$$

Equation (6.8) here acts as a global coupling for the system, which is otherwise governed by sequential transport between neighboring wells.

We shall now discuss the modeling of the contacts. We treat them as two "virtual" wells numbered by superscripts 0 and $N + 1$, respectively (Prengel et al. 1994). The $2N$ ordinary differential equations for $n_j^{(i)}$ ($i = 1, \ldots, N$; $j = 1, 2$) can now be solved numerically by a Runge–Kutta method, subject to appropriate boundary conditions for $n_j^{(0)}$ and $n_j^{(N+1)}$. In the following we will examine two different sets of boundary conditions, which in some respect represent two extremes from among a great variety of possibilities. The first set consists in "duplicating" the electron concentrations from the neighboring well

$$n_j^{(0)} = n_j^{(1)},$$

$$n_j^{(N+1)} = n_j^{(N)}. \tag{6.10}$$

The second set fixes the electron concentration to a given value, which we have arbitrarily chosen as twice the doping concentration (all electrons are assumed to be in the ground level):

$$n_j^{(0)} = 2 N_\text{D}^{(1)} l_1 \, \delta_{j1}$$

$$n_j^{(N+1)} = 2 N_\text{D}^{(N)} l_N \, \delta_{j1}. \tag{6.11}$$

This corresponds to a simplified treatment of an Ohmic contact as a carrier reservoir (Prengel et al. 1994, Bonilla et al. 1994b, Wacker et al. 1997, Schwarz and Schöll 1996). More sophisticated treatments of the contacts are also available. For instance, one may assume that the current density at the contacts is proportional to the local field (Steuer *et al.* 1999, 2000, Carpio *et al.* 2000). Alternatively, the actual potential distribution at the boundary could be taken into account within a transmission-type formalism (Aguado *et al.* 1997).

Equations (6.10) and (6.11) are the discrete versions of Neumann and Dirichlet boundary conditions, respectively. The essential difference between them is that (in case of a structurally "perfect" superlattice) the first one allows an exactly homogeneous field distribution $F^{(i)} = U/[(N+1)d]$ to be a stationary solution of the system, whereas the second one generally implies some inhomogeneities at the contacts.

For a given applied voltage U the homogeneous solution of a perfect superlattice can be calculated analytically (Prengel *et al.* 1994). The (particle) current density through the ith well, which, for a stationary state, equals the total current density through the superlattice j^{tot} and thus has to be independent of i, is given by

$$j^{\text{tot}} = j^{(i)} = \sum_{j,k} \left(n_j^{(i-1)} R_{r,jk}^{(i)} - n_k^{(i)} R_{l,jk}^{(i)} \right). \tag{6.12}$$

The resulting current–voltage characteristic is depicted in Fig. 6.2. It contains two

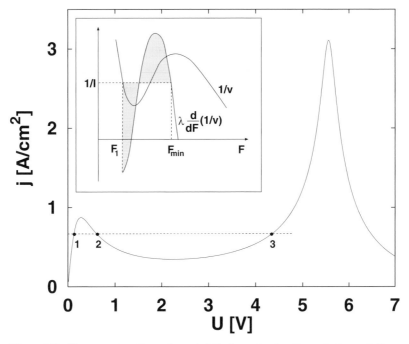

Figure 6.2. The current–voltage characteristic for a structurally perfect superlattice with $N = 40$ wells with a homogeneous field profile (doping concentration $N_{\text{D}} = 10^{14}$ cm^{-3}, barrier width $b = 15$ Å, and well width $l = 90$ Å). 1, 2, and 3 label the homogeneous steady states for a given current density. The inset shows the equal-areas rule which determines the minimum current I for domain formation such that the two shaded areas are equal. (After Schwarz and Schöll (1996).)

distinct peaks with a negative differential conductivity (NDC) regime in between the first peak and the subsequent minimum. The first peak is associated with sequential tunneling between equivalent levels of adjacent wells whereas for the second peak the lowest energy level of each quantum well is in resonance with the second level of the neighboring well, so resonant tunneling is the dominating transport mechanism.

6.2.2 The equal-areas rule for domains in superlattices

In spatially extended systems, N-shaped current-density versus field characteristics like the one shown in Fig. 6.2 generally lead to the formation of domain structures, cf. Fig. 2.5. Although the precise spatial profiles can only be computed numerically, some analytic insight can often be gained by the derivation of "equal-areas rules" that connect some gross features of the profiles with the control parameters, i.e. the current or doping density (Schöll and Landsberg 1988). Such equal-areas rules have been noted in the case of Gunn domains (Butcher 1965) as well as for current filamentation (Schöll 1987), as discussed in earlier chapters. Here we derive an equal-areas rule for the approximate (dimensionless) minimum current I for domain formation in superlattices.

If the intersubband relaxation time τ_{21} is fast compared with the inter-well tunneling processes, $n_2^{(i)} = 0$ holds in the steady state, and the system can effectively be described by only one variable per quantum well (which then can be chosen as either the charge density or the local electric field). In that case, neglecting the small diffusive contribution R_l, directed opposite to the field, eq. (6.12) reduces to

$$j^{\text{tot}} = j^{(i)} = n_1^{(i-1)}\left(R_{r,11}^{(i)} + R_{r,12}^{(i)}\right) \equiv n_1^{(i-1)} v\left(F^{(i)}\right) \tag{6.13}$$

and Fig. 6.2 directly shows the local inter-well transition probability per unit time, $R_r \equiv v$, times the (two-dimensional) doping concentration $l N_D$ as a function of the local field $F = U/[(N+1)d]$. Since the overall behavior of the system is essentially determined by the features of this characteristic function, especially by the existence of an NDC region, it is not surprising that phenomenological models that directly start by assuming such a current–field characteristic (Bonilla et al. 1994b) lead to similar results to ours for a perfect superlattice.

The spatially inhomogeneous steady state is given by substituting (6.13) into (6.7), which yields, with $n_2^{(i)} = 0$,

$$F^{(i)} = F^{(i+1)} + \lambda\left(1 - \frac{I}{v\left(F^{(i+1)}\right)}\right) \tag{6.14}$$

with dimensionless control parameter $I \equiv j^{\text{tot}}/(l N_D)$ and $\lambda \equiv N_D l e / \epsilon$. Equation (6.14) has the form of a noninvertible one-dimensional map $F^{(i)} = f\left(F^{(i+1)}; I, \lambda\right)$ equivalent to that derived from simple phenomenological models (Bonilla et al. 1994b). The fixed points of the map, $F = f(F; I, \lambda)$, yield the homogeneous solutions $I = v(F)$, i.e. for I chosen suitably, there exist three solutions $F_1 < F_2 < F_3$, where F_1 and F_3 correspond to the stable low- and high-field domain, respectively.

A stable inhomogeneous profile with a low-field domain at the cathode and a domain boundary at site i_B is given by

$$F_1 = F^{(i)} = f\left(F^{(i)}; I, \lambda\right) \quad \text{for} \quad 1 \le i \le i_B \tag{6.15}$$

$$F^{(i)} = f\left(F^{(i+1)}; I, \lambda\right) \quad \text{for} \quad i_B \le i \le N, \tag{6.16}$$

with $f'\left(F^{(i)}\right) \ge 0$ for $i > i_B$, i.e. the field increases monotonically toward the side of the high-field domain. The existence of such a sequence $(F^{(i_B)}, F^{(i_B+1)}, \ldots, F^{(N)})$ requires that (6.16) has a nonconstant solution for $i = i_B$. Since $f(F)$ is a nonmonotonic N-shaped function whose intersections with the line $F = F$ are the homogeneous fixed points, $F^{(i_B)} = f\left(F^{(i_B+1)}; I, \lambda\right)$ is possible only if the minimum of f lies at a value $\le F_{i_B} \equiv F_1$, which sets a lower bound to I. Therefore the minimum current I for domain formation is given by $f'(F^{(i_B+1)}) = 0$. Denoting $F_{\min} \equiv F^{(i_B+1)}$, from (6.16) it follows that

$$f(F_{\min}) = F_1 = f(F_1),$$

or

$$0 = \int_{F_1}^{F_{\min}} f'(F)\,dF = \int_{F_1}^{F_{\min}} \left[1 - \lambda I \frac{d}{dF}\left(\frac{1}{v(F)}\right)\right] dF, \tag{6.17}$$

or

$$\frac{F_{\min} - F_1}{I} = \lambda \int_{F_1}^{F_{\min}} \frac{-v'(F)}{v(F)^2}\,dF. \tag{6.18}$$

Equation (6.18) has the form of an equal-areas rule (see the inset of Fig. 6.2). It determines the minimum I where $F_{\min}$ and F_1 are given self-consistently by $f'(F_{\min}) = 0$ and $I = v(F_1)$, and $F_{\min} - F_1$ is the minimum field jump at the first quantum well of the domain boundary. With decreasing N_D ($\sim \lambda$), the minimum I grows until it reaches the maximum current $I_{\max}$ at the miniband peak. This criterion can also be used to discuss the minimum doping density ($\sim \lambda$) for which domain formation is possible:

$$N_{D,\min} \approx \frac{\epsilon/(el)}{I_{\max} \frac{d}{dF}\left(\frac{1}{v}\right)_{\max}}. \tag{6.19}$$

Similar analytic results have been obtained by Mityagin (1996), and by Wacker *et al.* (1997), who also proved the stability of the domains. Note that (6.19) resembles the condition (3.67) for traveling domains in *continuous* systems, which has also been applied to superlattice transport (Ignatov *et al.* 1985).

The build-up of a charge accumulation and the subsequent nucleation of field domains is a process of self-organized pattern formation, which is common in semiconductors with negative differential conductivity. However, there are essential differences compared with the classical Gunn domain instability, which is associated with a traveling triangular domain. For its existence it is required only that the $v(F)$ characteristic has one rising and one falling branch. In the superlattice, the high- and low-field domains correspond to the spatial coexistence of two stable states of the local $j(F)$ characteristic at the same current density, whereas the Gunn domains have a triangular field profile whose maximum does not attain a second stable state (Shaw et al. 1992). If the doping density is sufficiently high, the width of the domain boundary is of the order of the superlattice period. In this case, the domain boundary is localized in a specific quantum well and, in contrast to the Gunn diode, can not move continuously through the sample but can only jump from one quantum well to the next. As a result, there exist different stable stationary domain states (their number being approximately equal to the number of quantum wells) that arise from different locations of the boundary. Thus, unlike in continuous reaction–diffusion systems, there is a wide range of the control parameter I for which the velocity of the traveling front, corresponding to a domain boundary, is zero (Carpio et al. 2000).

6.2.3   Multistable current–voltage characteristics of imperfect superlattices

In order to simulate the current–voltage characteristic of a superlattice we calculate the temporal evolution of the system at a fixed voltage U from the full model (6.1)–(6.9) to obtain the stationary state which it reaches asymptotically. We then increase the external voltage by a small step ΔU (10 mV in the simulations presented here) using the previous stationary state as the new initial charge distribution. By performing a sweep-up and a sweep-down of the voltage in this way, we can thus obtain hysteresis effects and multistability (Kastrup et al. 1994).

The lowest curve of Fig. 6.3(a) (labeled "$\alpha = 0\%$") shows the simulated current–voltage characteristic of a "perfect" superlattice of $N = 40$ periods. In all the simulations presented in this section, the mean doping concentration of the GaAs layers is $N_D = 6.7 \times 10^{17}$ cm^{-3} and the widths of the wells and barriers are $l = 90$ Å and $b = 15$ Å, respectively. Those values have been chosen to match typical experiments (Grahn 1995a).

The current–voltage characteristic consists in a successive sequence of branches whose number is slightly less than the number of superlattice periods. Both sweep-up (the upper sequence of branches) and sweep-down (the lower sequence) are depicted. The current branches are a result of the fact that the field breaks up into a high-field domain on the anode side, where resonant tunneling between different energy levels of adjacent wells dominates, and a low-field domain on the cathode side of the superlattice with transport between equivalent energy levels as the relevant transport mechanism. The domain boundary between them, which is formed by a charge accumulation, extends over just two or three superlattice periods. For increasing applied voltage U the high-field domain expands toward the cathode well by well. Each jump in the current–voltage characteristic with increasing U corresponds to a shift of the charge accumulation that forms the domain boundary by one superlattice period toward the cathode. For sweep-down the same mechanism applies, but with the domain wall shifting in the opposite direction (Prengel *et al.* 1994). It can be shown analytically that the slope of the current branches depends linearly upon the spatial extent of the high-field domain, i.e. the index i enumerating the branches from right to left (Schwarz 1995). By simply reversing the sweep direction for an intermediate voltage, one can show both in the simulations (Prengel *et al.* 1994) and experimentally (Kastrup *et al.* 1994) that the branches of the sweep-up and sweep-down are connected two by two, forming parts of the same stable branch. Furthermore, by numerically computing directly the fixed points of the dynamic system regardless of their stability, it becomes apparent that those different stable branches are again joined to each other by unstable parts, thus forming a single continuous $j(U)$ characteristic (Schöll *et al.* 1996, Patra *et al.* 1998).

   Although simulations of a model that assumes a strictly periodic superlattice successfully explain the gross features of experimentally obtained current–voltage characteristics, the fine structure of the latter, however, exhibits a more irregular behavior, with various branches substantially varying in length. Our simulations show that this can be explained by taking into account fluctuations of the barrier and well widths and the doping along the various layers of the superlattice. The individual sequences of shorter and longer branches in the current–voltage characteristic for different superlattice samples from a single wafer are virtually identical (Helgesen and Finstad 1990, Wacker *et al.* 1995c). We can therefore conclude that *local* in-plane fluctuations in sample parameters do not play an essential role in the irregularity of

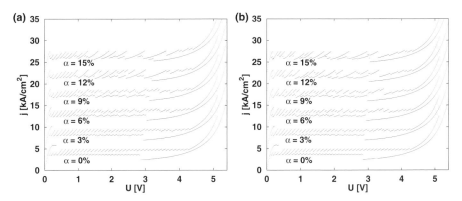

Figure 6.3. Calculated current–voltage characteristics for various degrees of doping fluctuations α (40 superlattice periods, doping concentration $N_D = 6.7 \times 10^{17}$ cm^{-3}, Neumann boundary conditions). In (a) and (b) two different random sequences have been used to simulate the effect of doping fluctuations. Consecutive characteristics are shifted horizontally by 4.5 kA cm^{-1} each. (After Schwarz and Schöll (1996).)

the current–voltage characteristics. On the other hand, interface roughness exists on a scale of some few nanometers (Etemadi and Palmier 1993), which is small compared with the sample's dimensions, which are typically of the order of 100 μm (Kastrup *et al.* 1994); they are effectively averaged out. This justifies our use of a one-dimensional transport model.

To simulate the effects of doping fluctuations we generate a random set $\{e_i\}$ of N values $e_i \in [-1, 1]$. The individual doping concentration of the ith well is then expressed as $N_D^{(i)} = N_D(1 + \alpha e_i)$ with a scaling parameter α that allows us to control the degree of disorder without changing the specific sequence of higher and lower doped wells. It is convenient to use α, which is proportional to the standard deviation of the random sequence, to characterize the degree of disorder.

Figures 6.3(a) and (b) show the current–voltage characteristics for two different random sequences $\{e_i\}$. In both figures sweep-up and sweep-down are depicted for increasing disorder α. Note that consecutive characteristics are shifted horizontally by 4.5 kA cm^{-1} each. For both realizations $\{e_i\}$ the general picture is the same: With increasing degree of disorder some of the current branches grow in length while in turn others shrink. For high values of α the latter can disappear altogether such that the adjacent "longer" branches extend over that voltage interval in the characteristic. Unlike in the case of a structurally perfect superlattice ($\alpha = 0$), for a sufficient degree of disorder some branches of the sweep-up and sweep-down are connected even at higher voltages (>1 V), an effect that is also found in measured current–voltage characteristics (Grahn 1995a).

Although Figs. 6.3(a) and (b) give very similar impressions for characteristics with corresponding α, they differ substantially in detail. Whereas in both cases the specific sequences of "longer" and "shorter" branches remain the same for different degrees of disorder (they just become more pronounced with increasing α) they are completely different for Figs. 6.3(a) and (b). The detailed shape of a current–voltage characteristic thus depends sensitively on the individual realization $\{e_i\}$ of the fluctuations.

On the other hand, one can in turn use the shape of the current–voltage characteristic to draw conclusions about the local defects of a sample. Since the charge accumulation which forms the domain wall shifts from the cathode to the anode with increasing applied bias, it effectively "scans" the superlattice structure well by well. Local perturbations of the periodicity of the superlattice lead to a deviation in the length of the corresponding current branch. If the perturbations of the structure are located apart from each other by more than the spatial extent of the domain wall, their individual influences can clearly be distinguished in the current–voltage characteristic. This provides a resolution of approximately two or three superlattice periods due to the spatial extent of the domain wall. It is then possible to directly deduce their positions within the structure from the index numbers of the perturbed current branches. This applies not only to deviations of the doping concentration but also to fluctuations of the widths of the individual layers of the superlattice sample (Wacker *et al.* 1995c).

Figure 6.4 explains how the current–voltage characteristic changes in the regime of domain formation if the donor density in a single quantum well is either increased (b) or decreased (c) with respect to the unperturbed case (a). It is evident that there is a direct correlation between the local doping density in the kth well and the peak current of the $(k + 1)$th branch of the current–voltage characteristic (counting from the right) corresponding to the location of the domain boundary at the $(k + 1)$th well. One can even infer quantitative information about the local doping density from the peak currents, i.e. the positions of the saddle-node bifurcations which mark the end of a stable branch. It can be shown analytically (Patra *et al.* 1998) under some simplifying assumptions that the maximum current reached on a particular stable branch is proportional to the doping density in the quantum well next to the well at which the domain boundary is located. This makes it possible to determine the local values of the ratio $N_D^{(i)}/N_D$ with good precision from measured current–voltage characteristics and thereby characterize the quality of a sample. This procedure has been tested by applying it to numerically simulated characteristics with random doping

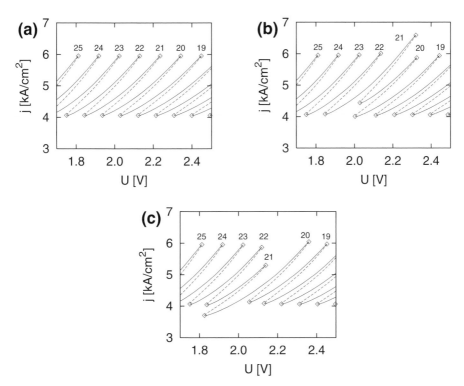

Figure 6.4. Part of the current–voltage characteristics in the regime of domain formation ($N_D = 7.9 \times 10^{17}$ cm^{-3}, Dirichlet boundary conditions). The doping density in the 20th well is (b) increased by 10%, and (c) decreased by 10%. The unperturbed characteristic is displayed in (a). The branches are numbered according to the location of the domain boundary at the kth well. Full and broken lines denote stable and unstable states, respectively. (After Patra *et al.* (1998).)

fluctuations (Schwarz *et al.* 1998) and to measured data (Patra *et al.* 1998).

Even though the full continuously connected current–voltage characteristic, consisting in alternating stable and unstable branches, manifests only slight changes if moderate disorder is introduced, the characteristic which is found under voltage sweep-up or sweep-down may change qualitatively. Individual branches are shifted and their lengths are changed according to the doping densities in the corresponding quantum wells; thus, some stable branches might be missed out completely during a voltage sweep-up or sweep-down if the doping densities in two adjacent quantum wells differ sufficiently. This is demonstrated in Fig. 6.5 by comparing voltage sweep-up and sweep-down (a) with the full stable branches (b). For a given degree of disorder α, the effect of branches being missed out is most pronounced at large doping or high voltage.

If several perturbations are located close to one another, due to the nonlinear nature of the system they interact and can no longer be distinguished in the current–voltage characteristic. Also, since different types of disorder (fluctuations of the doping concentrations, of the well widths, and of the barrier widths) have very similar impacts on the shape of the characteristic, they can not be identified therefrom. Instead, complementary experiments such as optical measurements are necessary. Nonetheless, it seems possible to judge the degree of disorder of a sample (and thus the quality of its growth process) from the overall appearance of its current–voltage characteristic by comparison. Simple macroscopic global measurements of the current–voltage characteristic can thus be used to quantify the microscopic growth-related disorder of a sample (Wacker *et al.* 1995c).

In fact, specifically tailored current–voltage characteristics can be obtained by intentionally introducing structural imperfections into the superlattice. This has

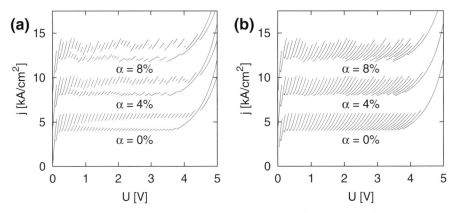

Figure 6.5. Current–voltage characteristics for superlattices with various degrees of doping fluctuations α; (a) characteristics for voltage sweep-up and sweep-down, and (b) stable parts of the full connected current–voltage characteristics ($N_D = 7.9 \times 10^{17}$ cm^{-3}, Dirichlet boundary conditions; the vertical scale is shifted for each α). (After Patra *et al.* (1998).)

been demonstrated by simulating a superlattice with a single wider barrier (Schwarz
et al. 1996a), see Fig. 6.6. For both directions of bias, with increasing voltage we
first obtain a sequence of short current branches with a low current density (a). It is
followed by a single long ascending branch leading to another sequence of notably
longer current branches at about $j = 5$ kA cm^{-2}. The lengths and current densities
of the latter branches are identical to those obtained in simulations of "perfect" or
weakly disordered superlattices of the same doping density (Schwarz et al. 1996b).
The characteristics of both directions of bias differ in the voltage at which the long
ascending branch is located, i.e. the number of branches in the "lower" and "upper"
sequences. For positive U the lower current sequence consists of eight and the higher
one of 29 branches, whereas for reverse bias the ratio is 30 to seven.

This asymmetric behavior can be explained with the help of the spatial field
distribution across the sample, which for positive bias is displayed in Fig. 6.6(b)
for voltages U up to 2 V. For very small voltages U the electric field is distributed
virtually homogeneously across the sample (corresponding to the initial sharp rise
of the current–voltage characteristic). At approximately 0.1 V this homogeneous
field distribution breaks up, and a high-field domain forms across the 32nd and 33rd
barriers. This can easily be understood if one bears in mind that, for stationary
states of the system, the current across each barrier must be identical. To satisfy
this condition the electric field across the thicker 32nd barrier, $F^{(32)}$, must be higher
than that elsewhere. However, the total voltage U is not yet high enough for the
high-field domain to extend over more than two superlattice periods. For the resulting
"inverse" field domain boundary, a local carrier depletion, i.e. an electron density
in the 33rd well smaller than the doping density, is required due to Gauss' Law.

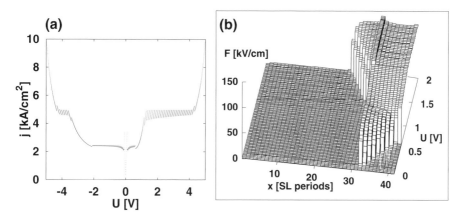

Figure 6.6. (a) The simulated current–voltage characteristic of a superlattice with
40 periods and $N_D = 6.7 \times 10^{17}$ cm^{-3}, with Dirichlet boundary conditions
(sweep-up). The 32nd barrier of the superlattice has been chosen 20% wider than the
others. (b) Simulated spatial profiles of the electric field F across the superlattice
(SL) structure for positive voltages. The SL periods are numbered starting from the
cathode. (After Schwarz et al. (1996a).)

As a result, the current density, which equals the local electron density times the field-dependent transport coefficient summed up over all transport processes, and thus the total current through the sample are small compared with the case of a "perfect" or weakly disordered superlattice of the same doping density, in which only charge accumulation occurs. The simulations show that, with a further increase of U, the high-field domain expands toward the anode period by period, creating the small discrete branches in the characteristic. When the anode has been reached, the current is no longer limited by electron depletion; it grows with rising U up to the value of the branches in the characteristic of a "normal" superlattice. The high-field domain now expands toward the cathode across the remaining superlattice periods the same way it would throughout a sample without the wider barrier. The remaining current–voltage characteristic is therefore effectively that of a "perfect" superlattice with a smaller number of periods (30) and a respective voltage offset. For negative bias, the same mechanism applies. However, the thicker barrier is located at a different position with respect to the cathode, resulting in the asymmetry of the $I(U)$ characteristic. The measured current–voltage characteristic of a superlattice that had intentionally been grown with one thicker barrier confirmed the model prediction (Schwarz et al. 1996a).

Finally, we discuss the influence of the boundary conditions. In Fig. 6.3 Neumann boundary conditions are used, whereas in Figs. 6.4–6.6 Dirichlet boundary conditions are employed. Except for the first and last few branches, for which the domain boundary is located close to one of the contacts, the characteristics are virtually identical if the same parameters are used (Schwarz and Schöll 1996). This again confirms that the shape of the current–voltage characteristic at a certain voltage U is determined by the local properties of the superlattice structure around the domain boundary, whereas those of the other wells and barriers, which are located within either the high- or the low-field domain, are of negligible influence. For the simulation of the *stationary* current–voltage characteristics of doped superlattices the boundary conditions thus play only a minor role. This changes drastically when self-sustained oscillations are considered.

6.3   Self-sustained oscillations

In the previous section we have considered solely the steady-state current–voltage characteristics of a superlattice, i.e. the fixed points of the dynamic system. For high doping concentrations of some 10^{17} cm^{-3} these correspond to electric-field-domain states, whereas for low doping a stable uniform field distribution forms (Schöll et al. 1996). In an intermediate doping regime undamped self-sustained oscillations of the current are found in a range of voltages in simulations (Wacker et al. 1995b, Bonilla 1995), which has also been observed experimentally (Kastrup et al. 1995, Grahn et al. 1995). In a simple model with only one variable per well, one can show that the homogeneous field distribution becomes unstable when the doping concentration N_D

exceeds a "critical" value for voltages within the NDC region (cf. Fig. 6.2) (Wacker *et al.* 1997) and self-generated oscillations arise (Schwarz and Schöll 1996, Kastrup *et al.* 1997).

We will first explain the mechanism of the oscillations and then discuss how disorder can affect them (Patra *et al.* 1998). In particular, we shall present bifurcation scenarios not only in terms of their dependencies on the bias voltage and mean donor density but also by taking the degree of disorder into account as a third, equally important system parameter. The various regimes of stationary multistable field domains and self-sustained oscillations are investigated in the framework of the simple model used in the previous sections of this chapter. Here we use Dirichlet boundary conditions, modeling Ohmic contacts as a carrier reservoir created by heavily doped boundary layers. Neumann boundary conditions have been used elsewhere (Wacker *et al.* 1995c, Schöll *et al.* 1996).

6.3.1 Bifurcation scenarios of spatio-temporal patterns

In order to gain insight into the general features of the spatio-temporal scenarios, we shall first briefly review the case of a structurally perfect sample. The various regimes of spatio-temporal behavior found from the simulations for a superlattice

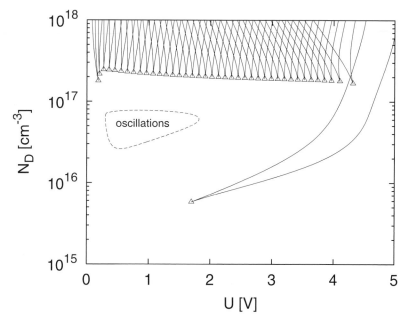

Figure 6.7. A phase diagram of spatio-temporal instabilities as a function of the external voltage U and doping density N_D for a "perfect" superlattice (Dirichlet boundary conditions). Saddle-node bifurcations are marked by solid lines, Hopf bifurcations by dashed lines, and cusp points as triangles ($\triangle$). (after Patra *et al.* (1998).)

without doping fluctuations are summarized in the phase diagram of Fig. 6.7. The current–voltage characteristics corresponding to cross-sections for various fixed N_D are depicted in Fig. 6.8.

The spatially homogeneous, N-shaped current–voltage characteristic is stable only at low doping (Fig. 6.8(a)). For higher doping, a smeared-out domain-like field distribution forms, and a bistable Z-shaped current–voltage characteristic arises (Fig. 6.8(b)), indicating a transition between a smeared-out field domain for high current and lower voltage, and an almost homogeneous field distribution for low current and higher voltage. The positions of saddle-node bifurcation points where a stable and an unstable steady state merge are indicated by diamonds; they form two lines in the lower part of the phase diagram in Fig. 6.7. At a minimum value of N_D the two lines end in a cusp point where the $j(U)$ characteristic changes from N-shaped to Z-shaped.

At higher doping, spatio-temporal instabilities lead to self-generated current oscillations associated with the build-up of space charges (Wacker *et al.* 1995*b*). Here, the current–voltage characteristic contains a regime of limit-cycle oscillations (Fig. 6.8(c)). This regime is confined by supercritical Hopf bifurcation points, marked as a closed dashed line in Fig. 6.7. Note that subcritical Hopf bifurcations can be found for different boundary conditions (Schöll *et al.* 1996).

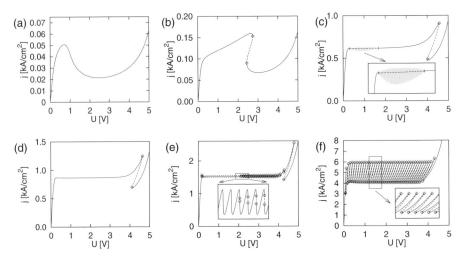

Figure 6.8. Current–voltage characteristics for doping densities of (a) $N_D = 3.2 \times 10^{15}$ cm^{-3}, (b) $N_D = 10^{16}$ cm^{-3}, (c) $N_D = 7 \times 10^{16}$ cm^{-3}, (d) $N_D = 10^{17}$ cm^{-3}, (e) $N_D = 2 \times 10^{17}$ cm^{-3}, (f) $N_D = 7.9 \times 10^{17}$ cm^{-3}. Throughout the following, stable stationary states are marked as solid lines, unstable ones as dashed lines, saddle-node bifurcation points as diamonds ($\diamond$), and Hopf bifurcation points as crosses ($+$). Oscillations are indicated by shaded areas within minimum and maximum current density. The insets show enlarged sections of the respective current–voltage characteristics. (After Patra *et al.* (1998).)

A typical oscillation is depicted in Fig. 6.9. The field profile in Fig. 6.9(a) shows that an inhomogeneous field distribution forms and moves toward the anode. The charge accumulation associated with it (Fig. 6.9(b)) grows in order to satisfy the global coupling condition (6.8), and eventually, near the anode, vanishes. Then the whole cycle is repeated. The corresponding oscillation of the current density, j, is almost perfectly sinusoidal. The oscillation exhibits a square-root dependency of the amplitude upon the distance from the bifurcation point, as shown in the inset of Fig. 6.9(c) for fixed N_D and increasing U. The frequency is not affected by this bifurcation and is determined near the bifurcation point by the imaginary part

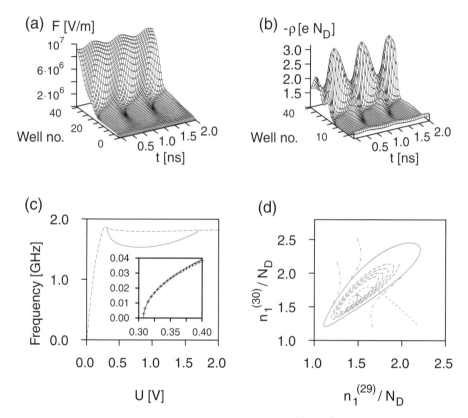

Figure 6.9. Limit-cycle oscillation ($N_D = 5 \times 10^{16}$ cm^{-3}, Dirichlet boundary conditions). (a) The temporal evolution of the field profile F. (b) The temporal evolution of the charge density ρ ($U = 1$ V). (c) The frequency of the oscillation versus U (solid line). The frequency corresponding to the imaginary part of the largest eigenvalue is shown as a dashed line. The inset depicts the amplitude of the current density j (in kA cm^{-2}) versus U (in volts) near one of the Hopf bifurcation points (+). (d) A phase portrait of n_1^{30} versus n_1^{29} (+: unstable fixed point, solid line: limit cycle, dashed line: trajectory lying in the center manifold, dotted line: trajectory starting outside the center manifold; the other trajectories start near the limit cycle). (After Patra *et al.* (1998).)

of the corresponding eigenvalue (Fig. 6.9(c)). Similar results are obtained for fixed U and variable N_D. In comparing these results with those obtained from other models (Kastrup *et al.* 1997), one should observe that the frequencies depend strongly on the choice of boundary condition, and here we assume heavily doped boundary layers.

The dynamics is fast in the directions orthogonal to the center manifold, in which the limit cycle as well as the corresponding unstable fixed point are embedded (Fig. 6.9(d)). Note that, in contrast to the formation of stationary field domains, the superlattice must contain a minimum number of quantum wells if oscillations are to be allowed (Wacker *et al.* 1997).

At the highest doping densities, the number of carriers available is sufficient to provide the space charge necessary to form a stable, stationary boundary between a low-field and a high-field domain. As the domain boundary is shifted from the anode to the cathode with increasing voltage, the current–voltage characteristic exhibits small modulations in the form of sequences of N-shaped (Fig. 6.8(e)) or Z-shaped (Fig. 6.8(f)) branches corresponding to different locations of the domain boundary, as has been shown experimentally (Kastrup *et al.* 1994). The various stable branches are connected by unstable parts, as depicted in the insets. Thus the current–voltage characteristic consists in a single continuously connected curve, along which stable and unstable parts alternate. The changes of stability are caused by saddle-node bifurcations, which show up as a complex pattern of intersecting lines in the upper part of Fig. 6.7. This diagram allows one to determine the positions of the individual current branches as well as, for given parameters U and N_D, the number of multistable states. The two lines of saddle-node bifur-cation points belonging to a particular unstable branch merge in a cusp point. Thus, the associated value of N_D is the minimum doping density necessary for multistability and hysteresis of current branches. Since the cusp points lie at slightly different values of N_D, in the respective range of N_D values hysteresis may set in only above a certain threshold voltage, as can be seen in the inset of Fig. 6.8(e). The overall behavior shown in Fig. 6.8 for the structurally perfect model is similar to that found in other models (Kastrup *et al.* 1994, Wacker *et al.* 1997). Note that the various scenarios can also be realized experimentally by varying the electron density by optical irradiation instead of varying the doping (Ohtani *et al.* 1998).

6.3.2   Effects of fluctuations in doping on oscillations

The main effect of disorder in the regime of limit-cycle oscillations is to shift and deform their locations in parameter space. For moderate disorder, it is determined by the location of the Hopf bifurcation points (Fig. 6.10, dashed) as in the perfect superlattice (Fig. 6.7). The change in amplitude and frequency of the oscillations for fixed voltage and fixed donor density is only slight.

Unfortunately, the total effect of a sequence of random fluctuations can not be predicted from a superposition of the effects of the individual single perturbations as in the regime of multistable stationary field domains (Patra *et al.* 1998). In contrast to the latter case, in which the charge accumulation forming the domain boundary is largely confined to a single quantum well, oscillations involve changes of the charges in a large number of quantum wells, resulting in a nonlinear interaction. Furthermore, the shape of the regime of oscillations looks very different for different realizations e_i for the same α, and thus gives no reliable indication of the global degree of disorder. The regime of oscillations generally tends, however, to become larger with increasing disorder.

For strong disorder, the regime of oscillations may thus extend to doping densities high enough for multistable stationary field domains associated with saddle-node bifurcation to form. If a Hopf bifurcation and a saddle-node bifurcation coincide, this is called a Takens–Bogdanov point (Guckenheimer and Holmes 1983). In a superlattice with strong doping fluctuations, two Takens–Bogdanov points and one cusp point are generated in a single codimension-three bifurcation by appropriately

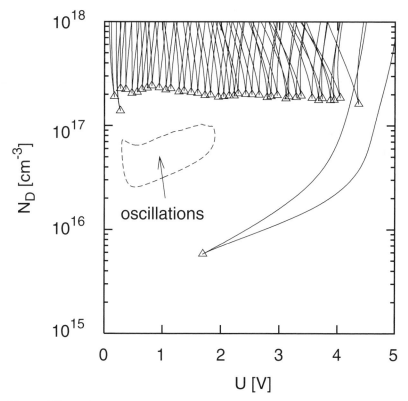

Figure 6.10. A phase diagram of spatio-temporal instabilities as a function of the mean donor density N_D and voltage U for a superlattice with doping fluctuations $\alpha = 8\%$ (Dirichlet boundary conditions). (After Patra *et al.* (1998).)

adjusting three control parameters: N_D, U, and α. The two Takens–Bogdanov points are connected by a curve of Hopf bifurcation points; if α is increased slightly, this curve merges with the closed curve of Hopf bifurcations found also in superlattices with moderate disorder or none at all (Fig. 6.10). In the phase diagram of the mean doping N_D versus voltage U for fixed α (Fig. 6.11(a)) this gives a single open curve (dashed), which encloses the shaded area of limit-cycle oscillations. Two pairs of full lines corresponding to saddle-node bifurcations of domain states are also shown, each ending in a cusp point (triangle). The inset shows the phase diagram near a Takens–Bogdanov point in more detail.

Since we are discussing a codimension-three bifurcation scenario, it is not sufficient to merely consider the location of the bifurcation points as a function of two parameters. Therefore, the same Takens–Bogdanov points (crosses) are also depicted in the (α, U)-plane for fixed N_D in Fig. 6.11(b). One of the cusp points from (a) is not shown in (b) since it does not depend on the degree of disorder; consequently, there are two saddle-node bifurcation lines extending down to $\alpha = 0$. It can be seen that the Hopf bifurcation line (dashed) connecting the two Takens–Bogdanov points exists only above a minimum value of α. With increasing α, at first the regime of oscillations (shaded) is limited by two Hopf bifurcation points. When the value of the control parameter α is slightly increased, a Hopf bifurcation and a saddle-node bifurcation merge in the first Takens–Bogdanov point. At higher α the regime of oscillations is limited at lower U by a Hopf bifurcation, whereas at higher U it ceases due to a global bifurcation: A homoclinic bifurcation in which a saddle-point collides with a limit cycle, forming a saddle-loop and subsequently disappearing into the "blue sky" (see Section 1.2.3). Here the amplitude of the limit cycle remains finite whereas the frequency tends to zero.

For higher α, we find different bifurcation scenarios. First, the regime of oscillations is no longer bounded from above by a homoclinic (saddle-loop) bifurcation but rather by a saddle-node bifurcation on a limit cycle (see Fig. 6.11(b) for $\alpha \geq 15.4\%$), cf. the classification in Section 1.2.3. For even higher α, a saddle-node bifurcation line originating from the cusp point (triangle) crosses the branch of Hopf bifurcation points (see the inset of Fig. 6.11(b)). The main regime of oscillations is now limited by two saddle-node bifurcations on the limit cycle (Fig. 6.12); in addition, there is a small oscillatory regime (left-hand inset) starting at the Hopf bifurcation and being destroyed in a nearby saddle-loop bifurcation. Note that, if the temporally averaged current is monitored, as is usually done in experiments, the current–voltage characteristic is expected to exhibit sharp transitions when the various stationary and oscillatory regimes are entered.

The saddle-loop bifurcation and the saddle-node bifurcation on a limit cycle result in a sharp decrease of the oscillation frequency to zero when the voltage approaches one of the bifurcation points (see the inset of Fig. 6.12). Therefore, if a superlattice is operated near one of these bifurcation points, the frequency of the oscillation can be controlled by just changing the applied external voltage, U, slightly. In a different model (Kastrup *et al.* 1997) another explanation for the increase of the frequency with bias is given by discussing the space available for the charge monopole to travel. This

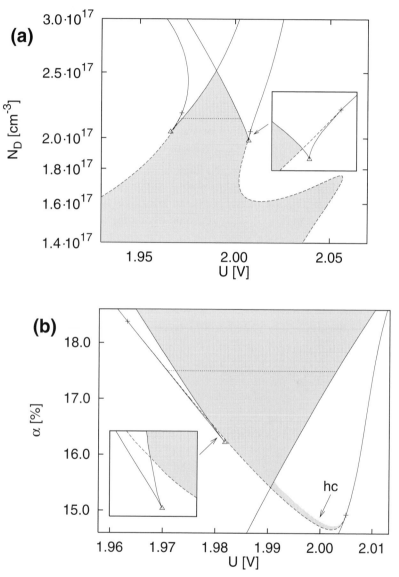

Figure 6.11. Part of the phase diagram near the two Takens–Bogdanov points (marked as crosses, +) in a superlattice with strong disorder. The regime of oscillations is shaded. Hopf bifurcations are denoted by dashed lines, saddle-node bifurcations by full lines, and homoclinic bifurcations by the label "hc". The triangles denote cusp points. (a) N_D versus U for fixed $\alpha = 17.5\%$. (b) α versus U for fixed $N_D = 2.14 \times 10^{17}$ cm^{-3}. The dotted, horizontal line marks the intersection of the slices of parameter space presented in the two figures. (After Patra *et al.* (1998).)

explanation is not applicable to the present model for superlattices of the short length chosen here but becomes relevant only for $N \geq 200$ superlattice periods. Voltage tuning of limit-cycle oscillations has indeed been observed experimentally in doped superlattices (Kastrup *et al.* 1997). Such behavior as well as the experimentally found discontinuous switching between various oscillatory modes (Zhang *et al.* 1997) can be consistently explained also within the present model.

In conclusion, the basic mechanism of the current oscillations is the periodic formation of a charge accumulation due to NDC that drifts across the superlattice and reduces the high-field region of the structure. We note that the choice of boundary conditions has a much greater influence upon the oscillations than it does upon the current–voltage characteristics. This is expected because the latter are just the stable fixed points of the dynamic system, whereas the shape of the oscillations depends on the global behavior of the system in a larger portion of the phase space. Similar results with respect to the dependency upon boundary conditions have been noted in the case of Gunn domains (Böer and Döhler 1969) and recombination domains (Böer and Voss 1968*a*) in bulk semiconductors. However, unlike in those continuous systems,

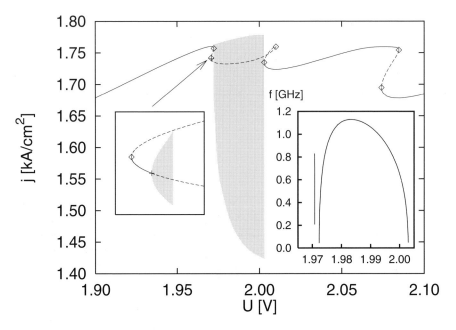

Figure 6.12. The current–voltage characteristic corresponding to the dotted line in Fig. 6.11 ($N_\mathrm{D} = 2.14 \times 10^{17} \ \mathrm{cm}^{-3}$ and $\alpha = 17.5\%$). The main regime of limit-cycle oscillations (shaded) is bounded on both sides by saddle-node bifurcations on the limit cycle. The left-hand inset shows the region near the Hopf bifurcation (+) on an enlarged scale. The small oscillatory regime (shaded) is bounded on the right-hand side by a saddle-loop bifurcation. The right-hand inset depicts the variation of the frequency f with the voltage U. (After Patra *et al.* (1998).)

in superlattices, at high enough currents, the traveling wavefronts may even move upstream against the direction of drift of the individual carriers, due to the interplay between the front end the dynamics at the injecting contact (Carpio *et al.* 2000).

6.4   The dynamics of domain formation

In this section we study the dynamics of field-domain formation in n-doped semi-conductor superlattices (Prengel *et al.* 1995, 1996, Kastrup *et al.* 1996, Shimada and Hirakawa 1997). Dynamic simulations of the switching of the applied voltage from 0 into the domain regime show that two formation mechanisms can occur: The formation of the domain boundary *in place* in the NDC regime of the drift-velocity versus field characteristic, and well-to-well hopping starting from the cathode in the bias regime of positive differential conductivity. The simulations are in good agreement with experimental results. Calculations for various barrier widths show that the formation times become shorter for thinner barriers, as expected due to the increase of the drift velocity.

Although static field domains in superlattices have been investigated intensively (Grahn 1995*b*), the dynamics of their formation and of the switching process has received considerably less attention. The dynamics of the switching between different stable stationary branches of the current–voltage characteristic, which is associated with the relocation of the domain boundary (Amann *et al.* 2001), has only recently been investigated experimentally by applying voltage pulses (Luo *et al.* 1998*a*). In the following we present simulations of the dynamic response of a superlattice when an applied bias voltage is switched on instantaneously.

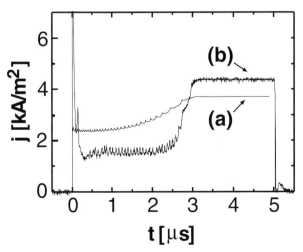

Figure 6.13. (a) The calculated response of the current to an applied voltage step from 0 to -1 V exhibiting well-to-well hopping spikes. (b) The experimental response of the current to an applied voltage of -1 V (two-dimensional doping density $N_{2d} = 1.5 \times 10^{11}$ cm^{-2}). (After Prengel *et al.* (1995).)

6.4.1   Well-to-well hopping of the domain boundary

The simulations refer to a superlattice consisting in $N = 40$ quantum wells of width $l = 9$ nm and depth $\Delta E_c = 982$ meV, which are weakly coupled by barriers of width $b = 4$ nm. The doping density is allowed to fluctuate randomly from well to well by 1% around a mean value $N_D = 1.67 \times 10^{17}$ cm^{-3}. The coupled rate equations (6.1) for the electron densities $n_k^{(i)}$ are solved with a sequential tunneling rate that includes resonant tunneling with a Lorentzian shape of width $\Gamma = 3.3$ meV and a "nonresonant" component in the WKB approximation (Kastrup *et al.* 1996).

In Fig. 6.13 we show the calculated (a) and measured (b) time dependencies of the current density when the voltage is switched from 0 to -1 V. The switching process in Fig. 6.13(a) is realized numerically by calculating the evolution of the electric field starting from a homogeneous field distribution of the appropriate strength. Figure 6.13(b) shows the corresponding experimental result obtained by recording the time-resolved response of the current to a voltage pulse of -1 V (the sharp turn-on and turn-off current spikes are cut off). The good quantitative agreement both of the time scale and of the detailed shape can clearly be seen. As the simulations show, the small current spikes are due to well-to-well hopping of a charge monopole that forms the boundary between the low- and high-field domains. To emphasize this, we show in Fig. 6.14 the velocity $v_{coc} = dk_{coc}/dt$ of the center of charge $k_{coc} = (\sum_{k=1}^{N} k\rho^{(k)})/(\sum_{k=1}^{N} \rho^{(k)})$ (where $\rho^{(k)} = n_1^{(k)} + n_2^{(k)} - N_D^{(k)}$) together with the current density for the time interval 0.5 μs $< t < 2.5$ μs. The velocity v_{coc} exhibits a sharp peak whenever the current density does, thus revealing a correlation between the (experimentally unobservable) motion of the charge monopole and the (measurable) current.

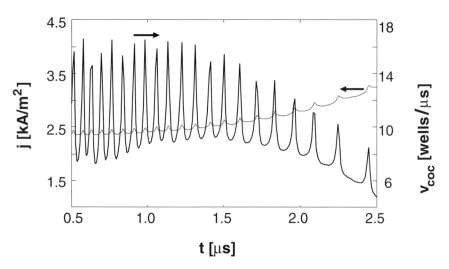

Figure 6.14. The calculated current density j (thin line) and velocity v_{coc} of the center of charge (thick line) for an enlarged section of Fig. 6.13(a) (after Prengel *et al.* (1995)).

The mechanism of the domain formation is thus the following: Starting from a homogeneous field distribution, a charge monopole forms at the cathode and moves toward the anode by well-to-well hopping, until a stable position is reached. This very behavior can be found in the spatial-field-distribution versus time plot of Fig. 6.15(b). However, a look at Fig. 6.15(a) shows that there can be modifications for lower voltages (0.6 V in this case). Here, a smaller charge accumulation moves much faster from the cathode toward the anode, but, at the same time, a charge accumulation at the anode is extending in the opposite direction. When they finally coalesce, there arises a much larger charge-accumulation layer, which forms a sharp domain boundary. Thus, the domain boundary forms "in place", near the location which ensures a stable field distribution. This applies to the regime of strong negative differential velocity of the drift-velocity versus field characteristic $v(F)$.

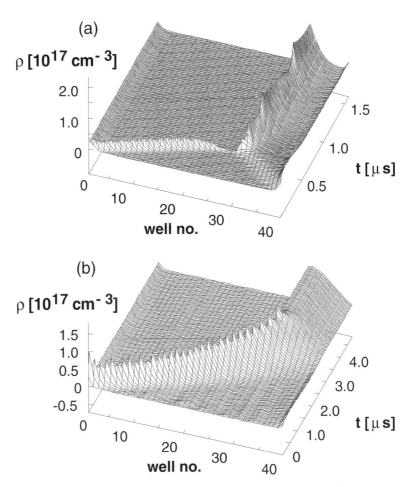

Figure 6.15. The calculated temporal evolution of the charge-density distribution for an applied voltage of (a) 0.6 V and (b) 1.0 V (after Prengel *et al.* (1995)).

The domain-formation time is interpreted as the time it takes for the current to cross a certain threshold level (35 μA for the experiment, 3.5 kA m^{-2} in the simulation). The times determined in this way are shown in Fig. 6.16 (full squares for the measured times, circles for the calculated times). The quantitative agreement is satisfactory, although the curves differ in some details.

Two important facts should be noted here. First, beyond 1 V the measured formation time decreases almost linearly with the applied voltage. In this voltage regime the domain boundary has to propagate from the cathode to its final position so that the formation time decreases the closer the final position of the boundary is to the cathode. This shows that the origin of the decreasing formation time is the decreasing propagation distance of the domain boundary at larger absolute voltages. Second, below 1 V the formation time increases with increasing voltage, indicating that the formation mechanism changes somewhat, as mentioned above and shown in Fig. 6.15(a). This seems to be related to the strong negative differential mobility of the $v(F)$ characteristic, a view that is supported by the observation that both the calculated maximum formation time and the minimum in the $v(F)$ characteristic occur at $\approx$1.5 V.

The same phenomenon of well-to-well hopping is also found in the regime of lower doping, where self-sustained current oscillations occur (Prengel *et al.* 1996). After the transient well-to-well hopping, a fine resolution of the self-generated domain oscillations reveals small periodic current modulations superimposed upon the main oscillations, indicating that also in this regime transport is dominated by well-to-well hopping of the space charge located in the domain boundary.

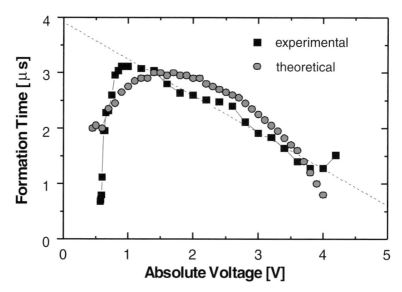

Figure 6.16. Experimental (full squares) and theoretical (circles) formation times versus applied voltages. The dashed line shows the linear regression for the experimental data points between 1 and 3 V. (After Prengel *et al.* (1995).)

6.4.2   Influences of the barrier width and doping density

Figure 6.17(a) shows the calculated domain formation times for three barrier widths of the superlattice: 40 Å (as described above), 38 Å, and 35 Å. One clearly sees that domain formation becomes much faster for thinner barriers, as expected, since the barrier width determines the tunneling time and thus the velocity of the well-to-well hopping of the charge-accumulation layer. In Fig. 6.17(b) the influence of varying doping density is shown to be weak.

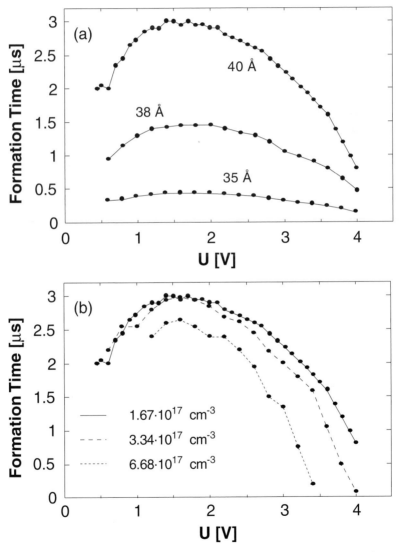

Figure 6.17. Calculated formation times (a) for various barrier widths and (b) for various doping densities (after Prengel *et al.* (1995)).

Chapter 7

Spatio-temporal chaos

7.1   Introduction

Spatio-temporal chaos is a feature of nonlinear spatially extended systems with large numbers of degrees of freedom, as described for instance by reaction–diffusion models of activator–inhibitor type. In this chapter we shall study a model system of this type that was introduced in Chapter 2 for layered semiconductor heterostructures, and used for a general analysis of pattern formation in Chapter 3. Here we shall investigate in detail the complex spatio-temporal dynamics including a codimension-two Turing–Hopf bifurcation, and asymptotic and transient spatio-temporal chaos. Chaos control of spatio-temporal spiking by time-delay autosynchronization is applied. The extensive chaotic state is characterized using Lyapounov exponents and a Karhunen–Loève eigenmode analysis.

7.2   Spatio-temporal spiking in layered structures

For understanding spatio-temporal chaos it is important to study first the elementary spatio-temporal patterns, which may eventually – in the course of secondary bifurcations – evolve into chaotic scenarios. One such elementary pattern, which has been observed experimentally in various different semiconductor devices exhibiting SNDC, e.g. layered structures such as p–n–p–i–n diodes (Niedernostheide *et al.* 1992*a*) and p–i–n diodes (Symanczyk *et al.* 1991*a*), or in impurity-impact-ionization breakdown (Rau *et al.* 1991, Spangler *et al.* 1992), and in electron–hole plasmas (Aliev *et al.* 1994), is the *spiking mode* of current filaments. The properties of localized spiking structures have been studied theoretically in detail by Kerner and Osipov (1982, 1989) in general reaction–diffusion systems. Spiking in a simple chemical reaction–diffusion model – the Brusselator – has been reported by De Wit *et al.* (1996). Stationary and traveling localized structures in layered semiconductor p–n–p–n structures have also been modeled (Gorbatyuk and Rodin 1992*c*, Niedernostheide *et al.* 1992*a*, 1992*b*, 1994*a*). A specific model for the heterostructure hot-electron diode that exhibits spatio-temporal spiking has been derived and analyzed in one spatial dimension (Wacker and Schöll 1992, 1994*a*, 1994*b*, Bose *et al.* 1994). A general analysis that specifies conditions for complex spatio-temporal dynamics in terms of competing spatial and temporal instabilities has been performed for two-dimensional spatial domains (Bose *et al.* 2000).

In this section we consider a simple, generic model (Wacker and Schöll 1994*b*) that displays such spiking behavior, independently from the specific microscopic transport mechanism that is effective in a particular device. It is a two-component reaction–diffusion system of activator–inhibitor type with global coupling, and has already been introduced in Section 2.4.5. Such activator–inhibitor systems are widely used in the description of active media, not only in semiconductor transport but also in chemical reaction systems (Engel *et al.* 1996). In the following we shall briefly re-derive this model and then study its dynamic spatio-temporal behavior.

7.2.1   A globally coupled reaction–diffusion model

In layered semiconductor structures the charge transport is impeded by conduction-band discontinuities or potential barriers, which leads to charge accumulation in the accompanying potential wells. This inhomogeneous space-charge distribution determines the electric-field distribution which in turn strongly influences transport. Therefore there is a condition for self-consistency, which often has more than one solution for a fixed applied bias. This indicates an important mechanism for bistability and SNDC arising from vertical transport in layered semiconductor structures.

The simplest example is the heterostructure hot-electron diode (HHED), which we have introduced in Chapter 2. This device essentially consists in a GaAs and an adjacent $Al_xGa_{1-x}As$ layer, both of which are undoped or only slightly n-doped, so that strong electric fields can arise. The layers are coated with highly doped contact layers. Owing to the conduction-band discontinuity, the AlGaAs barrier forms an obstacle to the transport of electrons if a bias is applied perpendicular to the layers. As shown in Figs. 7.1(a) and (b), in a certain range of voltage U across the sample, two states with different currents are possible (Hess *et al.* 1986, Belyantsev *et al.* 1986). If the electric field in the GaAs layer is small, the mean energy of the electrons is low and only a small tunneling current through the barrier is possible (Fig. 7.1(a)). Most of the voltage drops in the AlGaAs layer in this case. If the voltage is distributed differently, such that the electric field in the GaAs layer becomes high, thermionic emission sets in, resulting in a second stationary state with high current (Fig. 7.1(b)). In a certain range of voltage, bistability between these two states is found (Wacker and

Schöll 1991). They can be distinguished by the interface charge density ρ_s between the layers given by $\rho_s = \epsilon_2 \mathcal{E}_2 - \epsilon_1 \mathcal{E}_1$, where $\mathcal{E}_1$ and $\mathcal{E}_2$ are the respective electric fields, and ϵ_1 and ϵ_2 are the permittivities. The subscripts 1 and 2 denote the GaAs and AlGaAs layers, respectively.

As appropriate dynamic variables one may identify ρ_s and the voltage U across the structure. Let the voltage be applied in the z-direction. Then the interface between the layers extends in the (x, y)-plane. The dynamic behavior of $\rho_s(x, y, t)$ is given by the continuity equation (2.57):

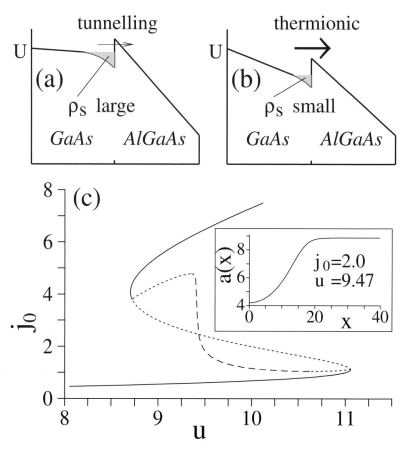

Figure 7.1. Bistability in the heterostructure hot-electron diode (HHED) for fixed sample voltage U (a) and (b). The low- and the high-current state are characterized by a large and a small surface charge density ρ_s between the layers, respectively. (c) The current–voltage characteristic for the generic model for $L_x = 40$ and $\mathcal{T} = 0.05$. The full line depicts homogeneous stable states. Inhomogeneous stable states of the shape given in the inset are marked by the dashed line. The dotted lines refer to states that are unstable. The stability is determined for $\alpha = 2$. (After Wacker and Schöll (1994b).)

$$\frac{\partial \rho_s(x, y, t)}{\partial t} = j_1 - j_2 - \left(\frac{\partial j_x^\parallel}{\partial x} + \frac{\partial j_y^\parallel}{\partial y} \right), \tag{7.1}$$

where j_1 and j_2 denote the current densities in the GaAs and the AlGaAs layer, respectively, and $j^\parallel = \mu_\parallel |\rho_s| \mathcal{E}_\parallel$ is the interface current density in the two-dimensional electron gas between the two layers. As shown in Chapter 2, Section 2.4.5, eq. (7.1) can be expressed as

$$\frac{\partial}{\partial t} \rho_s(x, y, t) = j_1 - j_2 + \frac{\mu_\parallel}{\epsilon_1/L_1 + \epsilon_2/L_2} \nabla \cdot (|\rho_s| \nabla \rho_s), \tag{7.2}$$

where L_1 and L_2 are the layer widths.

The dynamic equation for U has been derived in Section 1.5.2, see also (2.63):

$$\frac{dU}{dt} = \frac{1}{C + C_{\text{ext}}} \left[I_0 - \int_A dx \, dy \left(\frac{C}{C_1} j_1 + \frac{C}{C_2} j_2 \right) \right], \tag{7.3}$$

where I_0 is the total current, $C_1^{-1} = \int_0^{L_1} dz \, (A\epsilon(z))^{-1}$ and $C_2^{-1} = \int_{L_1}^{L} dz \, (A\epsilon(z))^{-1}$ are the inverse intrinsic capacitances of the two layers, $C = 1/(C_1^{-1} + C_2^{-1})$ is the total intrinsic capacitance, and C_{ext} is the external parallel capacitance.

In Section 2.4.5 various physical models for the current densities $j_{1/2}(\rho_s, U)$ have been discussed (Wacker and Schöll 1992, 1994a). All of them yield an S-shaped current–voltage characteristic in a certain range of sample parameters, and *spatio-temporal spiking*, whereby a current filament temporarily switches on and off. In a simplified generic model, see (2.64) and (2.65), the constitutive equations in terms of dimensionless variables $u(t) = U/U_s$ and $a(x, t) = \rho_s/V_s$ are given by (Wacker and Schöll 1994b)

$$\frac{\partial}{\partial t} a(x, t) = f(a, u) + a''(x, t), \tag{7.4}$$

$$\frac{d}{dt} u(t) = \alpha (j_0 - \langle j(a, u) \rangle), \tag{7.5}$$

where we restrict ourselves to one transverse coordinate x (assuming a small lateral extension L_y in the y-direction), and $\langle j \rangle$ denotes the spatial mean value of the current density j over x:

$$\langle j \rangle = \frac{1}{L_x} \int_0^{L_x} j(x, t) \, dx,$$

L_x is the transverse device width, and $f(a, u)$ and $j(a, u)$ are given by

$$f(a, u) = \frac{u - a}{(u - a)^2 + 1} - Ta, \tag{7.6}$$

$$j(a, u) = u - a. \tag{7.7}$$

We shall now demonstrate that the equations (7.4) and (7.5) can be considered an activator–inhibitor system. The homogeneous steady states (denoted by a superscript

∗) are obtained by setting the temporal and spatial derivatives in (7.4) and (7.5) equal
to zero:

$$a^* = \frac{j_0}{\mathcal{T}(j_0^2 + 1)}, \quad u^* = j_0 + a^*. \tag{7.8}$$

From this we obtain the voltage versus current relation

$$u(j_0) = j_0 + \frac{j_0}{\mathcal{T}(j_0^2 + 1)}. \tag{7.9}$$

For $\mathcal{T} < \frac{1}{8}$ there is a regime of negative differential resistance

$$\frac{du}{dj_0} = 1 - \frac{j_0^2 - 1}{\mathcal{T}(j_0^2 + 1)^2} < 0, \tag{7.10}$$

where $j_0(u)$ exhibits bistability, i.e. we obtain an S-shaped current–voltage characteristic. We use $\mathcal{T} = 0.05$ throughout the following. In the range for which $du/dj_0 < 0$,
i.e. on the middle branch of the S-shaped $j_0(u)$ characteristic (Fig. 7.1(c)), the
following holds:

$$f_a \equiv \frac{\partial}{\partial a}\left(\frac{u - a}{(u - a)^2 + 1} - \mathcal{T}a\right)\bigg|_{(a^*,u^*)} = -\mathcal{T}\frac{du}{dj_0} > 0. \tag{7.11}$$

Hence, for fixed $u = u^*$, small homogeneous fluctuations δa from the homogeneous
steady state a^* will increase in time because

$$\frac{\partial \delta a}{\partial t} = f_a\, \delta a. \tag{7.12}$$

Therefore a acts as an activator. From

$$f_u \equiv \frac{\partial f}{\partial u}\bigg|_{(a^*,u^*)} = -\frac{j_0^2 - 1}{(j_0^2 + 1)^2} < \mathcal{T}\frac{du}{dj_0} < 0 \tag{7.13}$$

we find that u inhibits the increase of a. The production of the inhibitor u increases
with the activator concentration a and saturates with increasing u. Thus we have
derived an activator–inhibitor system with global coupling due to $\langle a \rangle$ in a form
common for various physical systems (Elmer 1992, Schimansky-Geier et al. 1991,
Mikhailov 1994).

Note that such global couplings as $\langle u - a \rangle$ are essential for stabilizing current
filaments, cf. Section 3.4. They are generally introduced in semiconductor charge
transport through the external circuit, which imposes a restriction on the total current.
They have been shown to give rise to a competition between several current filaments
similar to Ostwald ripening in equilibrium thermodynamics, i.e. if several filaments
are coupled via this nonlocal circuit interaction, the initially largest filament will
eventually survive (Schimansky-Geier et al. 1991, Kunz and Schöll 1992, Schwarz
et al. 2000a).

Current filamentation in S-shaped NDC semiconductor systems with *two* global constraints has also been studied (Gorbatyuk and Rodin 1992c, Gorbatyuk and Niedernostheide 1996, Meixner *et al.* 1998b). In modern thyristor-like devices driven by a spatially distributed gate electrode, an additional global constraint, besides the main supply circuit, is imposed by the control circuit and leads to nonlocal coupling between the device cross-sections. It provides means for the total stabilization of uniform states with NDC and essentially determines the filament width.

7.2.2 Elementary spatio-temporal patterns

In addition to the homogeneous stationary solutions, we obtain inhomogeneous solutions. In the following we use a transverse length $L_x = 40$ and Neumann boundary conditions for $a(x, t)$ (i.e. $a'(0, t) = a'(L_x, t) = 0$).

Figure 7.1(c) shows the stationary current–voltage characteristic $j_0(u)$. The stability of the homogeneous states is determined by a linear mode analysis. Observing the similarity of eq. (7.5) to (3.23), we can use the results of Section 3.2, setting $1/(RC) = \alpha$ and $j_U \equiv 0$. On the middle branch of the S-shaped characteristic a Hopf bifurcation occurs at $k = 0$, cf. (3.37), if $f_a - \alpha = 0$. Spatially homogeneous limit-cycle oscillations are possible for

$$f_a > \alpha, \tag{7.14}$$

which requires $\alpha < \frac{1}{8} - \mathcal{T}$. Additionally we find an instability against spatial fluctuations $\delta\rho_s(x, t) \sim \delta a \sim \cos(n\pi x/L_x)$, cf. (3.72), if

$$f_a > \left(\frac{n\pi}{L_x}\right)^2. \tag{7.15}$$

The mode with the longest wavelength ($n = 1$) becomes unstable first, and the instability arises on most of the middle branch of the homogeneous characteristic (independently from α). It leads to the bifurcation of an inhomogeneous spatial structure in the form of a half-filament as shown in the inset of Fig. 7.1(c). The branch corresponding to the current filament is also plotted on the current–voltage characteristic of Fig. 7.1(c), as a dashed curve. If the profile is symmetrically extended to negative x-values, i.e. if the system's size is doubled, a full filament in the usual sense is recovered without changing the physics. For an infinite one-dimensional system an equal-areas rule (Schöll 1987, Schöll and Landsberg 1988) for the existence of a current filament can be derived from the constitutive equation (7.4) in the steady state by integrating over da using the boundary conditions at $x = +\infty$ and $x = -\infty$ in the standard way, see eq. (3.7), as was illustrated in Chapter 4 for SNDC due to impurity breakdown. Here this yields

$$\int_{a_1}^{a_3} da \left(\frac{u_{co} - a}{(u_{co} - a)^2 + 1} - \mathcal{T}a\right) = 0. \tag{7.16}$$

For the coexistence voltage u_{co} at which two stable phases with a_1 and a_3 from the lower and upper branches of the $j_0(u)$ characteristic, respectively, coexist with a planar interface. we find $u_{co} = 9.41$, which is located at the almost vertical part of the filamentary branch of the characteristic on the left-hand side of the bistability range in Fig. 7.1(c).

The stability of the inhomogeneous branch may be determined numerically. We find that the inhomogeneous branch with positive differential conductance $dj_0/du > 0$ is unstable. It can be shown that this holds generally (Bass *et al.* 1970, Wacker and Schöll 1995), see Section 3.4. The inhomogeneous branch with $dj_0/du < 0$ is stable for $\alpha = 2$, which is typical for the HHED without external capacitance. The dynamic simulation shows that the system always tends to the homogeneous or inhomogeneous stationary state, so that no oscillatory behavior is found. Note that the stationary profiles do not depend upon α, but their stability in general does.

Considering lower values of α (i.e. applying an external capacitance), a Hopf bifurcation is found on the middle homogeneous branch for $\alpha < 0.075$. This can lead to homogeneous relaxation oscillations of the type described by Wacker and

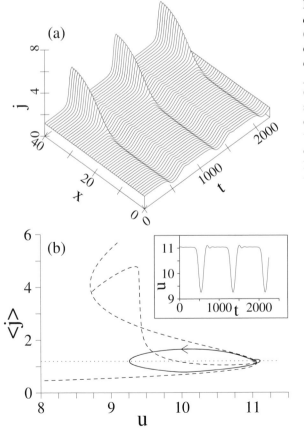

Figure 7.2. Spiking for $\alpha = 0.05$ and $j_0 = 1.2$. (a) The spatio-temporal evolution of the current density $j(x, t) = u(t) - a(x, t)$. (b) The phase portrait of the spatially averaged current density $\langle j \rangle$ versus the voltage u. The dotted line depicts $\langle j \rangle = j_0$ (the load line). The inset shows the voltage as a function of time. (After Wacker and Schöll (1994*b*).)

Schöll (1991) for the HHED. Additionally, another type of oscillatory behavior is found, as is shown in Fig. 7.2 for $\alpha = 0.05$, $L_x = 40$, and $j_0 = 1.2$. The cycle can be described as follows: We start at the stationary operating point on the middle branch of the homogeneous characteristic with a slight inhomogeneity in $a(x)$. Since the operating point is unstable against spatially inhomogeneous fluctuations, a current filament begins to nucleate. This filament does not attain a stationary state but disappears immediately and the initial state is reached again. We call this behavior *spiking*.

This behavior sets in at $j_0 = j_{0C} = 1.138$ when the homogeneous stationary state becomes unstable against inhomogeneous fluctuations. For small $j_0 - j_{0C} > 0$ the time interval between the spikes is very long because the growth of the inhomogeneity is very slow. Upon increasing the control parameter j_0 the repetition frequency of the spikes becomes larger, but their individual shape remains unchanged. For $j_0 \geq j_2 \approx 1.295$ the spiking behavior disappears and the system tends toward the stable inhomogeneous filamentary state. It is interesting to note that the homogeneous stationary state is stable against *homogeneous* fluctuations in this range of j_0. The Hopf bifurcation leading to homogeneous relaxation oscillations does not occur until $j_0 = 1.328$ for the parameter $\alpha = 0.05$. So spatio-temporal oscillations (spiking) may occur without an oscillatory instability of the *homogeneous* system. The inhomogeneous branch is stable for $j_0 > j_1 \approx 1.245$, at which a subcritical Hopf bifurcation occurs. Thus, in the range $1.24 < j_0 < 1.295$, bistability between a stationary filamentary distribution and spiking appears.

Similar behavior can be found for various parameters (α and L_x). For $\alpha = 0.035$ we find spiking like in Fig. 7.2 for $1.138 < j_0 < 1.247$. At this value the homogeneous state becomes unstable against homogeneous fluctuations via a Hopf bifurcation and so we find distinct intermediate oscillations (which are almost homogeneous) between the spikes. The interaction of the frequency of the intermediate oscillations with the growth of the spikes leads to period doubling at $j_0 = 1.251$, as shown in Fig. 7.3. Upon further increasing j_0 a period-doubling scenario leading to chaos is found. Figure 7.4 displays typical chaotic behavior.

To analyze this situation we have plotted the first-return map in Fig. 7.4(c). This was obtained by taking the value $da(n) = \log_{10} |a(0, t_n) - a(L_x, t_n)|$ for the times t_n when $u(t)$ reaches a local minimum. The return map consists in three discontinuous parts. First there is a nearly straight line for $da < -0.5$, for which the inhomogeneity grows in each cycle. This corresponds to the almost homogeneous oscillations. Increasing da leads to a sudden increase up to $da \approx 0.5$ on a second branch of the map when the spike appears. Afterwards there is a reinjection mechanism on the third branch, reducing da significantly toward the first branch. Increasing the control parameter j_0 leads to longer series of intermediate oscillations until the homogeneous oscillatory state becomes stable at $j_0 = 1.328$.

Note that the spatial inhomogeneity becomes very small between the spikes. Therefore spatial fluctuations due to noise should play a significant role in determining the growth of the spikes in real systems. This leads to irregular time intervals Δt between the spikes and might be a stronger effect than the deterministic chaotic mechanism discussed above. For stochastic fluctuations general relations for the mean value of Δt and its standard deviation as functions of the control parameter have been derived (Wacker and Schöll 1992, Wacker 1993).

Some systematics can be brought into the spatio-temporal dynamic behavior by noting that several instabilities are competing in the system, depending upon the values of L_x, j_0, and α (Bose *et al.* 2000).

■ An oscillatory instability of the uniform steady state (Hopf bifurcation) for

$$\alpha < \alpha_u(j_0) \equiv f_a. \tag{7.17}$$

■ A spatial instability of the uniform steady state (cf. Section 3.3.2, eq. (3.72)), leading to a stationary filament for

$$\left(\frac{\pi}{L_x}\right)^2 < \alpha_u(j_0). \tag{7.18}$$

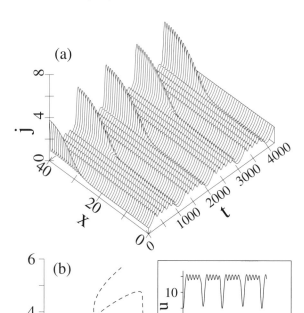

Figure 7.3. Period doubling of spiking for $\alpha = 0.035$ and $j_0 = 1.258$. (a) The spatio-temporal evolution of the current density $j(x, t)$. (b) The phase portrait of the spatially averaged current density $\langle j \rangle$ versus the voltage u. The inset shows the voltage as a function of time. (After Wacker and Schöll (1994*b*).)

▓ An oscillatory instability of the stationary filament (cf. Section 3.4.2) for

$$\alpha < \alpha_f(j_0, L_x) \equiv \lambda_1, \tag{7.19}$$

where $\lambda_1 > 0$ is the critical eigenvalue of the filament (Alekseev *et al.* 1998).

For sufficiently large L_x, e.g. for $L_x = 40$, (7.18) is satisfied for $j_0 > j_{0C}$, and the uniform steady state is always unstable against spatial fluctuations. Two cases may arise, viz.

$$\alpha_f(j_0, L_x) < \alpha_u(j_0), \tag{7.20}$$

in which case there is bistability between a stationary filament and uniform relaxation oscillations for $\alpha_f < \alpha < \alpha_u$, and

$$\alpha_u(j_0) < \alpha_f(j_0, L_x), \tag{7.21}$$

in which case, for $\alpha_u < \alpha < \alpha_f$, the uniform state is spatially unstable and the stationary filament is temporally unstable, but no oscillatory instability of the uniform state arises. Thus the system has no stationary attractor, and mixed spatio-temporal modes are expected. If, additionally, the conditions for temporal, (7.17), and spatial,

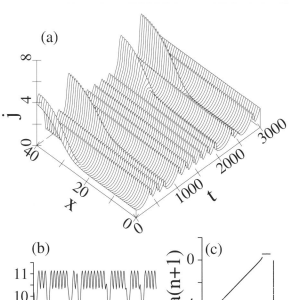

Figure 7.4. Chaotic spiking for $\alpha = 0.035$ and $j_0 = 1.31$. (a) The spatio-temporal evolution of the current density $j(x, t)$. (b) The voltage as a function of time. (c) The return map for the quantity $da(n) = \log_{10} |a(0, t_n) - a(L_x, t_n)|$ at times t_n when $u(t)$ reaches its minimum. (After Wacker and Schöll (1994*b*).)

(7.18), instabilities of the uniform state coincide in a codimension-two bifurcation point,

$$\alpha = \alpha_u = \left(\frac{\pi}{L_x}\right)^2, \tag{7.22}$$

then its vicinity is the regime in which spatio-temporal spiking and complex dynamics including spatio-temporal chaos are likely to occur. Of course, the linear stability analysis can not determine whether the resulting bifurcations are subcritcal or supercritical. The analysis carries over to two spatial dimensions, but then the condition (7.19) becomes different for corner filaments and edge current layers (Bose *et al.* 2000).

7.2.3   Chaotic bifurcation scenarios

We shall now develop a comprehensive phase diagram of the various dynamic regimes which occur for particular values of the parameters, and provide an analysis of the bifurcation scenarios (Bose *et al.* 1994). Depending on the parameter α, i.e. the magnitude of C_{ext}, on the external current j_0, and on initial conditions, various spatio-temporal patterns are found in the time-dependent simulation, as shown in Fig. 7.5. Apart from homogeneous stable states, one can distinguish homogeneous oscillations (a), stable stationary filaments (b), and spiking, which may be periodic (c) or chaotic (d).

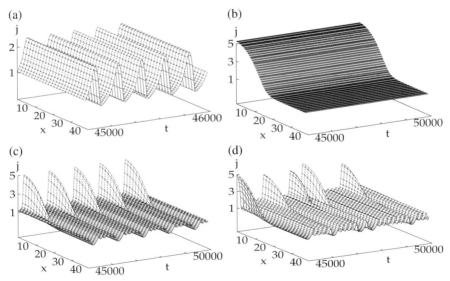

Figure 7.5. Spatio-temporal patterns $j(x, t)$ of the generic model for $\alpha = 0.035$. (a) Homogeneous oscillations ($j_0 = 1.40$). (b) A stationary filament ($j_0 = 2.0$). (c) Periodic spiking ($j_0 = 1.24$). (d) Chaotic spiking ($j_0 = 1.31$). (After Bose *et al.* (1994).)

A phase diagram of the (j_0, α)-plane indicating the regimes in which a particular behavior dominates is provided in Fig. 7.6. It can clearly be seen that the spiking behavior sets in at $j_0 = 1.138$, at which the homogeneous branch becomes unstable against fluctuations of the shape $\delta a(x) \sim \cos(\pi x/L_x)$. In the given parameter range, first an inhomogeneous distribution forms within a short time. This distribution breaks down immediately, a nearly homogeneous state is reached again and the same cycle is repeated. The period of the oscillations is mainly determined by the growth of the inhomogeneity, which is very sensitive to noise (Wacker 1993). Upon increasing j_0 the spiking disappears via one of two different scenarios, depending on the value of α.

For larger α the spiking mode vanishes via a subcritical Hopf bifurcation and the system switches to the stable filament (Bose *et al.* 1994). This is shown in Fig. 7.7, in which the minimum inhomogeneity $\delta a = \min_t |a(0,t) - a(L_x, t)|$ is plotted versus the control parameter. There is bistability between the spiking (s) and the stationary filamentary branch (f) in the voltage interval $1.24 < j_0 < 1.295$ for $\alpha = 0.05$, as mentioned above. Upon decreasing j_0, starting on the filamentary branch, a subcritical Hopf bifurcation at $j_0 = 1.24$ is found. The unstable limit cycle which bifurcates from there is denoted by (u). Its spatio-temporal structure can be visualized for a short time by starting with appropriate initial conditions (see Fig. 7.8(a)); it has the form of a breathing filament. The dotted line (u) between the two stable branches schematically marks the boundary between the two basins of attraction of s and f. Dynamically, the boundary between the attractor basins represents a periodic repeller. In a spatially averaged $(\langle j \rangle, u)$ phase portrait the periodic repeller represents an unstable limit cycle,

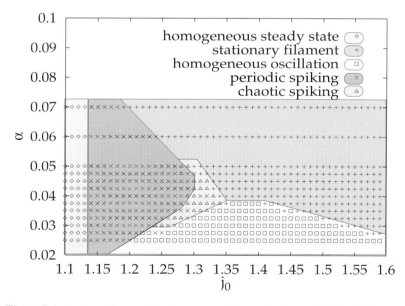

Figure 7.6. A phase diagram of various asymptotic spatio-temporal states in the (j_0, α) control parameter space (after Bose *et al.* (1994)).

whereas the stationary filament and the periodic spiking correspond to a stable fixed point and a stable limit cycle, respectively. They are shown in Fig. 7.8(b) together with two transient trajectories (dotted) corresponding to two different initial conditions near the unstable limit cycle. One tends to the spiking attractor and the other to a stationary filament.

For lower values of α, with increasing j_0 the spiking gives way to spatially homogeneous periodic relaxation oscillations. These are generated by a Hopf bifurcation of the intermediate homogeneous state corresponding to the NDC branch, as discussed in the context of Fig. 7.3 above. For values of α between these extremes there occurs chaotic behavior, which seems to be caused by the interplay of these effects.

We shall now study the chaotic behavior of spatio-temporal spiking in more detail. Figure 7.9 depicts the Fourier spectrum of $u(t)$ for the dynamic states shown in Figs. 7.5(c) and (d). The sharp spectrum in Fig. 7.9(a) reflects periodic spiking. The broad-band noise in Fig. 7.9(b) indicates that the irregular spiking behavior is indeed due to deterministic chaos, not to quasiperiodicity.

Only the spikes and their amplitudes are considered in the next plot. Starting with a small cosine-shaped inhomogeneity, the model is simulated for various values of j_0, discarding transients. In Fig. 7.10 the minima of $u(t)$ corresponding to the current spikes are plotted versus the control parameter j_0. Period-doubling scenarios occur on

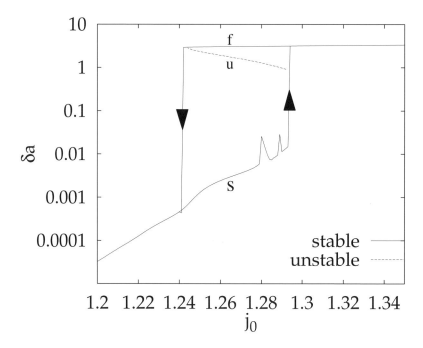

Figure 7.7. Hysteresis behavior between spiking (s) and the stable filament (f) with an unstable limit cycle (u) as separatrix. The various cycles are identified by the minimum δa of $|a(0, t) - a(L_x, t)|$ in time ($\alpha = 0.05$). (After Bose *et al.* (1994).)

both sides of the j_0 range shown. Several periodic windows can be identified in the chaotic regime. It is interesting to note the abrupt crisis-like blow-up of the chaotic band at $j_0 = 1.301$ and $j_0 = 1.317$, where an unstable periodic orbit appears to collide with the boundary of the chaotic attractor.

In Fig. 7.11(b) the period-doubling scenario for a different value of α is shown. Also the chaotic band "blows up" at $j_0 = 1.28$. We can identify several periodic windows in the chaotic regime, for example there is a period-3 window around $j_0 = 1.264$. Additionally, an intermittency scenario occurs with decreasing j_0 at $j_{0c} \approx 1.267$. To determine the type of intermittency we have examined the distribution of the laminar lengths l, i.e. the lengths of the time intervals during which spiking is periodic, and the scaling behavior of the mean laminar length $\langle l \rangle$ as functions of the control parameter j_0. Both the characteristic U-shaped laminar length distribution in Fig. 7.12(a) and the scaling of $\langle l \rangle^{-2} \sim (j_{0c} - j_0)$ in Fig. 7.12(b) indicate type-I intermittency.

(a)

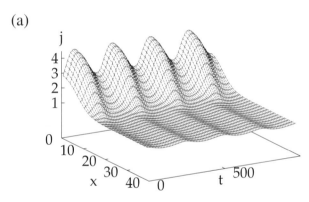

(b)

Figure 7.8. Dynamic behavior for $\alpha = 0.05$ and $j_0 = 1.27$. (a) The spatio-temporal evolution of the current density $j(x, t)$. (b) The phase diagram of $\langle j \rangle$ versus u, showing two different trajectories starting near the unstable pulsating filament (unstable limit cycle u) and tending toward the stable stationary filament (fixed point marked ×) and the stable spiking state (s), respectively. (After Bose *et al.* (1994).)

As further characterization of the chaotic behavior we have calculated the Lyapunov exponents. In general the Lyapunov exponents of an n-dimensional ordinary differential equation (ODE) are defined by the temporal development of an infinitesimal n-sphere. After some time t this n-sphere will become an n-ellipsoid. The ith Lyapunov exponent λ_i is then determined by the ith principal axis $p_i(t)$ as

$$\lambda_i = \lim_{t \to \infty} \frac{1}{t} \ln\left(\frac{p_i(t)}{p_i(0)}\right).$$

There are several numerical difficulties in the computation of the Lyapunov exponents directly following this definition. To avoid these Benettin *et al.* (1980) and

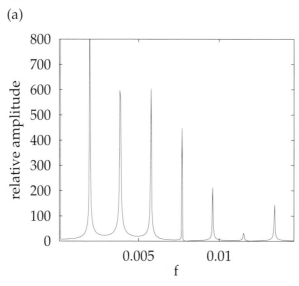

(a)

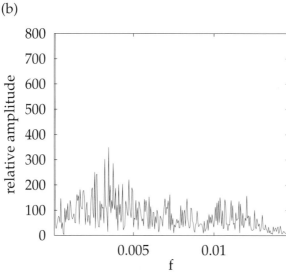

(b)

Figure 7.9. Fourier spectra of $u(t)$ for $\alpha = 0.035$ and (a) $j_0 = 1.24$ and (b) $j_0 = 1.31$. The Fourier transform was taken in the interval $t = 50\,000$ to $t = 100\,000$. (After Bose *et al.* (1994).)

independently Shimada and Nagashima (1979) introduced a method using not only
the phase space but also the tangent space. A fiducial trajectory (the center of the
n-sphere) is described by the system of ODEs and the trajectories of points on the
surface of the sphere are described by the linearized system. In particular, the principal
axes are defined by the action of the linearized equations on an initially orthonormal
vector frame anchored to the fiducial trajectory.

Still two problems arise. The principal axis vectors diverge in magnitude and the
vectors also tend to fall along the direction of the most rapid growth, so that they
become indistinguishable by numerical methods. However, these two problems can
be eliminated by the re-orthonormalization of the vector frame from time to time.
Here the Gram–Schmidt orthonormalization procedure may be used. If s is the time
between two orthonormalizations, the Lyapunov exponents are now given by

$$\lambda_i = \lim_{k \to \infty} \frac{1}{ks} \sum_{l=1}^{k} \ln \alpha_l^{(i)},$$

where $\alpha_l^{(i)}$ is the growth of the ith principal axis vector between the $(l-1)$th and
the lth re-orthonormalization. Since the method described holds only for ODEs, we
discretize $v(x, t)$ at N equally spaced points. In the following we use $N = 25$.

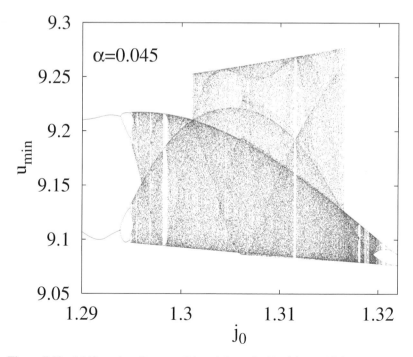

Figure 7.10. A bifurcation diagram of the minima of $u(t)$ with $u < 10$ for
$\alpha = 0.045$ as a function of j_0 (after Bose *et al.* (1994)).

Figure 7.11(a) shows the two largest Lyapunov exponents (including zero) in the interval $1.24 \leq j_0 \leq 1.33$. Our system displays periodic or chaotic spiking for these

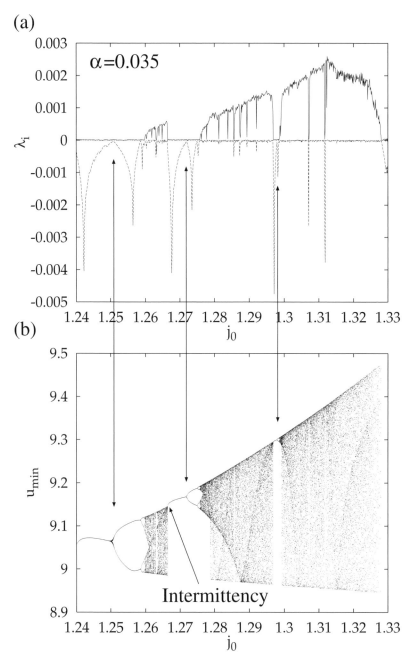

Figure 7.11. (a) The Lyapunov spectrum and (b) the bifurcation diagram for $\alpha = 0.035$ (after Bose *et al.* (1994)).

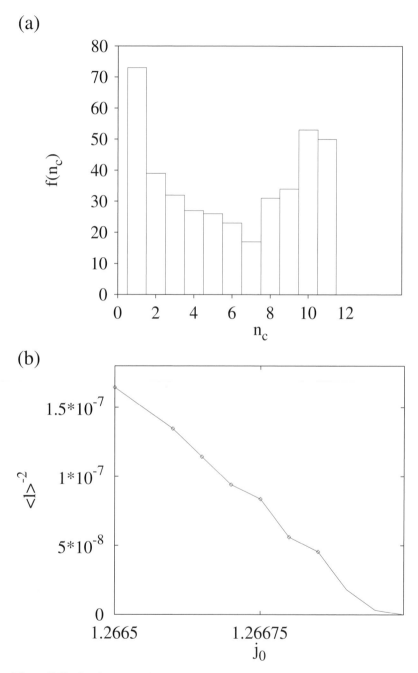

Figure 7.12. Spatio-temporal intermittency. (a) The distribution of the laminar lengths $f(n_c)$ for $j_0 = 1.2666$, where the laminar lengths are described by the number of periodic cycles n_c. (b) The scaling of the mean laminar length $\langle l \rangle$ as a function of the control parameter j_0 ($\alpha = 0.035$). (After Bose *et al.* (1994).)

parameters. For the periodic case the maximum exponent is zero, whereas chaotic behavior is characterized by at least one positive exponent. At the period-doubling bifurcations the second Lyapunov exponent also vanishes. The Lyapunov spectrum closely matches the bifurcation diagram in Fig. 7.11(b).

According to Kaplan and Yorke (1979) the fractal dimension of a chaotic attractor d_f can be estimated as

$$d_f = j + \frac{1}{|\lambda_{j+1}|} \sum_{i=1}^{j} \lambda_i,$$

where λ_i are the Lyapunov exponents in decreasing sequence, and j is defined by

$$\sum_{i=1}^{j} \lambda_i > 0, \qquad \sum_{i=1}^{j+1} \lambda_i < 0.$$

For the chaotic regime of our system $j = 2$ holds, so we need the three largest exponents in order to calculate the dimension. With the same parameters as those in Fig. 7.11(a), we get a fractal dimension slightly larger than 2, e.g. 2.023 for $j_0 = 1.266$ and 2.094 for $j_0 = 1.313$. This shows that, although spatio-temporal degrees of freedom are excited in our system, chaotic spiking is associated with a *low-dimensional* chaotic attractor. This is corroborated by a further analysis applying an expansion of the spatio-temporal data in terms of optimally adapted eigenmodes: a so-called Karhunen–Loève decomposition. This technique will be discussed in a general context in Section 7.3.4 below.

In conclusion, the transition from periodic to chaotic behavior occurs either via period doubling or by an intermittent scenario. Because the chaotic behavior can be described by only a few degrees of freedom, this model does not exhibit high-dimensional spatio-temporal chaos.

7.2.4   Comparison with experiments

Bistability between spiking and a stationary filament occurs in a range of control parameters and leads to hysteretic scenarios if the external current is swept up and down. Note that spiking has been found only if the operating point (i.e. the intersection of the homogeneous $j(u)$ characteristic with $j = j_0$) is located at low j_0 near the right-hand end of the SNDC range in Fig. 7.1, which is related to the condition (7.21) $\alpha_u(j_0) < \alpha_f(j_0, L_x)$ for a mixed spatio-temporal instability in the vicinity of the codimension-two bifurcation $\alpha = \alpha_u = (\pi/L_x)^2$ discussed in Section 7.2.2. In Section 7.3 we shall further elaborate on the question of when spatio-temporal spiking appears in a slightly different type of reaction–diffusion system, namely with *local* coupling, and find that its occurrence is closely related to a codimension-two Turing–Hopf bifurcation and the mixing of spatial (Turing) and temporal (Hopf) symmetry-breaking instabilities.

Experimentally, the spatio-temporal spiking behavior has been observed in various semiconductor devices exhibiting S-shaped current–voltage characteristics. These observations seem to be in good agreement with the spiking behavior of our generic model. Thus, it is plausible that it should be possible to describe these devices by a generic reaction–diffusion system of the type (7.4) and (7.5) for the relevant dynamic variables.

Unfortunately, the spatio-temporal behavior of the heterostructure hot-electron diode has not yet been investigated experimentally, due to the small time and length scales involved, and because of problems due to the large power dissipation of the magnitude $P/A \sim 10 I_s U_s/A = 400$ kW cm^{-2} in this structure. However, Spangler et al. (1992) investigated the impact-ionization impurity breakdown in n-GaAs. They observed that large-amplitude voltage oscillations occur just above the impact-ionization threshold. These oscillations are strictly periodic and rise in frequency up to about 5 MHz with increasing average current. They result from repetitive ignition and extinction of a filament that proceeds with increasing average current into a stable filamentary current flow. Brandl and Prettl (1991) have called this behavior *flashing filaments*.

The stochastic appearance of spikes in the pre-breakdown regime of impurity-impact ionization in p-Ge was mentioned by Rau et al. (1991). They reported that the first occurrence of spontaneous oscillations is connected to the onset of breakdown, but still appears in the positive-differential-resistance region. Here the oscillation seems to be stochastic. With increasing bias voltage it becomes more and more regular without changing its shape or amplitude and the frequency increases. By comparison with stationary filamentary states, the authors conclude that these oscillations are the result of the *firing* of a filament, not the result of the breakdown of the whole sample during a short period of time.

In a semiconductor gas-discharge system filaments switching on and off and moving in an irregular manner have been observed experimentally and described in terms of a two-layer model (Dohmen 1991). Symanczyk et al. (1991a) reported observing spontaneous current and voltage oscillations close to the first jump in the current–voltage characteristic of Si p–i–n diodes. By measuring the spatial distribution of the surface potential, which is connected with the local current density, they proved that these oscillations are caused by current filaments switching on and off. Spiking has also been observed in GaAs p–i–n diodes (Niedernostheide 1995).

Multifilamentary current distributions and current oscillations in p–n–p–i–n diodes exhibiting SNDC due to filaments switching periodically on and off have been reported by Niedernostheide et al. (1992a). The static current–voltage characteristics have been explained in detail by a transport model (Gorbatyuk and Niedernostheide 1996), see Section 2.4.7. We shall now briefly review recent experiments on silicon p^{+}–n^{+}–p–n^{-}–n^{+} diodes (Niedernostheide et al. 1996b) and draw a comparison with the predictions following from the reaction–diffusion model (7.4) and (7.5).

The main bifurcation parameters are the external dc driving voltage V_s and a capacitance C connected parallel to the sample and determining the temporal evolution of the device voltage. By means of spatially and temporally resolved measurements of the recombination radiation, it has been shown that there is a strong correlation between the *global* current and voltage oscillations and the *local* dynamics of the current-density distribution in the form of spiking filaments. This makes it possible to provide a systematic analysis of the spatio-temporal behavior of the current density as a function of the two parameters V_s and C.

The device is connected to a dc voltage source V_s via a load resistor of resistance R_0 in such a way that the p^+ layer is positively biased with respect to the n^- layer. Thus, the middle n^+–p junction is reverse-biased and causes a low-current branch in the current–voltage characteristic $I(V)$ (Fig. 7.13). When the device's voltage reaches values of ≈ 50 V, the electric field in the space-charge region of the n^+–p junction is sufficiently large that impact ionization takes place, leading to an autocatalytic increase in the number of charge carriers. The charge carriers are immediately separated in the high-field zone and cause an additional carrier injection when they reach the outer p–n junctions. This leads to the regime of negative differential resistance in the $I(V)$ characteristic, as shown in Fig. 7.13. For sufficiently large currents, the differential resistance becomes positive again, leading to an S-shaped $I(V)$ characteristic. From other measurements (Niedernostheide *et al.* 1992a, 1994a) it is known that the current-density distribution in the low-current branch of the $I(V)$

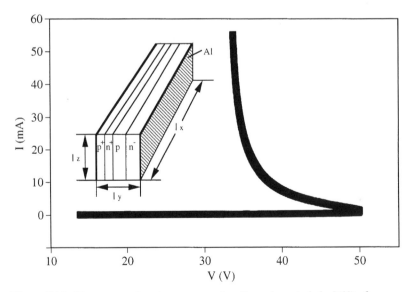

Figure 7.13. The measured stationary current–voltage characteristic $I(V)$ of a p–n–p–n diode; $R_0 = 20$ kΩ. The inset shows the typical geometry of a diode. All experimental results presented refer to a sample with $l_x = 3.6$ mm, $l_y = 0.87$ mm, and $l_z = 0.84$ mm. (After Niedernostheide *et al.* (1996b).)

characteristic is uniform; on proceeding along the branch with negative differential resistance, increasing the current, the current-density distribution contracts to a current filament. The contraction phase is typically completed for a current of ≈20 mA; then a current filament with a well-defined width is stable.

The system's behavior changes drastically when a sufficiently large capacitance ($C \geq 5$ nF) is connected parallel to the device. In this case one observes spontaneously arising oscillations of the current and voltage of the device. A typical scenario for increasing the voltage V_s can be described as follows: When V_s exceeds a critical value $V_{s,c1}$, small-amplitude oscillations with a frequency f of about 4.3 kHz appear at $V_{s,c1}$. The amplitude of the oscillations rises continuously from zero with increasing V_s, indicating a supercritical Hopf bifurcation. At a second critical value, $V_{s,c2}$ large-amplitude oscillations are superimposed on the small-amplitude oscillations. These oscillations vanish when the voltage V_s exceeds a third critical value, $V_{s,c3}$. When the voltage is decreased from large values, the large-amplitude oscillations appear at a value $V_{s,c4} < V_{s,c3}$. Thus, there is a hysteretic behavior indicating a subcritical bifurcation. When the voltage is further decreased to $V_{s,c2}$, the small-amplitude oscillations appear instead of the large-amplitude oscillations, and at $V_{s,c1}$ the oscillations vanish completely. Note that, within experimental error, no hint of hysteretic behavior has been found either for the transition at $V_{s,c2}$ or for that at $V_{s,c1}$.

In order to gain insight into the spatio-temporal behavior of the current-density distribution, the recombination radiation was measured by using a gated high-sensitivity near-infrared video camera with minimal gate times, down to 10 ns. In all measurements one of the x–y-planes of the sample (cf. Fig. 7.13) was focussed onto the photocathode of the camera. Since the recombination radiation of silicon is rather weak, it is necessary to integrate about several hundreds to thousands of single shots. For that purpose the self-sustained voltage oscillations have been used to generate a trigger pulse, the delay and width of which could be adjusted in such a way that it was possible to image the radiation distribution at any stage of the period with a suitable gate time.

In Figs. 7.14(a)–(c) time traces of the device voltage $V(t)$, the sample current $I(t)$, and the I–V phase portrait for a typical small-amplitude oscillation are depicted. Figures 7.14(d) and (e) show the measured distributions of the recombination radiation at two different times. They correspond to the times at which the current trace reaches its maximum and minimum, respectively. The white lines mark the imaged ($l_x \times l_y$) area of the sample. The main direction of current flow is from the right to the left. At the top of Fig. 7.14(d) the sequence of the multilayered structure is indicated schematically. In both camera records the light intensity is distributed uniformly. However, in Fig. 7.14(e) the light intensity is located mainly near the anode, whereas in Fig. 7.14(d) a nearly uniform emission near the cathode can also be recognized and the intensity of light near the anode is greater than that in Fig. 7.14(e). Thus, one may conclude that the light-intensity distribution performs a homogeneous relaxation oscillation. Taking the light-intensity distribution as a first approximation for the current-density distribution, this in turn implies that the small-amplitude oscillations are connected with homogeneous oscillations of the current density.

Figure 7.15 elucidates the case of large-amplitude oscillations. While $V(t)$ still changes quite slowly in the course of time (Fig. 7.15(a)), the sample current $I(t)$ (Fig. 7.15(b)) exhibits sharp peaks with a width of about 5 µs. The I–V phase portrait (Fig. 7.15(c)) reveals that the oscillation covers a large area in this plane and, in particular, encloses the branch with the negative differential resistance of the stationary $I(V)$ characteristic (cf. Fig. 7.13). The camera records depicted in Figs. 7.15(d)–(f) indicate that the peak appearing in the time trace $I(t)$ is caused by a spiking filament. Figure 7.15(d) shows the light-intensity distribution when

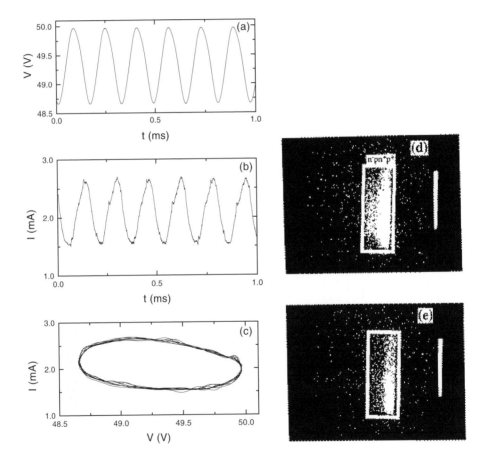

Figure 7.14. Measured time traces of the device voltage $V(t)$ (a) and the sample current $I(t)$ (b), and the corresponding phase portrait of I versus V (c); the camera records show the emitted light-intensity distributions at the maximum (d) and minimum (e) of $I(t)$, indicating that this is a homogeneous relaxation oscillation. The parameters are $R_0 = 10$ kΩ, $C = 0.02$ µF, gate time 10 µs, and number of integrated single shots 33 167. (After Niedernostheide *et al.* (1996*b*).)

the sample current is maximal. Apparently, a localized current filament has been formed. The strong current through the filament causes a rapid partial discharge of the capacitance C. The record shown in Fig. 7.15(e) was taken 10 μs after the current peak. The filament is still observable; however, the light intensity is drastically reduced in comparison with Fig. 7.15(d), indicating that the filament fades away. 60 μs after the generation, the filament is completely extinguished (Fig. 7.15(f)). It rises again with the next current peak. When the number of integrated single shots is increased, one finds that, during the low-current phases, in which no filament is present, a weak uniform light-intensity distribution is observable, indicating a uniform low-current-density state. Thus, these data imply that the periodic large-amplitude oscillations emanate from a periodically spiking filament that fades away immediately

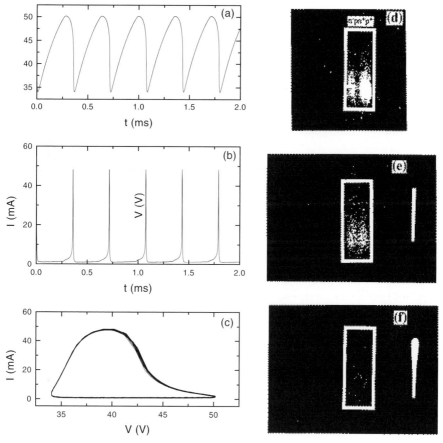

Figure 7.15. Measured time traces of the device voltage $V(t)$ (a) and the sample current $I(t)$ (b), and the corresponding phase portrait of I versus V (c); the camera records show the emitted light-intensity distributions at the maximum (d) of $I(t)$ and with delays of 10 μs (e) and 60 μs (f) with respect to (d) and indicate that this is a spiking filament. The parameters are $R_0 = 20$ kΩ, $C = 0.02$ μF, gate time 1 μs, and number of integrated single shots 27 850. (After Niedernostheide *et al.* (1996*b*).)

after it has been nucleated; during the recovery time the current density relaxes to a uniform state from which a new spiking filament is generated in the next cycle.

By using the results that the small-amplitude and large-amplitude oscillations correspond to homogeneous relaxation oscillations and spiking filaments, respectively, we are now able to provide a systematic classification of the spatio-temporal behavior of the current density connected with these oscillations in the device under consideration. In Fig. 7.16 the bifurcation points, determined by evaluating time series of the sample current as described above, in the C–V_s control-parameter space are indicated by various symbols. The lines are introduced to guide the eye and to separate the regions in which different behaviors have been observed. Clearly, one can distinguish five different regions. For low values of V_s the system realizes stable uniform current-density distributions, which become unstable in favor of uniform relaxation oscillations or a spiking filament when V_s exceeds $V_{s,c1}$. The former are stable only in a small voltage interval and also transform to spiking filaments at $V_s = V_{s,c2}$. Periodically spiking filaments can be stabilized in a wide parameter range. In the region that is bounded by the lines through the triangles and the rhombi, bistability of spiking and static filaments has been found. For even larger voltages only the latter are stable. Note that the width of the bistable region where a static and a spiking filament can be stabilized at the same value of V_s decreases with increasing C.

The transition from the homogeneous relaxation oscillation to the periodically spiking filament takes place in a very small voltage interval. A detailed analysis of the time traces $I(t)$ at this transition reveals that the transition consists in two stages.

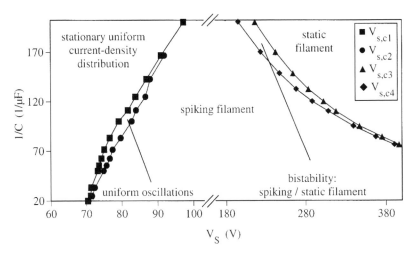

Figure 7.16. Classification of the experimentally observed spatio-temporal behavior in the C–V_s control parameter space for $R_0 = 20$ kΩ. The symbols mark experimentally determined points of bifurcation from uniform current-density distributions to uniform oscillations or spiking filaments (squares), from uniform oscillations to spiking filaments (circles), and from spiking to static filaments (triangles) or vice versa (rhombi). (After Niedernostheide *et al.* (1996*b*).)

At voltages just below the appearance of spikes, the amplitude of the homogeneous oscillations exhibits small fluctuations, indicating that the spatially uniform oscillation becomes temporally irregular. With increasing voltage V_s the amplitude fluctuations of the small-amplitude oscillations increase and the onset of the spiking regime is characterized by the occasional appearance of large-amplitude current spikes in an irregular manner, as shown in the time trace of $I(t)$ in Fig. 7.17. The duration of the phases in which the current density oscillates homogeneously varies statistically. With increasing V_s, the mean duration of these phases decreases and, finally, a periodic signal develops. The strong correlation between the appearance of the peaks of $I(t)$ and the nucleation of a spiking filament in the case of periodic large-amplitude oscillations suggests that the peaks appearing irregularly in $I(t)$ are connected with complex nonperiodic spatio-temporal spiking of a filament.

We shall now draw a comparison with the model (7.4) and (7.5). First of all we note that the type of the homogeneous oscillations (Fig. 7.5(a)) is in good qualitative agreement with the experimental data in Fig. 7.14. The spiking behavior depicted in Fig. 7.5(c) exhibits a spatio-temporal current distribution, which is in agreement with the behavior of the observed recombination radiation in Fig. 7.15 if we assume that the intensity of the radiation is related to the local current density.

Nevertheless, the temporal behaviors both of the current and of the voltage in the case of spiking differ with respect to the following points. The shape of the oscillations is different from that observed in the experiment, in which the current exhibits extremely sharp peaks, while the voltage changes more smoothly. These differences may be caused by the detailed forms of the functions $f(a, u)$ and $j(a, u)$. Second, the theoretical results exhibit long phases between the spikes, during which both $\langle j \rangle$ and u are almost constant. During these phases the phase trajectory is very close to the homogeneous fixed point of the model equation, so that the growth of the inhomogeneity takes a long time. This growth process is extremely sensitive to noise

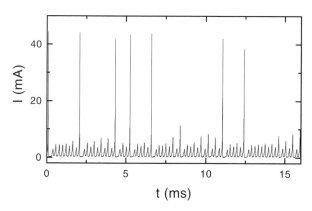

Figure 7.17. The measured nonperiodic time trace $I(t)$ near the transition from homogeneous relaxation oscillations to a periodically spiking filament. The parameters are $R_0 = 20\,\mathrm{k\Omega}$, $C = 0.02\,\mathrm{\mu F}$, $V_s = 76.75\,\mathrm{V}$. (After Niedernostheide *et al.* (1996*b*).)

or spatial inhomogeneities inside the sample, since these destroy the homogeneity of the fixed point.

Now we compare the bifurcation scenarios of the generic model (Fig. 7.6) and the experiment (Fig. 7.16). Here we have to identify $\alpha \sim 1/C$ and $j_0 \sim V_s$. The experimentally observed transition from the spiking to the static filamentary state with increasing bias V_s (the right-hand side of Fig. 7.16) is in good agreement with the simulation of the generic model. In both cases a certain range of bistability between these two states is found. The theory (Bose *et al.* 1994) predicts that the static filament becomes unstable via a subcritical Hopf bifurcation. In the experiment one finds a spiraling out of the $j(t)$ signal from the old fixed point on decreasing the voltage over $V_{s,c4}$, which is in good agreement with there being a subcritical Hopf bifurcation.

The left-hand side of the experimental bifurcation diagram (Fig. 7.16) shows that the spiking behavior is generated from the homogeneous steady state via a small regime of homogeneous oscillations that occur for a lower bias. This is in contrast to the result from the generic model, in which the spiking behavior is generated directly from the stationary homogeneous steady state, which becomes unstable against pattern formation. We suggest that this difference may be related to the following two aspects.

One important aspect could be the fact that the internal capacitance C_2 of the device is not a fixed quantity but strongly depends on the voltage and the state in which the diode is operated. As soon as the autocatalytic process becomes effective the sample current strongly increases and causes changes in the voltage drops across the p–n junctions, leading to an increase of the junction capacitances. Because the mean sample current increases with increasing V_s, the capacitance C_2 should effectively increase with increasing V_s. With a fixed external capacitance C, the quantity $\alpha \approx C_2/C$ should then increase with the external current j_0. This means that the path through the bifurcation diagram might not be a horizontal line at fixed α with increasing j_0 but could rather be a more complicated curve that successively passes through the various regimes in Fig. 7.6. By following such a curve we can successively find a homogeneous stationary state, homogeneous oscillations, spiking, and a stationary filament with increasing j_0, which is completely analogous to the experimentally observed behavior. Additionally, the functions $f(a, u)$ and $j(a, u)$ may look different for the p–n–p–n diode investigated experimentally. This may lead to a deformation of the phase diagram and may explain the different widths of the regimes, while the topological properties should be conserved. At the transition between the homogeneous oscillations and the periodic spiking the generic model exhibits chaotic behavior. This could be connected with the irregular oscillations observed experimentally at this transition point (Fig. 7.17).

In conclusion, all experimentally observed current-density patterns in the parameter range investigated can be consistently explained by the simple reaction–diffusion model. It has been shown that, besides uniform relaxation oscillation, periodically spiking filaments may spontaneously appear in a semiconductor device. The experimental results give a clear indication that, in a certain parameter range, also irregularly spiking filaments with relatively quiescent phases of spatially uniform but temporally irregular oscillations occur close to the transition to stable uniform oscillations. As a further important result, bistability between a spiking and a static filament causes hysteresis when the bias voltage source is swept up and down. This feature is a clear-cut distinction from the scenario of circuit-limited oscillations which was proposed for the oscillations in p-Ge (Rau *et al.* 1991). The trajectories of these relaxation-type oscillations cross through the filamentary branch and would stay there if the filament were stable. We emphasize that spiking oscillations in other semiconductor devices may also appear without any external capacitance. However, the possibility of varying an external capacitance provides a convenient method for a systematic analysis of various bifurcation scenarios of spiking filaments in semiconductor devices.

7.2.5 Chaos control

During the last decade, the nonlinear dynamics of high-dimensional, spatially ex- tended systems (Mikhailov 1994, Busse and Müller 1998) on the one hand, and the control of low-dimensional temporal chaos (Schuster 1999) on the other, has stimulated a large amount of work. Only recently have there been attempts to combine these two issues and thus to extend the knowledge acquired from control of low-dimensional chaos to dynamic systems with large numbers of degrees of freedom. The main difficulty arises from the fact that the standard control techniques (Ott *et al.* 1990, Pyragas 1992, Socolar *et al.* 1994), which we have reviewed in Chapter 1, seem to work only for unstable periodic orbits with very few unstable directions. High-dimensional systems exhibiting spatio-temporal chaos do not normally have this property. Therefore – depending on the particular system under consideration – a va- riety of approaches were proposed in the literature (Petrov *et al.* 1994, Battogtokh and Mikhailov 1996, Ding *et al.* 1996, Bleich *et al.* 1997, Münkel *et al.* 1997, Grigoriev *et al.* 1997).

In this section we shall apply and extend methods of time-delayed feedback control that have been developed for low-dimensional systems, viz. ordinary dif- ferential equations (Pyragas 1992) and iterated maps (Socolar *et al.* 1994), to the chaotic spatio-temporal spiking attractor of the semiconductor system introduced above (Franceschini *et al.* 1999). We shall demonstrate that the unstable periodic orbits (UPOs) corresponding to spatio-temporal spiking can be stabilized by applying an appropriate global feedback. It should be noted that, although our system of partial differential equations has infinitely many degrees of freedom, it possesses a low-dimensional chaotic attractor, as we have seen in Section 7.2.3 (Bose *et al.* 1994). We shall critically evaluate analytic approximations for the limits of control originally developed for low-dimensional temporal chaos (Just *et al.* 1997) and show that the delay time can be extrapolated with high accuracy, while the theoretical limit for the control of UPOs in terms of the product of the period and the largest Lyapunov exponent is not reached. If the global feedback is modified by a spatial filter, we achieve stabilization of various spatial patterns.

In order to stabilize UPOs of the chaotic dynamic system (7.5) and (7.4) we shall employ a continuous feedback as suggested by Pyragas (1992) for ordinary differential systems to synchronize the current state of the system with a time-delayed version of itself ("time-delay autosynchronization", TDAS). However, for the distributed system (7.5) and (7.4) it has not been possible to obtain chaos control by simply applying this feedback to the temporal variable u. Rather, the most efficient control of the spatio-temporal dynamics is achieved when the spatially averaged variable $a(x, t)$ is used for the construction of the control signal $\epsilon(t)$ and this signal is fed back into the same spatio-temporal variable $a(x, t)$, i.e. if eq. (7.4) is replaced by

$$\frac{\partial a(x, t)}{\partial t} = f(a, u) + \frac{\partial^2 a}{\partial x^2} + \epsilon(t), \tag{7.23}$$

$$\epsilon(t) \equiv K(\langle a \rangle(t - \tau) - \langle a \rangle(t)) \equiv K\xi(t),$$

where K is the control amplitude, and τ is the delay time (Fig. 1.11). Note that the feedback is a spatially homogeneous signal that applies the same perturbation to every point of the distributed system. Nevertheless, it turns out that it is capable of stabilizing the extremely inhomogeneous spatio-temporal spikes representing the UPOs of the system. Physically, it may be realized by an external control circuit using a lateral gate electrode located in the active layer. Since the dimensionless control variable $\langle a \rangle = u - \langle j \rangle$ is given in terms of the voltage drop u and the integral current $\langle j \rangle$ through the device, i.e. electric quantities that are easily accessible experimentally, it is straightforward to implement a control circuit that couples the delayed difference $\epsilon(t)$ back to the lateral gate potential. If this gate electrode is extended in the x-direction and imposes the same electric input along the whole gate, a global coupling to the distributed variable $a(x, t)$ is realized. A concrete setup of such a distributed lateral gate structure for a gate-driven p–n–p–n thyristor device has been described (Meixner et al. 1998a, 1998b), see Chapter 2, Section 2.4.8. (Note that the global couplings due to external circuits are used there in a different context, i.e. for control of the propagation of lateral current-density fronts, as discussed in Chapter 3.) Similar lateral gate configurations have also been realized experimentally in mesa-etched resonant-tunneling semiconductor structures, in which the charge density in the active quantum-well layer corresponding to the internal variable a can be readily controlled by tuning the gate potential.

A natural extension of the Pyragas scheme is given by a feedback that takes into account several previous states and for which the term "extended time-delay autosynchronization" (ETDAS) was proposed (Socolar et al. 1994). Using this extension, we obtain the control signal

$$\epsilon(t) = K\left((1 - R)\sum_{m=1}^{\infty} R^{m-1}\langle a\rangle(t - m\tau) - \langle a\rangle(t)\right) \equiv K\xi(t), \qquad (7.24)$$

$$0 \leq R < 1,$$

where R controls the contributions of previous states, cf. (1.37); small values of R indicate a short time memory, whereas larger values give a stronger weight to all previous states. With increasing R orbits of higher orders and orbits with larger Lyapunov exponents are likely to be stabilized (Bleich and Socolar 1996, Just *et al.* 1999). The case $R = 0$ corresponds to the simple TDAS scheme.

In our computer simulations we choose as initial conditions the unstable state corresponding to the middle branch of the S-shaped current–voltage characteristic perturbed by small inhomogeneous random fluctuations ($\delta a(x)$ and δu) of less than 1%. The bifurcation parameter j_0 is chosen in the regime of chaotic spiking (Fig. 7.18). The initial condition thus falls within the basin of the chaotic spiking attractor. We use a forward Euler algorithm with a spatial discretization in 25 grid points.

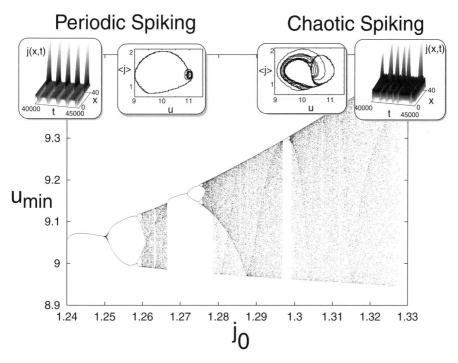

Figure 7.18. The bifurcation diagram of spatio-temporal spiking of current filaments. The minima of the normalized voltage $u(t)$ with $u < 10$ are plotted versus the control parameter j_0 which is the normalized driving current. The insets show the spatio-temporal evolution of the current density $j(x, t) = u(t) - a(x, t)$ and the $(u, \langle j\rangle)$ phase portraits of periodic ($j_0 = 1.245$) and chaotic spiking ($j_0 = 1.31$). Other parameters are $L_x = 40$, $\alpha = 0.035$, and $\mathcal{T} = 0.05$. All quantities are dimensionless. (After Franceschini *et al.* (1999).)

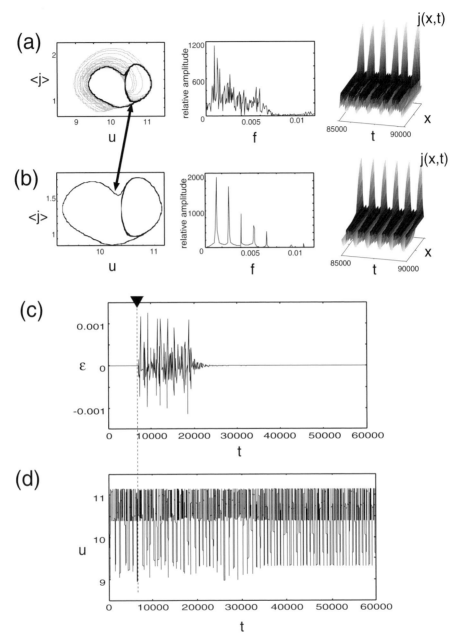

Figure 7.19. The phase portrait of the attractor in the $(u, \langle j \rangle)$ phase plane, power spectrum of $u(t)$, and spatio-temporal distribution of the current density $j(x, t)$ for (a) the uncontrolled and (b) the controlled system. The thick trajectory in the phase portrait of the chaotic attractor in (a) represents the controlled UPO from (b). In (c) and (d) are shown the control signal $\epsilon(t)$ and the voltage $u(t)$ versus time, respectively. Control is switched on at $t = 7000$. The parameters are $j_0 = 1.302$, $\tau = 732.4$, $K = 0.000\,548$, $R = 0.2$, and $N = 8$. The other numerical parameters are the same as those in Fig. 7.18. (After Franceschini *et al.* (1999).)

Figure 7.19 shows the stabilization of such a chaotic spiking mode and elucidates the dynamics of the uncontrolled (Fig. 7.19(a)) and the controlled system
(Fig. 7.19(b)) by comparing the phase portraits of the attractor projected onto the
$(u, \langle j \rangle)$ phase plane, the power spectra of the voltage $u(t)$, and the spatio-temporal
distributions of the current density $j(x, t)$. The simulations were performed for 10^5
time units. The data of the last 2×10^4 time units were used to plot the phase portrait
and to compute the power spectrum. Figure 7.19(a) illustrates the embedding of the
UPO in the chaotic attractor by showing the phase portraits both of the controlled
UPO (thick curve) and of the uncontrolled system. The delay time is equal to the
period of the stabilized periodic orbit. The large loop in the phase plane corresponds
to the localized spike of the current density $j(x, t)$ whereas the smaller loop reflects
the uniform small-amplitude relaxation oscillation in between two spikes; note that
altogether it represents a period-one orbit. We have used a slightly modified ETDAS
algorithm by truncating the series in (7.24) for $m > N$; the memory thus comprises
the N previous states at times $t - \tau, t - 2\tau, \ldots, t - N\tau$, and neglects further states
in the past. N is chosen such that R^N is smaller than 10^{-6}. Thus, with increasing R,
more previous states have to be taken into account.

Figures 7.19(c) and (d) show the time series of the control signal $\epsilon(t)$ and the
dynamic variable $u(t)$. Control is switched on at $t = 7000$. After about 20
UPO periods the control signal vanishes and the chaotic oscillations of $u(t)$ become
periodic.

Stabilization of periods of higher order is shown in Fig. 7.20. For a fixed bifurcation
parameter j_0 the unstable period-one, period-two, and period-four orbits can be
stabilized with a subsequently increasing memory amplitude R. We have been able to
achieve control of the period-one orbit with $R = 0$, but higher orbits (of periods >2)
require $R \neq 0$. In (b) and (c) the additional frequency peaks at a half and a quarter of
the fundamental frequency can be seen in the power spectra, especially in the higher
harmonics. Denoting the periods of the controlled UPO by τ_1, τ_2, and τ_4, respectively,
we find that the periods of higher order τ_2 and τ_4 are not exact integer multiples of τ_1,
but there is a discrepancy of about 1% between $n\tau_1$ and τ_n. This deviation is significant
since control fails for $\tau_2 = 2\tau_1$ or $\tau_4 = 4\tau_1$ however one adjusts the coupling constant
K.

By continuous variation of τ and K and simultaneous monitoring of the control
signal $\epsilon(t) = K\xi(t)$, the control parameters τ and K can be adjusted to their optimal
values (Pyragas 1992). A satisfactory stabilization is achieved when the control
signal becomes negligibly small after some transient time. Therefore we compute the
temporally averaged asymptotic control signal $\langle |\xi| \rangle_t$ for each simulation. We proceed
according to the following steps. First, the delay time τ is varied for a fixed value of
the estimated control amplitude K. If K is chosen on the right order of magnitude,
$\langle |\xi| \rangle_t$ as a function of τ exhibits sharp resonant minima, as depicted in Fig. 7.21; the
positions of the minima indicate the unknown periods T_i of the UPOs (Pyragas 1992).
Next, τ is fixed to one of these UPO resonances, and K is optimized. Generally, the

control signal vanishes in a finite control domain of K values (Fig. 7.22). In Fig. 7.22 we have also plotted the largest nonzero Lyapunov exponent λ which is positive for the uncontrolled chaotic attractor ($K = 0$) and negative in case of successful stabilization of the UPO. An optimum value of K is indicated by a minimum Lyapunov exponent. Note, however, that a negative Lyapunov exponent alone is not sufficient for successful control of a UPO since it might also be due to an artificially induced limit cycle that exists only with a nonvanishing control signal such as that shown in Fig. 7.22 for the K range above K_{max}. It is the result of unintentional synchronization of the system with the control signal giving rise to a limit cycle of period $\Theta \neq \tau$ that is not a UPO of the chaotic attractor.

By iterating the procedure described above, an optimal choice of both τ and K can be obtained, although this may be quite cumbersome due to the extremely sharp

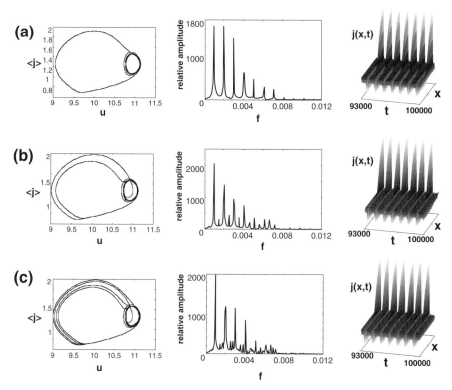

Figure 7.20. The stabilized periodic attractors in the $(u, \langle j \rangle)$ phase plane, power spectrum of $u(t)$, and spatio-temporal distribution of the current density $j(x, t)$ for the controlled system with the bifurcation parameter $j_0 = 1.262$: (a) the period-one, (b) the period-two, and (c) the period-four orbit. The control parameters are (a) TDAS: $\tau = 985.9$ and $K = 0.000\,513$; (b) ETDAS: $\tau = 1949.2$, $K = 0.000\,312$, $R = 0.1$, and $N = 6$; (c) ETDAS: $\tau = 3909.5$, $K = 0.000\,288$, $R = 0.2$, and $N = 10$. The other numerical parameters are the same as those in Fig. 7.18. (After Franceschini *et al.* (1999).)

resonances in τ. Empirical schemes to improve the estimate of the delay time by use of a self-adaptive control have been suggested for special cases (Kittel *et al.* 1995). A very promising technique to extrapolate the unknown UPO period T on the basis of repeated application of an analytic approximation formula was proposed for temporal chaos control in ordinary differential equations by Just *et al.* (1998). If the delay time τ and the UPO period T do not coincide, the system responds with a periodic signal of period Θ for not too large a delay mismatch. An expansion in terms of the delay mismatch $\tau - T$ yields

$$\Theta(K, \tau) = T + \frac{K}{K - \kappa} (\tau - T) + \mathcal{O}((\tau - T)^2). \tag{7.25}$$

Θ depends on the control parameters K and τ and on a system parameter κ, and obeys the constraint $\Theta(K, \tau = T) = T$. The parameter κ comprises all details about the coupling of the control force and is unknown, as is the UPO period T. It is therefore sufficient to determine Θ for two sets of the parameters K and τ. Then T and κ can be computed from the system of two nonlinear equations (7.25) using Newton's method. In the next iteration we use the computed approximation of T as the new delay τ in the subsequent simulation. Again we expect the system to synchronize with a new period Θ. Using this set of data and the better of the two initial guesses,

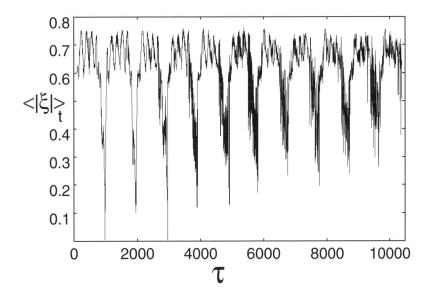

Figure 7.21. The mean asymptotic control signal $\langle|\xi|\rangle_t$ as a function of the delay time τ for $j_0 = 1.262$ and fixed $K = 0.000\,479$ (TDAS). For each value of τ the control signal $|\xi|$ has been averaged over the last 20 000 time steps of a simulation of 2×10^5 time units. The other numerical parameters are the same as in Fig. 7.18. (After Franceschini *et al.* (1999).)

we can again solve the system (7.25) and thus obtain a better approximation of T. The algorithm therefore makes available the possibility of a recursive approximation of the unknown UPO period T. The above procedure can be applied to our system exhibiting spatio-temporal chaos. For the bifurcation parameter $j_0 = 1.262$ (cf. Figs. 7.21 and 7.22), $K = 0.0006$, and initial guesses of the unknown period $\tau_1^{(0)} = 500$ and $\tau_2^{(0)} = 1200$ we have determined Θ and numerically solved (7.25) for $T^{(1)}$. After only four simulations with different values of τ and three iterations of (7.25) we obtain the UPO period $T^{(3)} \approx 983.2$. In spite of the large initial delay mismatch this result

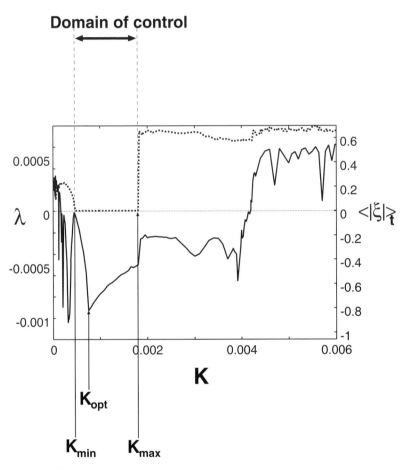

Figure 7.22. The largest nonzero Lyapunov exponent λ (full line) and mean asymptotic control signal $\langle |\xi| \rangle_t$ (broken line) as functions of the control amplitude K for $j_0 = 1.262$ and fixed $\tau = 985.9$ (TDAS). Optimal control (K_{opt}) is achieved when the control signal vanishes and $\lambda(K)$ reaches its minimum. K_{min} and K_{max} mark the boundaries of the domain of control. For each value of K the control signal $|\xi|$ has been averaged over the last 20 000 time steps of a simulation of 2×10^5 time units. The other numerical parameters are the same as those in Fig. 7.18. (After Franceschini *et al.* (1999).)

is in excellent agreement with the true UPO period $T = 985.9$ (the relative deviation is less than 0.3%!) and is reproducible for even worse initial guesses and other values of j_0. This indicates that the algorithm is very robust and efficient not only for purely temporal chaos but also for high-dimensional systems.

Once we have adjusted the delay time τ to the UPO period T we can vary the control amplitude K in a wide interval (from K_{min} to K_{max}) and still achieve control of the UPO. This domain of control is characterized by a vanishing control signal (and thus by $\langle |\xi| \rangle_t = 0$) and by a negative Lyapunov exponent $\lambda(K)$ (Fig. 7.22). At the lower boundary of this K domain the stabilized orbit decays via a period-doubling bifurcation ("flip instability" corresponding to a torsion of nearby orbits by π during one cycle) while at the upper boundary a Hopf bifurcation occurs (Just *et al.* 1999, Just 1999). The Hopf bifurcation adds an incommensurate frequency to the stabilized UPO limit cycle and thus gives rise to a two-torus preceding the onset of chaos via quasiperiodicity.

Figure 7.23 presents an overview of the bifurcations with respect to K. For small values of K we find chaotic behavior characterized by a positive Lyapunov exponent and a nonvanishing control signal, as expected for too weak control. As we increase K, an inverse period-doubling cascade sets in, which finally leads to the stabilized UPO in the domain of control (a). At each period-doubling bifurcation the Lyapunov exponent vanishes (Schuster 1988). In the period-doubling regime the control signal remains positive. Within the domain of control optimal stabilization is achieved at $K = K_{opt}$ for which the negative Lyapunov exponent $\lambda(K)$ is minimum (b). Further increase of K leads to a Hopf bifurcation associated with a rise of the control signal $\langle |\xi| \rangle_t$ to slightly positive values at the upper boundary of the control domain (c) followed by a sharp increase as chaos sets in (d). The Lyapunov exponent λ becomes zero at the Hopf bifurcation as a signature of the two-torus, which possesses two vanishing Lyapunov exponents (note that a second Lyapunov exponent is zero throughout the whole range of K), and jumps to positive values only as chaos occurs. For a very large control amplitude the controlled variable a is driven too hard and decouples from the other variable; therefore the control of UPOs fails.

Figure 7.24 presents simulations for the points (a), (b), (c), and (d) of Fig. 7.23. Figures 7.24(a) and (c) correspond approximately to the respective bifurcation points, whereas Fig. 7.24(b) represents the stabilized reference UPO at $K = K_{opt}$, and Fig. 7.24(d) shows the onset of chaos just above the Hopf bifurcation. The power spectra in Figs. 7.24(a) and (c) clearly reveal the appearance of the subharmonic frequency and the higher incommensurate frequency associated with period-doubling and Hopf bifurcation, respectively (marked by arrows). The onset of chaos in Fig. 7.24(d) exhibits intermittent behavior whereby the dynamics seems to oscillate between the two-torus and the chaotic attractor. These two dynamic states are plotted separately both in the phase portrait and in the power spectrum.

For different parameters j_0 one can also observe more complex bifurcation scenarios. In Fig. 7.22, for example, the Hopf bifurcation is concealed by a regime in which a synchronized limit cycle (negative Lyapunov exponent) is enforced by the control ($\langle|\xi|\rangle_t \neq 0$). A closer inspection shows that, at least in some range of $K > K_{max}$, bistability between the forced limit cycle and a controlled UPO occurs, depending upon the initial conditions. This can be understood by noting that the basin of attraction of the controlled UPO becomes smaller as the boundaries of the control domain are approached.

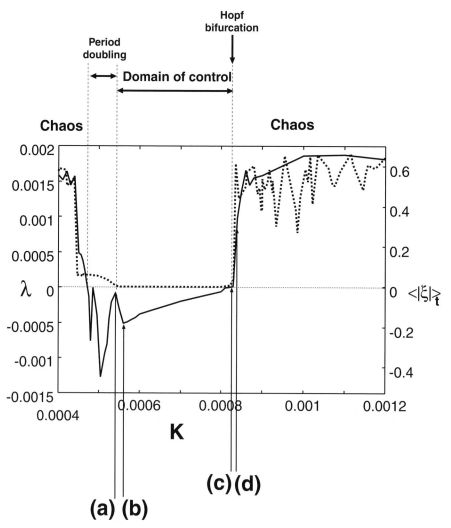

Figure 7.23. The same as Fig. 7.22 for $j_0 = 1.302$, $\tau = 732.4$, $R = 0.2$, and $N = 8$. The following values of K are marked: (a) period-doubling bifurcation (flip instability), (b) optimal UPO control, (c) Hopf instability (torus), and (d) the onset of chaos. For each value of K the control signal $|\xi|$ has been averaged over the last 20 000 time steps of a simulation of 2×10^5 time units. (After Franceschini *et al.* (1999).)

Up to now we have considered only a spatially homogeneous feedback $\epsilon(t)$ at every point of the distributed system. The idea of using a spatially inhomogeneous feedback arises from the observation that there exist various locally stable or even unstable spatially inhomogeneous modes in the uncontrolled system, e.g. spiking current filaments located in the center of the sample, but these are observed only for special initial conditions. Note that, in the absence of control, a random initial

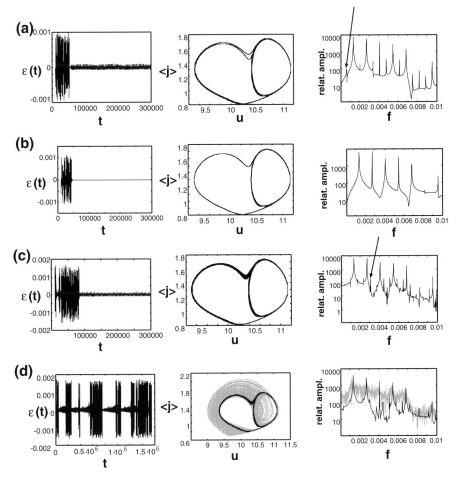

Figure 7.24. The control signal $\epsilon(t)$, phase portraits $(u, \langle j \rangle)$, and power spectrum of $u(t)$ for various values of the control amplitude K marked in Fig. 7.23: (a) period-doubling (flip instability) at the lower boundary of the control regime, (b) the reference UPO at optimum $K = K_{\text{opt}}$, (c) Hopf instability at the upper boundary of the control regime, and (d) the onset of chaos beyond the Hopf bifurcation via intermittency: quasiperiodic (black lines) and chaotic (gray lines) spiking oscillations alternate. The arrows in (a) and (c) indicate the additional frequencies with respect to (b). The parameters are $j_0 = 1.302$, $\tau = 732.4$, $R = 0.2$, and $N = 8$, with (a) $K = 0.000\,54$, (b) $K = 0.000\,56$, (c) $K = 0.000\,834$, and (d) $K = 0.000\,838$. (After Franceschini *et al.* (1999).)

distribution always gives rise to a (spiking) filament that is pinned to the boundary (cf. Fig. 7.25(a)); this dominant spatial mode results from the attraction exerted upon filaments by Neumann boundary conditions, see Section 3.4. By tailoring appropriate spatio-temporal control signals $\epsilon(x, t)$ that resemble the spatial profile of the desired modes, one can expect to stabilize some of them from random initial conditions. Thus chaos control in spatio-temporal systems could be used for pattern selection, which would lead to promising applications as distributed semiconductor memory devices and neural networks.

We illustrate this idea for a spatial mode $j(x, t)$ whose profile is symmetric and exhibits a peak (spike) at the center of the sample. This mode is unstable and can be observed only if the initial distribution of $j(x, t)$ is perfectly symmetric and the system is noise-free. In order to stabilize this symmetric spiking mode we apply a spatially modulated control signal $\epsilon_{inhom}(x, t)$ that favors the symmetry considered and can be

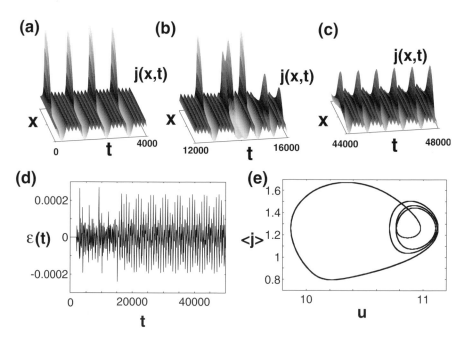

Figure 7.25. Suppression of the dominant spatial mode in favor of an unstable one by spatial filtering. In (a)–(c) the spatio-temporal distribution of the current density $j(x, t)$ is plotted for various time regimes. (a) An asymmetric spiking filament at the boundary, induced by random initial conditions. (b) The transient regime of coexistence of both spiking modes after control is switched on at $t = 2000$. (c) The symmetric spiking mode prevails over the asymmetric one and forms the asymptotic pattern. (d) The control signal $\epsilon(t)$. (e) A phase portrait of the stabilized limit cycle corresponding to (c). The parameters are $j_0 = 1.262$, $K = 0.0002$, $R = 0$, and $\tau = 985.5$, and the asymptotic period of the response signal is $\Theta = 778.3$. The other numerical parameters are the same as those in Fig. 7.18. (After Franceschini *et al.* (1999).)

constructed from the homogeneous control signal $\epsilon(t)$ in the following way:

$$\epsilon_{\text{inhom}}(x, t) = \epsilon(t)\left[1 - \frac{1}{2}\cos\left(\frac{2\pi}{L_x}(x)\right)\right],$$ (7.26)

$$\epsilon(t) \equiv K(\langle a\rangle(t - \tau) - \langle a\rangle(t)).$$

Figure 7.25 shows the results of a simulation with this spatially inhomogeneous feedback control. For random initial conditions the dominant spatial mode is always given by a chaotically spiking profile $j(x, t)$ with the spike located at the boundary (Fig. 7.25(a)). After control has been switched on, there is a transition period during which the dominant mode (that favored by the boundary conditions) coexists with the symmetric mode (which is favored by the control signal) (Fig. 7.25(b)). Finally the symmetric mode wins and the corresponding orbit is stabilized asymptotically (Fig. 7.25(c)). The phase portrait in Fig. 7.25(e) suggests that the controlled orbit is periodic though not a UPO of the chaotic attractor since the control signal $\epsilon(t)$ does not vanish after some transient time (Fig. 7.25(d)). Thus the period of the stabilized limit cycle Θ does not coincide with the delay time τ. Nevertheless, we stress that, by means of a weak perturbation, we have succeeded in suppressing the dominant *spatial* mode in favor of an unstable one and replacing the *temporally* chaotic sequence of spikes by a periodic sequence.

Let us now discuss the limits of control of spatio-temporal chaos. A rough estimate of the efficiency of the ETDAS scheme for temporal chaos in ordinary differential equations has been given by Just *et al.* (1999) and Just (1999): Only those UPOs which obey the constraint

$$\lambda\tau \leq 2\frac{1 + R}{1 - R}$$ (7.27)

can be stabilized. Thus the simple TDAS scheme ($R = 0$) is expected to be able to stabilize UPOs whose product of the Lyapunov exponent λ and the period τ does not exceed 2. In practice we have been able to control UPOs in our spatially extended system with global feedback only for lower values.

Empirically it has been found that both TDAS and ETDAS work successfully for UPOs with $\lambda\tau \leq 1.2$. If, on the other hand, $\lambda\tau > 1.2$, the TDAS scheme starts to fail while ETDAS still works successfully if the Lyapunov exponent of the uncontrolled system is not too large. The best results obtained (Franceschini *et al.* 1999) were the stabilization of a high-period UPO with a small Lyapunov exponent for $j_0 = 1.262$ ($\lambda = 4.8 \times 10^{-4}$, $\tau = 3909.5$, and thus $\lambda\tau \approx 1.9$) and of a highly unstable UPO for $j_0 = 1.305$ ($\lambda = 2.58 \times 10^{-3}$, $\tau = 730.3$, and $\lambda\tau \approx 1.9$). Both orbits could be controlled only with ETDAS.

In parameter regions where the largest Lyapunov exponent is greater than 2.4×10^{-3} (i.e. $1.309 \leq j_0 \leq 1.314$) we have not been able to stabilize any UPO at all, although we extrapolate periods on the order of $\tau \approx 700$, and (if the Lyapunov exponent of the UPO and of the uncontrolled system do not differ significantly) $\lambda\tau < 2$

would still hold. This could possibly be improved by making use of a technique developed by Zoldi to calculate UPOs explicitly using a damped-Newton method (Zoldi and Greenside 1998, Zoldi *et al.* 2000). It seems that it might be possible to stabilize some of these UPOs by means of more sophisticated control techniques, e.g. with a spatially filtered control signal.

Figure 7.26 summarizes the empirical findings for various values of λ and τ by showing the domains of control in the diagram of $\langle |\xi| \rangle_t$ as a function of K for various ETDAS realizations. The control domain comprises all values of K for which $\langle |\xi| \rangle_t$

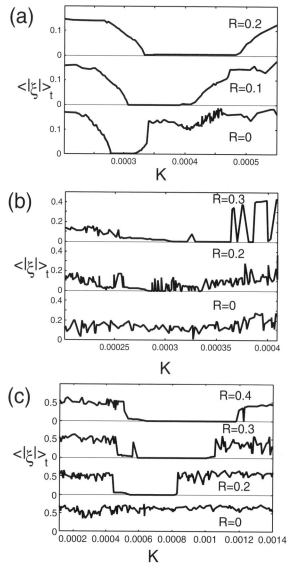

Figure 7.26. Domains of control for various values of $\lambda \tau$ and R. UPO stabilization is achieved if the mean control signal $\langle |\xi| \rangle_t$ vanishes in some interval of K. (a) $j_0 = 1.262$, $\tau = 1949.9$ (period two), and $\lambda \tau \approx 1.2$. (b) $j_0 = 1.262$, $\tau = 3909.5$ (period four), and $\lambda \tau \approx 1.9$. (c) $j_0 = 1.302$, $\tau = 732.4$ (period one), and $\lambda \tau \approx 1.4$. For each value of K the control signal $|\xi|$ has been averaged over the last 20 000 time steps of a simulation of 2×10^5 time units. The other numerical parameters are the same as those in Fig. 7.18. (After Franceschini *et al.* (1999).)

vanishes. In Fig. 7.26(a), for example, we have $\lambda = 6.2 \times 10^{-4}$, $\tau = 1949.2$ for the period-two UPO, and thus $\lambda\tau \approx 1.2$. For this uncritical value of $\lambda\tau$ both the TDAS ($R = 0$) and the ETDAS ($R > 0$) scheme work in a finite K interval, as is evident from the vanishing control signal. It should be noted that with increasing R the control domain is shifted to larger values of K and becomes wider, in good agreement with the theoretical predictions for temporal chaos (Just *et al.* 1999). For the period-four UPO at the same value of j_0 (Fig. 7.26(b)) $\lambda\tau \approx 1.9$, and control has no longer been achieved by the TDAS scheme ($R = 0$), whereas the ETDAS scheme with $R > 0$ still works, although large fluctuations in $\langle|\xi|\rangle_t$ are present. They are a result of the shrinking attractor basin, which prevents stabilization for some realizations of the random initial conditions, even though for $R = 0.2$, for example, the theoretical limit of $\lambda\tau = 3$ is nowhere near having been reached. In Fig. 7.26(c) we consider a UPO with $\lambda\tau \approx 1.4$ for a different bifurcation parameter j_0; here the Lyapunov exponent is larger ($\lambda = 1.88 \times 10^{-3}$) and the simple TDAS scheme does not allow stabilization even of the period-one UPO. On the other hand, there is still a large control domain for the feedback schemes with longer memory ($R > 0$), and the small shift of the lower stability boundary (flip instability) and the stronger shift of the upper stability boundary (Hopf instability) with increasing R agree well with the theoretical predictions (Just *et al.* 1999).

In conclusion, we have seen that a variety of spatio-temporal UPOs embedded in a chaotic attractor of a distributed system can be stabilized using an extended time-delay autosynchronization algorithm (Franceschini *et al.* 1999). These UPOs correspond to spiking current filaments. We have critically evaluated numerical techniques and analytic approximations originally developed for *temporal* chaos (Just *et al.* 1998, 1999) and found that several properties of chaos control in low-dimensional temporal systems carry over to the *spatio-temporally* chaotic reaction–diffusion system. We have confirmed that the delay time of global time-delayed feedback control (which is adjusted to the UPO period T) can be extrapolated with high accuracy from the periodic response of the system. We have also gained insight into the mechanism of spatio-temporal chaos control by analyzing the bifurcations at the boundaries of the control domain. While the gross features agree with the case of temporal chaos, we have found a higher sensitivity to noise, which appears to be due to the smaller attractor basins and multistability of the spatio-temporal patterns associated with the larger number of degrees of freedom. The approximate theoretical limit for the control of temporal UPOs, which is given by $\lambda\tau \leq 2$ for the Pyragas feedback, and by $\lambda\tau \leq 2(1 + R)/(1 - R)$ for the extended time-delay autosynchronization scheme, has not been reached in the simulations. Nevertheless, the ETDAS scheme represents a significant improvement on the simple Pyragas feedback. If the global feedback is modified by a spatial filter, one can achieve pattern selection and stabilization of otherwise unstable spatial modes corresponding to various locations of the spikes within the sample.

These findings offer promising potential applications since the feedback can be readily realized for distributed bistable semiconductor systems by gate control circuits. Thus it should be possible to build stable, tunable microwave oscillators on that principle.

7.3   Transient spatio-temporal chaos

In this section we introduce a second length scale into the two-component reaction–diffusion system (7.4) and (7.5) given by the diffusion of the inhibitor u which now becomes spatially dependent. This system is found to exhibit transient spatio-temporal chaos before reaching a periodic attractor.

7.3.1   Multiple spiking in a locally coupled reaction–diffusion model

Deterministic chaos is not only displayed in the form of chaotic attractors in the long-time behavior of dissipative dynamic systems, but has also been found in connection with chaotic repellers in the transient phase (Tél 1990, Grebogi *et al.* 1983). Of particular interest is transient chaos in spatially extended systems, since such systems generically exhibit long transients that may practically preclude the observation of the asymptotic attractors (Mikhailov 1994). Therefore the dynamic behavior may be dominated by spatio-temporal chaos even if the system asymptotically displays stable motion, e.g. in terms of a periodic or fixed-point attractor. Such behavior has been found in some extended systems in which the transient times may grow exponentially with the system's size. It has been proposed that this feature might be of relevance for certain regimes of turbulence (Shraiman 1986, Crutchfield and Kaneko 1988). Transient irregular dynamics in extended systems have been studied mainly for chains of coupled maps (Crutchfield and Kaneko 1988, Politi *et al.* 1993) and for the Kuramoto–Sivashinsky (KS) equation (Shraiman 1986, Hyman *et al.* 1986), but a spatially continuous reaction–diffusion system that exhibits this kind of behavior has recently been presented and analyzed (Wacker *et al.* 1995*a*). Such systems are of particular interest for the description of various nonlinear active media associated, for example, with chemical reactions (Swinney and Krinsky 1991) and semiconductor charge transport (Niedernostheide *et al.* 1992*b*, 1994*a*, Wacker and Schöll 1994*b*, Kerner and Osipov 1994). The system considered here exhibits transient spatio-temporal chaos and an exponential increase of the transient times with the system's size. Thus, the transient chaotic dynamics will determine the behavior observed of a sufficiently large system even on a long time scale.

We consider the following dimensionless reaction–diffusion system which is a variant of the system (7.4) and (7.5) discussed in Section 7.2. It can be derived for charge transport in layered semiconductors using several simplifications (Wacker and

Schöll 1994*b*, Wacker 1993). The basic assumption is no longer that the contacts are ideal planar Ohmic contacts, which resulted in an x-independent voltage across the device, but rather that they are modeled by a contact layer of finite extent b in the z-direction with constant conductivity σ (Radehaus *et al.* 1987). The linear contact layer and the SNDC layer are located at $-b < z < 0$ and $0 < z < L$, respectively. The potential distribution $\phi(x, z)$ is fixed by the constant external voltage U_0 at $z = -b$, and by the transversally dependent voltage $U(x)$ at $z = 0$ (Fig. 7.27). In the Ohmic contact layer $j = \sigma \mathcal{E}$ holds. Since the dielectric relaxation time is small for a good Ohmic conductor, charge accumulations will immediately relax, and we can assume that the potential distribution within the Ohmic layer satisfies Poisson's equation $\Delta \phi \approx 0$, and is approximately given by the *Ansatz*, cf. Radehaus *et al.* (1987),

$$\phi(x, z) = U(x) + (U(x) - U_0)\frac{z}{b} - \frac{\partial^2 U}{\partial x^2}(q_1 z + q_2 z^2 + q_3 z^3), \tag{7.28}$$

with $q_1 = b/3$, $q_2 = \frac{1}{2}$, and $q_3 = 1/(6b)$ (Wacker 1993). For continuity reasons the z-component of the total current density has to satisfy

$$\sigma \mathcal{E}_L + \epsilon_L \frac{d\mathcal{E}_L}{dt} = j_N + \epsilon_N \frac{d\mathcal{E}_N}{dt} \tag{7.29}$$

at $z = 0$, where subscripts L and N denote the linear and the nonlinear layer, respectively. With $\mathcal{E}_L = -(\partial \phi / \partial z)$, $\mathcal{E}_N \approx U/L$, $\epsilon_L/b \ll \epsilon_N/L$, and $j_N b/\sigma \gg U$ one obtains the approximate relation

$$\frac{\partial U}{\partial t} = \frac{L}{\epsilon_N}\left(\frac{I_0}{A} - j_N + q_1 \sigma \frac{\partial^2 U}{\partial x^2}\right) \tag{7.30}$$

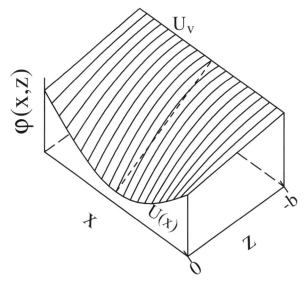

Figure 7.27. The potential distribution in the Ohmic contact layer for given $U_v = U_0$ at $z = -b$ and $U(x)$ at $z = 0$.

with the effective current density $I_0 = A\sigma U_0/b$. The dimensionless form of this equation together with eq. (7.4) then yields the dynamic system, cf. eqs. (1.38) and (1.39),

$$\frac{\partial a(x,t)}{\partial t} = f(a,u) + \frac{\partial^2 a}{\partial x^2}, \tag{7.31}$$

$$\frac{\partial u(x,t)}{\partial t} = \alpha[j_0 - (u-a)] + D\frac{\partial^2 u}{\partial x^2}, \tag{7.32}$$

with

$$f(a,u) = \frac{u-a}{(u-a)^2 + 1} - \mathcal{T}a. \tag{7.33}$$

The essential feature which distinguishes this system from the previous reaction–diffusion system (7.4) and (7.5) studied in Section 7.2 is that the global coupling $\langle a \rangle$ is replaced by a local coupling, and diffusion of u is added. Thus a second length scale, the diffusion length $\sqrt{D}$ of u, is introduced. While the diffusion length of a, which is unity in our dimensionless notation, determines the width of the current filament wall, i.e. the transition layer, the second diffusion length limits the width of the filament itself. Therefore now stable current-density distributions with several filaments become possible. In systems with only one diffusion length these multiple filaments can be shown to be generally unstable both in finite and in infinitely extended one-dimensional systems by invoking a generalized oscillation theorem of Sturmian theory (Schöll 1983, 1987). In contrast, reaction–diffusion systems with *two* diffusing variables exhibit a Turing instability (Turing 1952) of spatially periodic patterns for sufficiently large D. As we will see in the following, this can result not only in stationary multiple filaments, but also in multiple spiking patterns, depending on the parameters. Systems of this kind have been studied thoroughly (Ohta *et al.* 1990, Radehaus *et al.* 1990, Mikhailov 1992, 1994, Kerner and Osipov 1994, Niedernostheide *et al.* 1992a, Willebrand *et al.* 1992). Gorbatyuk and Rodin (1990, 1994, 1995) showed theoretically that gate-turn-off thyristors (GTO) can exhibit a Turing instability. During the switch-off process the injection channel in any element of the integrated p–n–p–n structure is intentionally compressed along one transversal dimension by the gate control. Under certain conditions the current distribution simultaneously becomes unstable along the other transverse direction with respect to the mode whose wavelength is close to the thickness of the device. For a typical GTO geometry this instability leads to the formation of spatially periodic arrays of current filaments (up to ten) and subsequent destruction of the device due to the avalanche local overheating. The theoretical results are in good agreement with experiments (Gorbatyuk *et al.* 1989).

Under current control, i.e. with a fixed control parameter j_0, the model (7.31) and (7.32) has a unique spatially homogeneous fixed point

$$a^* = \frac{j_0}{\mathcal{T}(j_0^2 + 1)}, \tag{7.34}$$

$$u^* = j_0 + a^*. \tag{7.35}$$

It is the same as that for the model (7.4) and (7.5), cf. Fig. 7.1. For $T = 0.05$ and $1.12 < j_0 < 4.09$ (which we will consider in the following) we find

$$\left. \frac{\partial}{\partial a} \left(\frac{u - a}{(u - a)^2 + 1} - Ta \right) \right|_{(a^*, u^*)} > 0. \tag{7.36}$$

Thus, a deviation in a from the homogeneous value a^* will grow in time for fixed $u = u^*$. Therefore a acts as an activator and u as an inhibitor in terms of nonlinear dynamics, as discussed in Section 7.2.1. As before, we use Neumann boundary conditions $\partial a / \partial x = \partial u / \partial x = 0$ for $x = 0, L$, where L is the size of the system. The equations can be solved with finite differences and a forward Euler algorithm. In terms of the semiconductor model the quantity $u - a$, which we plot, corresponds to the current density $j(x, t)$ which is the physical quantity of interest.

Note that the nonlinearity in eq. (7.31) is a specific feature of the reaction–diffusion system considered here, which distinguishes it from models with the simpler cubic nonlinearity that is often used (Mikhailov 1994, Ohta *et al.* 1990, Schimansky-Geier *et al.* 1995). As discussed in Section 7.2.2, in certain parameter regimes, complex spatio-temporal dynamics in the form of spiking is expected. We therefore suggest that this model is generic for a general class of noncubic activator–inhibitor systems.

For a wide range of the parameters T, α, D, and L the homogeneous fixed point successively exhibits a Hopf bifurcation (for homogeneous fluctuations) and a Turing instability (for inhomogeneous fluctuations) with increasing j_0. In part of this range periodic spatio-temporal spiking is found (Wacker and Schöll 1994*b*), whereby a spatially periodic pattern forms and vanishes periodically in time as shown in Fig. 7.28(a). The difference from Section 7.2 is that now patterns with multiple spikes occur. For the following simulations (Wacker *et al.* 1995*a*) we use $T = 0.05$, $\alpha = 0.02$, $D = 8$, and $j_0 = 1.21$ as parameters. For this case the Hopf and Turing instabilities occur at $j_0 = 1.185$ and 1.204, respectively. The system's size L is the only parameter which is changed. However, similar results are found when the other parameters are varied in a small range close to the values chosen here. We use the initial conditions $a(x, 0) = a^* + r(x)$ and $u(x, 0) = u^*$, where $r(x)$ is given by a random sequence from the interval $[-0.5, 0.5]$ at each position of the discretized spatial lattice. Then the periodic spiking behavior is found after a long transient irregular behavior. Figure 7.28 shows both a part of the transient (b) and the periodic behavior (a) for $L = 600$. The periodic pattern is characterized by three spikes that arise periodically at two different positions separated by half a period. (A spike with a maximum located at the boundary is counted as a half.) The transient behavior is dominated by similar spikes occurring at irregular positions. In order to visualize the dynamics more clearly we have plotted the position of the local maxima of $u(x, t) - a(x, t)$ exceeding the value 3 in the space–time plane (Fig. 7.28(c)). For different initial conditions we obtain similar results with different transient times.

Surprisingly, some simulations yield a slightly different periodic pattern consisting of 2.5 or 3.5 spikes within the length $L = 600$ and with a slightly dilated temporal period. Apart from this difference, the asymptotic patterns are nearly identical. Figure 7.28(d) shows that the same type of transient behavior appears for $L = 2000$, too. In Fig. 7.29 the temporal periods of the various stable asymptotic patterns observed for $L = 600$ and various initial conditions are plotted versus the control parameter j_0. It can be seen that multistability between spiking with various spatial periods and spatially homogeneous Hopf oscillations occurs in some ranges of j_0.

We now define the transient time τ for which the periodic state is reached to within a certain accuracy. For instance, we find $\tau = 30\,946$ for the simulation shown in Figs. 7.28(a)–(c). By simulating 994 different initial conditions for $L = 600$ we obtain a mean value $\langle \tau \rangle = 2.91 \times 10^4$ and a standard deviation $\sigma = 2.39 \times 10^4$. The distribution function $f(\tau)$ can be constructed by defining the number $A(t)$ of simulations with transient times $\tau \geq t$. $A(t)$ is a monotonic function, in contrast to the probability density $f(\tau)$, for which the problem of coarse-graining occurs for a finite number of data. The relation between these quantities is given by

$$A(t) \approx N \int_t^\infty d\tau \, f(\tau) \tag{7.37}$$

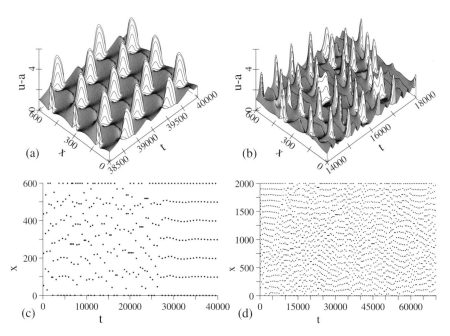

Figure 7.28. Spatio-temporal dynamics for a single run with system size $L = 600$: (a) asymptotic periodic spiking, (b) transient irregular spiking, and (c) positions of the spikes as a function of time t. In (d) is shown a part of the transient behavior for a different run with $L = 2000$. (After Wacker *et al.* (1995a).)

for large total numbers of simulations N. The result for $A(t)$ is given in Fig. 7.30(a) for $L = 600$. We find a scaling behavior $A \sim \exp(-t/s)$ with an inverse escape rate $s = 2.37 \times 10^4$, which is in good agreement with the value of σ expected for an exponential distribution $f(\tau) \sim \exp(-\tau/s)$. For $\tau < 5000$ there are only a few values. Thus, the mean value $\langle \tau \rangle$ is slightly larger than s and σ, in contrast to the result for transient chaos in a system of *ordinary* differential equations (Reznik and Schöll 1993). It has been checked that the influence of discretization error on the distribution is negligible by performing the simulation for $L = 600$ with finer time and space steps.

If one performs this procedure for various sizes, one always finds a similar exponential decay,

$$A(t) \sim \exp(-t/s(L)). \tag{7.38}$$

The results for $\langle \tau \rangle$ and σ are shown in Table 7.1. Except for $L = 400$ different asymptotic patterns have always been found for each system size. The different patterns observed are also given in Table 7.1. As expected, the number of spikes increases linearly with the system's size. In Fig. 7.30(b) we have plotted the mean value $\langle \tau \rangle$ and the standard deviation σ as functions of L. The logarithmic plot exhibits

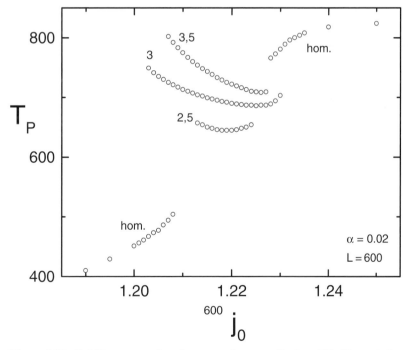

Figure 7.29. Stability ranges of spatio-temporal patterns for $L = 600$. The periods T_p of stable homogeneous oscillations (hom) and spiking with various numbers of periods (2.5, 3, and 3.5) are plotted versus the current density j_0. (After Meixner *et al.* (1997a).)

an exponential dependency on the system's length over about 1.5 decades with the exception of the points $L = 450$ and $L = 650$ which are significantly higher. The distribution of the periodic patterns given in Table 7.1 shows that, for these sizes, two different patterns arise with nearly equal probability, whereas for the other lengths one pattern clearly dominates. This indicates that the system has a natural intrinsic spatial period (wavelength) of roughly $\Delta x = 200$ into which it usually locks. Owing to the boundary conditions, this period can be reached only if L is an integer multiple

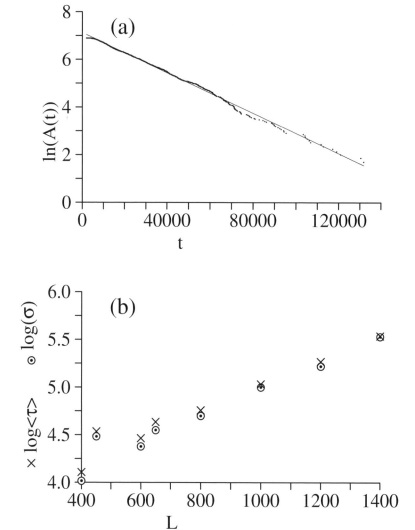

Figure 7.30. (a) A semilogarithmic plot of the number $A(t)$ of simulations with transient time $\tau \geq t$ as a function of t for $L = 600$. (b) A semilogarithmic plot of the mean transient times $\langle \tau \rangle$ (crosses) and their standard deviations σ (circles) as functions of the system's size L. (After Wacker *et al.* (1995*a*).)

of 100. For $L = 650$ and $L = 450$ this is not the case and the system experiences longer transient times before it reaches a periodic attractor. Such a feature has also been observed for the Kuramoto–Sivashinsky model (Hyman *et al.* 1986).

The exponential increase of the transient times can easily be understood. Assuming that the spikes occur almost independently from each other in the irregular regime, they on average emerge once in the spatio-temporal interval $b_x \times b_t$. Here b_x and $2b_t$ are the spatial and temporal periods of the periodic state, respectively. The exact periodic state is characterized by L/b_x spikes appearing at fixed positions at the same time. If all spikes appear within a certain range of attraction $\Delta x \times \Delta t$ around the exact positions, the pattern will develop toward the exact periodic state via some rocking transient. The probability for this event is given by $W = \lambda [\Delta x \, \Delta t / (b_x b_t)]^{L/b_x}$, where the factor λ reflects the various possible patterns which can be realized. If λ is not dependent on L we obtain an exponential law for the transient times $\tau \sim 1/W$. From Table 7.1 we conclude that essentially only one specific asymptotic pattern is realized for all multiple lengths of 100. Nevertheless, the relative number of simulations exhibiting different patterns increases slightly with L, which could result in some deviations from the exponential law.

Owing to the approximately exponential increase of the transient times with the system's length, a large system will practically not exhibit any periodic behavior. A typical simulation for $L = 2000$ is shown in Fig. 7.28(d). One can detect some phases for which almost periodic behavior appears in a certain range of x. However, the spatial extent of these regular phases decrease nearly linearly in time. This clearly demonstrates that the transition to a periodic state does not occur via the formation of

Table 7.1. *Results from numerous runs with different initial conditions for various sample sizes L. The mean value $\langle \tau \rangle$ and the standard deviations σ of the transient times found in various runs are shown. The asymptotic patterns observed in the runs are characterized by the number of spikes appearing at the same time. (A spike located at the boundary is counted as half a spike.) The distribution of these different patterns are also listed*

L	$\langle \tau \rangle$	σ	Distribution of the number of spikes			
400	12 787	10 356		210×2		
450	34 297	30 389		74×2	119×2.5	
600	29 102	23 931		976×3	18×3.5	
650	43 283	35 440		97×3	104×3.5	
800	57 165	49 986		193×4	7×4.5	
1000	107 524	100 067	3×4.5	187×5	6×5.5	
1200	184 730	165 787	14×5.5	174×6	10×6.5	
1400	345 674	338 066	22×6.5	150×7	16×7.5	1×8

a periodic nucleus, but instead the periodic state must be achieved within the whole sample at once.

In order to characterize the irregular behavior one can calculate the Lyapunov exponents in the transient state. Strictly speaking, the Lyapunov exponent is defined only in the limit of infinite sampling times. Nevertheless, we extract a local Lyapunov exponent by taking the gliding mean over the preceding 5000 time units only. This approximation gives a Lyapunov exponent that varies somewhat in time both in the transient and in the periodic state, but exhibits a distinct drop when the transient ends (Fig. 7.31). For the calculation of the Lyapunov spectrum the procedure of Benettin *et al.* (1980) and Shimada and Nagashima (1979) has been used; see Section 7.2.3. For $L = 600$ we have obtained five positive Lyapunov exponents, whereas for $L = 1200$ there are ten positive ones. This indicates that the number of Lyapunov exponents in a certain range of values is proportional to the system's size (constant *Lyapunov density*). Such a behavior can be expected if regions far from each other are essentially decoupled, so that they do not influence each other significantly. Thus the number of degrees of freedom is proportional to the system's size. This *extensive* spatio-temporal chaos will be characterized in more detail in Section 7.3.5.

In the transient regime the largest exponent is of the order of 2.5×10^{-3}. We have not found any significant dependency of the largest Lyapunov exponent on L. At the end of the transient the largest Lyapunov exponent drops to zero and the second Lyapunov exponent becomes negative, as expected for stable periodic behavior.

Generally, each repeller has at least one positive Lyapunov exponent of the order of its escape rate (Tél 1990). Since the escape rates ($1/s = 4.2 \times 10^{-5}$ for $L = 600$) are significantly smaller than the largest Lyapunov exponents we have found, we conclude

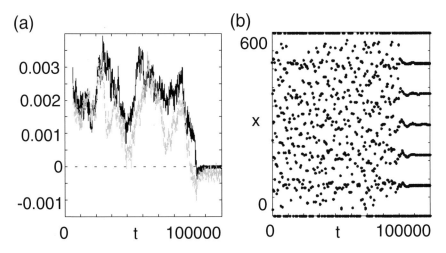

Figure 7.31. (a) Typical temporal evolution of the largest two Lyapunov exponents for $L = 600$ and (b) the corresponding spatio-temporal dynamics (after Wacker *et al.* (1995*a*)).

that those are indeed associated with the intrinsic chaotic structure of the repeller rather than with its unstable manifold.

7.3.2 Spreading of localized perturbations

While the Lyapunov exponents reveal only the *temporal* behavior of perturbations, the *spatio-temporal* properties of the system can be examined by considering the temporal spreading of a localized perturbation. To this end we add a perturbation to $a(x_d, t_d)$ at a distinct location x_d at a given time t_d. For $t > t_d$ we calculate the trajectories both for the unperturbed and for the perturbed initial condition at $t = t_d$. In order to visualize the deviations resulting from the perturbation we have plotted the logarithm of the difference between these two trajectories in Fig. 7.32. We find that the perturbation spreads with a constant velocity v that seems to be independent from the width L. From the numerical results it may be conjectured that a relation among v, the largest Lyapunov exponent λ_1, and the spatial period of the spikes Δx of the form $v \approx \lambda_1 \Delta x$ holds.

A constant propagation velocity has also been observed in a coupled map lattice (Politi *et al.* 1993). There are some points, especially in the simulation shown for $L = 1200$, where the propagation stops and the perturbation is reduced dramatically. Afterwards the residual perturbation grows again. During the initial stage the propagation is roughly parabolic rather than linear. This behavior might be due to pure diffusion. Only when the diffusive spreading slows down can the linear propagation be detected.

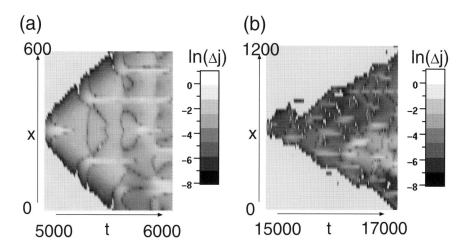

Figure 7.32. The temporal evolution of a localized perturbation in the transient phase for (a) $L = 600$ and (b) $L = 1200$. The perturbation is localized at $t = 15\,000$ in the center of the system. The grayscale corresponds to the logarithmic difference between values $j(x, t)$ for the perturbed and the unperturbed system. (After Wacker *et al.* (1995*a*).)

In conclusion, these results demonstrate that a simple generic reaction–diffusion system exhibits extensive transient spatio-temporal chaos in a certain regime of the parameter space. In this regime there exist positive local Lyapunov exponents during the transient time. A statistical analysis of the transient times for various system sizes shows that there is an exponential increase of the transient time with growing system size. More information about the spatio-temporal dynamics can be gained by analyzing the spreading of a perturbation, which is found to propagate with a constant velocity apparently independently from the system's size. This velocity is connected with the spectrum of Lyapunov exponents of the system and might be an interesting quantity for characterizing systems exhibiting spatio-temporal chaos.

7.3.3   The codimension-two Turing–Hopf bifurcation

In order to gain a deeper understanding of the complex spatio-temporal dynamics associated with spiking we shall now perform a systematic bifurcation analysis and point out the connection with a codimension-two bifurcation (Meixner *et al.* 1997*b*). To perform a linear stability analysis we linearize the dynamic system (7.31) and (7.32) around the spatially homogeneous fixed point (7.35) for small space- and time-dependent fluctuations $(\delta a, \delta u) \sim \exp(\lambda t + ikx)$ and obtain

$$\lambda \begin{pmatrix} \delta a \\ \delta u \end{pmatrix} = \begin{pmatrix} f_a - k^2 & f_u \\ \alpha & -\alpha - Dk^2 \end{pmatrix} \begin{pmatrix} \delta a \\ \delta u \end{pmatrix}. \tag{7.39}$$

This yields the characteristic equation

$$\lambda^2 - \Theta\lambda + \Delta = 0, \tag{7.40}$$

where

$$\Theta = f_a - \alpha - (D+1)k^2, \tag{7.41}$$

$$\Delta = \alpha T + (\alpha - f_a D)k^2 + Dk^4, \tag{7.42}$$

with

$$f_a = \frac{j_0^2 - 1}{(j_0^2 + 1)^2} - T, \tag{7.43}$$

$$f_u = -\frac{j_0^2 - 1}{(j_0^2 + 1)^2}. \tag{7.44}$$

The physical control parameter of the bifurcation scenarios is the current density j_0. To simplify the formalism, we introduce

$$\gamma = \frac{j_0^2 - 1}{(j_0^2 + 1)^2} \tag{7.45}$$

as the bifurcation parameter. The physical parameter j_0 can always easily be found once γ is fixed. Note, however, that the relation

$$j_0^2 = \frac{1 \pm \sqrt{1 - 8\gamma}}{2\gamma} - 1 \tag{7.46}$$

yields two positive values of j_0 in the range $0 < \gamma < \frac{1}{8}$ since $\gamma(j_0)$ is nonmonotonic (see Fig. 7.33). We have then

$$f_a = \gamma - T, \tag{7.47}$$

$$f_u = -\gamma, \tag{7.48}$$

and hence

$$\Theta(k) = \gamma - T - \alpha - (D + 1)k^2, \tag{7.49}$$

$$\Delta(k) = \alpha T + (\alpha - \gamma D + DT)k^2 + Dk^4. \tag{7.50}$$

As additional control parameters we will use the ratio α of the time scales of a and u (which can be varied by changing the external capacitor), the effective diffusion constant D (which is the squared ratio of the two intrinsic length scales of u and a), and the system's size L. The parameter $T = 0.05$ will be kept fixed throughout the following.

The roots of (7.40) yield the dispersion relation

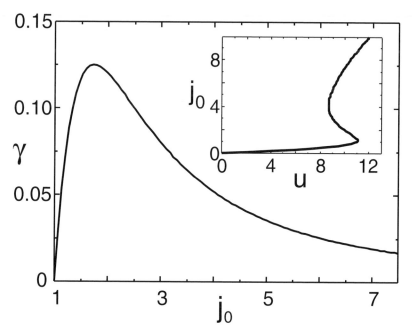

Figure 7.33. The bifurcation parameter γ as a function of the physical control parameter j_0. The inset shows the normalized static current–voltage characteristic for $T = 0.05$. (After Meixner et al. (1997b).)

$$\lambda_{1,2}(k) = \tfrac{1}{2}\left(\Theta \pm \sqrt{\Theta^2 - 4\Delta}\right).\tag{7.51}$$

Modes with $\mathrm{Re}\,\lambda > 0$ are unstable. The Neumann boundary conditions allow only cosine modes $\cos(kx)$, where $k = n\pi/L$ with integer n. For those unstable modes the real and imaginary parts of λ are plotted versus k for various values of α and j_0 in Figs. 7.34(a) and (b). It can be seen that instabilities occur in two ranges of k around $k = 0$ (Hopf modes; λ complex) and around a finite value $k \approx 20$ (Turing modes; λ real). We shall now investigate their occurrence and interactions in detail.

A Hopf bifurcation of a limit cycle occurs if $\Theta = 0$ (while $\Delta > 0$), i.e. if

$$\gamma = \mathcal{T} + \alpha + (D + 1)k^2.\tag{7.52}$$

When the bifurcation parameter γ (or equivalently j_0) is increased, the homogeneous steady state will first become unstable against the homogeneous mode $k = 0$, and will evolve toward a homogeneous limit cycle above the Hopf threshold of instability:

$$\gamma^{\mathrm{H}} = \alpha + \mathcal{T}.\tag{7.53}$$

At the bifurcation point, the frequency of these temporal oscillations is given by $\omega_c = \mathrm{Im}\,\lambda = \sqrt{\Delta(k = 0)}$, i.e.

$$\omega_c = \sqrt{\alpha \mathcal{T}}.\tag{7.54}$$

A Turing instability, leading to the bifurcation of stationary, spatially periodic patterns of wave vector k, appears if $\lambda(k) = 0$ at some finite k, i.e. if $\Delta = \Delta(k) = 0$. The graph of $\Delta(k)$ is a fourth-order parabola with a minimum value

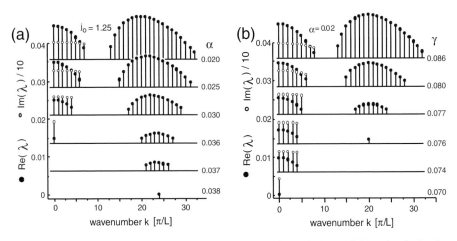

Figure 7.34. Dispersion relations for various values of (a) α and (b) γ (equivalently j_0). $\mathrm{Re}\,\lambda$ (full circles) and $\mathrm{Im}\,\lambda$ (open circles) are plotted versus the wave vector $k = n\pi/L$ in units of π/L for unstable modes only. In (a) $j_0 = 1.25$ and in (b) $\alpha = 0.02$ is fixed. ($L = 600$, $\mathcal{T} = 0.05$, and $D = 8$.) (After Meixner *et al.* (1997b).)

$$\Delta(k_{\mathrm{min}}) = \alpha\mathcal{T} - \frac{(\alpha - \gamma D + D\mathcal{T})^2}{4D} \tag{7.55}$$

at

$$k_{\mathrm{min}}^2 = -\frac{\alpha - \gamma D + D\mathcal{T}}{2D}. \tag{7.56}$$

The Turing instability first occurs at that critical wave vector $k_{\mathrm{c}} = k_{\mathrm{min}}$ for which the minimum $\Delta(k_{\mathrm{min}})$ becomes zero, i.e.

$$\alpha - \gamma D + D\mathcal{T} = \pm 2\sqrt{\alpha\mathcal{T}D}. \tag{7.57}$$

The root with the plus sign must be discarded because it leads to an imaginary wave number. Equation (7.57) determines the critical value of the control parameter γ corresponding to the Turing threshold of instability:

$$\gamma^{\mathrm{T}} = \left(\sqrt{\mathcal{T}} + \sqrt{\frac{\alpha}{D}}\right)^2. \tag{7.58}$$

The corresponding wave number is obtained by inserting (7.58) into (7.56), which gives

$$k_{\mathrm{c}}^2 = \sqrt{\frac{\alpha\mathcal{T}}{D}}. \tag{7.59}$$

In Fig. 7.35 the curves at which the Hopf and the Turing instabilities occur are plotted in the (α, j_0) control parameter space for fixed $\mathcal{T} = 0.05$ and $D = 8$. Below the full line (Hopf bifurcation) the homogeneous steady state is unstable with respect to temporal oscillations, and below the dashed line (Turing instability) it is unstable with respect to stationary spatially periodic patterns. In the overlap of these two instability regimes there is a competition between temporal and spatial symmetry-breaking instabilities. This can lead to an interaction of these two types producing particularly complex spatio-temporal patterns if the thresholds for the two instabilities occur close to each other. This is the case in the vicinity of degenerate points (marked by C_1, C_2) where the Hopf and the Turing bifurcation coincide: These are called *codimension-two Turing–Hopf points*, because two control variables are necessary to fix this bifurcation point in a generic system of equations.

At codimension-two Turing–Hopf points (C_1, C_2), we have $\gamma^{\mathrm{T}} = \gamma^{\mathrm{H}}$, in other words

$$\alpha + \mathcal{T} = \left(\sqrt{\mathcal{T}} + \sqrt{\frac{\alpha}{D}}\right)^2. \tag{7.60}$$

For a given α, we find the critical value of D:

$$D_{\mathrm{c}} = \left(\sqrt{\frac{\mathcal{T}}{\alpha} + 1} - \sqrt{\frac{\mathcal{T}}{\alpha}}\right)^{-2}. \tag{7.61}$$

If $D < D_{\mathrm{c}}$, it follows from (7.53) and (7.58) that $\gamma^{\mathrm{H}} < \gamma^{\mathrm{T}}$, and hence, with increasing γ, corresponding to increasing j_0 near C_1, or decreasing j_0 near C_2, the Hopf threshold is the first to be crossed and hence the Hopf instability will be the first to occur near criticality. If $D > D_c$, on the contrary, the first bifurcation will occur toward Turing patterns.

For fixed D, codimension-two Turing–Hopf points occur at

$$\alpha_c = \frac{4D\mathcal{T}}{(D-1)^2}, \tag{7.62}$$

which yields the two points (C_1, C_2) at $\alpha_c \approx 0.033$ marked in Fig. 7.35. If $\alpha < \alpha_c$, then, with increasing j_0, the Hopf bifurcation is the first to occur. For $\alpha > \alpha_c$ on the contrary, the Turing bifurcation appears first. This can be seen more clearly in the inset of Fig. 7.35 which shows the region near C_1 on an enlarged scale.

In summary, there is a one-parameter family of codimension-two Turing–Hopf points if we consider the control parameters α, D, and j_0 as adjustable parameters. The projections of this curve onto the three coordinate planes in (α, D, j_0) control parameter space is shown in Fig. 7.36.

Generally, in the vicinity of a codimension-two Turing–Hopf point, the bifurcation scenarios can be divided into two main groups. The first one gathers the dynamics resulting from the interaction between the Turing mode (k_c) and the Hopf mode (ω_c). It has been shown (Keener 1976, Kidachi 1980, Guckenheimer and Holmes 1983) using

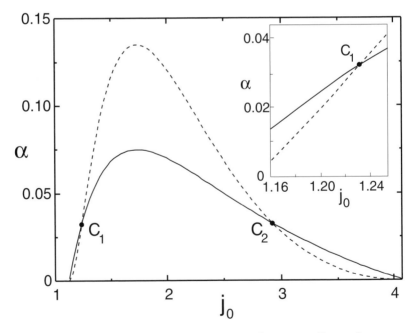

Figure 7.35. Instability regimes in the (α, j_0) control parameter diagram for $\mathcal{T} = 0.05$ and $D = 8$. The full and the dashed lines denote the Hopf bifurcation and the Turing instability, respectively. The codimension-two Turing–Hopf points are marked by C_1 and C_2. The inset shows the vicinity of C_1 on an enlarged scale. (After Meixner *et al.* (1997*b*).)

the amplitude-equations formalism (Cross and Hohenberg 1993) that the competition between these two modes can lead to three different solutions, which are the pure Turing structure, the pure Hopf oscillation, and a mixed mode (k_c, ω_c) consisting of a Turing pattern oscillating globally in time (De Wit *et al.* 1993). Depending on the specific parameters of the system (De Wit *et al.* 1996), the relative stabilities of these three solutions lead to two bifurcation scenarios: Either the mixed mode is always unstable while the Turing and Hopf states are bistable in a given domain of the control parameter, or the mixed mode is stable for values of parameters for which the Turing and Hopf modes are both unstable. The coupling between the Turing and the Hopf modes thus leads either to Turing–Hopf bistability or to a mixed mode with one wave number and one frequency.

The second main group of dynamics near the codimension-two Turing–Hopf point results from subharmonic instabilities (Hill and Stewart 1991, Cheng and Chang 1992, Fujimura and Renardy 1995, Lima *et al.* 1996) of the pure Turing or Hopf modes. Let us first consider the Turing mode with wave number k_c. Close to the codimension-two Turing–Hopf point, its subharmonic mode with wave number $k_c/2$ may have a complex linear eigenvalue with a small growth rate and a frequency $\omega(k_c/2)$. Near the codimension-two Turing–Hopf point, a resonant interaction between the two modes $(k_c, 0)$ and $(k_c/2, \omega(k_c/2))$ can give rise to a new stable mixed state called a subharmonic Turing mixed mode (Lima *et al.* 1996), or also in short a *sub-T* mode. This new mixed solution corresponds to a spatial structure with two wave numbers (k_c and $k_c/2$) oscillating in time with one frequency, $\omega(k_c/2)$. An analogous subharmonic instability can also occur if the system is originally in a pure Hopf state with frequency ω_c. If the control parameter is increased, a resonant interaction between this Hopf mode and the subharmonic mode $(k(\omega_c/2), \omega_c/2)$ can occur. The new stable solution resulting from this interaction is then a subharmonic

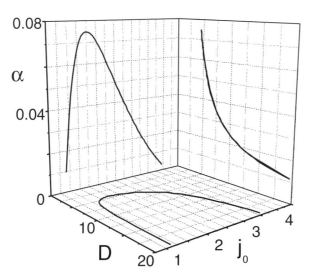

Figure 7.36. Projections of the curve of codimension-two Turing–Hopf points onto the three coordinate planes of the (α, D, j_0) control parameter space (after Meixner *et al.* (1997*b*)).

Hopf mixed mode (De Wit *et al.* 1996), or in short a *sub-H* mode, characterized by two frequencies (ω_c and $\omega_c/2$) and one wave number, $k(\omega_c/2)$. In the particular case in which $k(\omega_c/2)$ is on the order of the subharmonic wave number of the Turing mode, i.e. $k \approx k_c/2$, another mixed state with two wave numbers ($k(\omega_c/2)$ and $2k \approx k_c$) and two frequencies (ω_c and $\omega_c/2$) can be observed. This mixed solution is termed a subharmonic Turing–Hopf mode (De Wit *et al.* 1996), or also a *sub-HT* mode, since it results from the resonance near the codimension-two Turing–Hopf point of a sub-H mode and a Turing mode.

In short, the spatio-temporal dynamics near a codimension-two Turing Hopf point feature either bistability between steady structures and temporal oscillations or various mixed states ranging from the simple Turing–Hopf mixed mode to various types of subharmonic modes. Let us note that, in large systems, each of these solutions can undergo phase instabilities for given values of the parameters, giving rise to spatio-temporal chaos. The Benjamin–Feir and Eckhaus instabilities are the well-known phase instabilities of the pure Hopf and Turing modes, respectively (Cross and Hohenberg 1993). Conditions for which the Turing–Hopf mixed mode (De Wit *et al.* 1993) and the sub-T mode (Lima *et al.* 1996) become chaotic due to a phase instability have also been given.

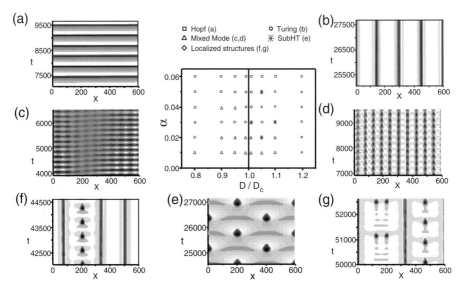

Figure 7.37. Regimes of different asymptotic spatio-temporal behaviors near the codimension-two Turing–Hopf bifurcation given by the line $D/D_c = 1$. The symbols in the (α, D/D_c) control parameter space denote various types of space–time patterns, which are illustrated by typical space–time plots of $j(x,t)$ as insets: (a) Hopf oscillations (squares), (b) Turing patterns (circles), (c) and (d) mixed modes (triangles), (e) subharmonic spatio-temporal spiking (asterisks), and (f) and (g) localized Turing–Hopf structures (diamonds). (After Meixner *et al.* (1997*b*).)

A systematic overview of the various dynamic behaviors in our semiconductor model can be gained by investigating the various regimes in a diagram of α versus D/D_c (Fig. 7.37). In physical units, this diagram corresponds to the ratio of time scales τ_a/τ_u versus the ratio of the squared length scales l_u^2/l_a^2 of the two variables a and u. By normalizing D by its critical value $D_c(\alpha)$ for each α we transform all codimension-two Turing–Hopf points onto the straight line $D/D_c = 1$. As discussed before, for $D/D_c > 1$ the Turing instability occurs first with increasing j_0, whereas for $D/D_c < 1$ the Hopf bifurcation occurs first. Therefore, on the right-hand side of the diagram, the Turing modes are expected to be dominant near criticality, whereas on the left-hand side this applies to the Hopf modes. For various points in the control parameter space of Fig. 7.37 scenarios as a function of j_0 have been simulated with random initial conditions. Typical space–time plots observed at these points for values of j_0 near the Turing–Hopf point C_1 (cf. Fig. 7.35) are shown as insets. It should, however, be noted that, depending upon the initial conditions, the asymptotic behavior obtained can be quite different. If j_0, and thus γ, is chosen to be above the first instability threshold, γ^H or γ^T, and D/D_c is sufficiently far from the Turing–Hopf point, then the only patterns found for Neumann boundary conditions are indeed pure Hopf oscillations (a) (for $D/D_c < 1$) or Turing structures (b) (for $D/D_c > 1$). If D/D_c approaches unity, the spatial and temporal modes may interact, leading to the various types of dynamics predicted theoretically. The Turing–Hopf mixed mode with one wave number and one frequency is recovered near the codimension-two line (d). In large systems, the temporal oscillations of the mixed mode might not be exactly in phase, as shown in (c). Some of the subharmonic mixed states have also been obtained in our semiconductor model. They are associated with a subharmonic instability of the pure modes (e), and can be explained by invoking a resonance of the subharmonic Hopf mode with frequency $\omega_c/2$ and wave vector $k(\omega_c/2)$ with a Turing mode with wave vector k_c if $k(\omega_c/2) \approx k_c/2$. Thus the spatio-temporal dynamics is characterized by two frequencies, ω_c and $\omega_c/2$, and two wave vectors, k_c and $k_c/2$. At each location of the system, the current density oscillates in time with the frequency $\omega_c/2$, and the spatial pattern has the wave vector $k_c/2$. The original frequency ω_c and wave vector k_c are still visible in the alternating spatial and temporal shift of the pattern by one period in space and in time. This is the periodic spatio-temporal spiking mode which we have discussed in detail earlier in Section 7.3.1 (Fig. 7.28(a)). Our interpretation in terms of two frequencies and two wave vectors is corroborated by a Karhunen–Loève decomposition of the relevant spatio-temporal eigenmodes in section 7.3.4. We have now gained a more profound understanding of its nature. It naturally appears close to codimension-two Turing–Hopf points.

More complex interactions occur at other points in the $(\alpha, D/D_c)$ diagram, (f) and (g), or when the bifurcation parameter j_0 is further increased. In particular, bistability between Turing and Hopf modes is obtained for several values of parameters. In the bistability regime, localized Turing–Hopf structures such as droplets of one state embedded in the other, and fronts between a Turing pattern and an oscillating region, are commonly observed. In Fig. 7.37(f) a spiking "droplet" is embedded in a Turing structure, and in (g) a one-period Turing structure is embedded in a complex oscillating state. For much larger j_0 in the vicinity of the second codimension-two Turing–Hopf point C_2 spatial coexistence of a Hopf oscillation and a Turing structure is obtained (Fig. 7.38). Depending upon the initial conditions, various localized structures occur: A Turing–Hopf front with a fixed boundary between the two phases (Fig. 7.38(a)), Turing domains embedded between two Hopf states (Fig. 7.38(b)), and alternating sequences of localized Hopf and Turing domains (Fig. 7.38(c)) are found. Similar behavior has been found in other two-component reaction–diffusion systems (Heidemann *et al.* 1993, Perraud *et al.* 1993, De Wit *et al.* 1996). Such localized, coexisting structures indicate that, in this range of parameters, there is bistability between Turing and Hopf modes, in contrast to the mode mixing which is effective at lower values of j_0, near the codimension-two point C_1. It may be conjectured that this difference in behavior can be explained by a similar argument to that used in Section 7.2.2 for the model with one diffusion length, (7.4) and (7.5), where it has been related to the different order of instability thresholds near the codimension-two bifurcation, see (7.20) and (7.21). It should also be noted that the point C_1, where spiking and mixed modes occur, is located at a j_0 value in the right-hand corner of the SNDC regime (cf. the inset of Fig. 7.33), whereas C_2, where localized structures are found, lies in the left-hand corner close to the coexistence value u_{co} of the model with one length scale (the condition for spatial coexistence of two stable homogeneous

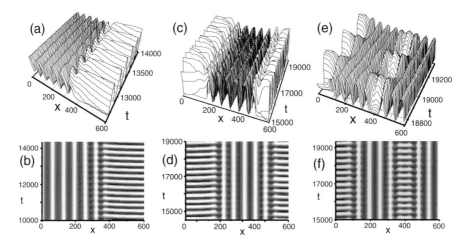

Figure 7.38. Localized structures near the codimension-two Turing–Hopf point C_2 for $\mathcal{T} = 0.05$, $\alpha = 0.02$, $D = 8$ (corresponding to $D/D_c = 0.67$), $j_0 = 3.1$, and various initial conditions: (a) a Turing–Hopf front, (b) a Turing domain embedded between two Hopf states, and (c) localized Turing–Hopf structures. The current density $j(x,t)$ is shown as a density plot and as a three-dimensional representation. (After Meixner *et al.* (1997*b*).)

phases is satisfied at a value of u_{co} on the left-hand side of the SNDC range of the $j_0(u)$ characteristic).

A detailed analysis of the bifurcation scenario as a function of j_0 is made in Fig. 7.39 for a given set of parameters α and D, summarizing the above explanations. The stability ranges of the various spatio-temporal patterns are schematically indicated. The Hopf oscillation persists throughout the range of j_0 values between the two Hopf bifurcations (cf. Fig. 7.35). In the immediate vicinity of the thresholds C_1 and C_2, the period of the oscillations is that given by the linear stability analysis (compare the values for ω_c and k_c found by linear stability analysis at the Hopf and Turing instabilities, respectively, and the corresponding temporal and spatial periods $\tau_c = 2\pi/\omega_c$ and $\Lambda_c = 2\pi/k_c$), whereas this period increases strongly near $j_0 = 1.25$. The subharmonic Hopf–Turing (sub-HT) spiking mode exists only for low values of j_0 near C_1, and localized structures and the pure Turing mode are found at higher values near C_2. Note that, since the localized structures are due to a Turing–Hopf bistability, they exist in the range of j_0 for which the pure Turing and Hopf modes are both observed. The Turing–Hopf mixed mode with one wave number and one

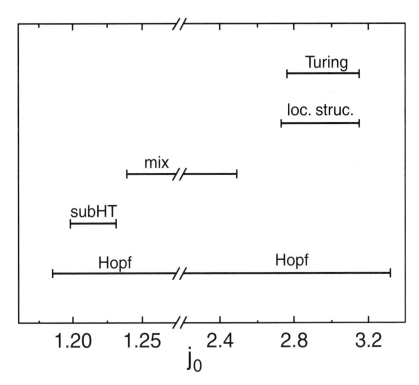

Figure 7.39. Stability regimes of various patterns as a function of j_0 for $T = 0.05$, $\alpha = 0.02$, and $D = 8$. Here mix stands for the Turing–Hopf mixed mode with one wave number and one frequency. Note that localized structures (loc. struc. are observed for values of parameters for which the Turing and Hopf modes are stable. (After Meixner *et al.* (1997*b*).)

frequency appears in a wide intermediate range between these two ends. It should be noticed that, near C_2, the wavelength of the Turing mode is that given by the linear stability analysis. If j_0 is then decreased, this wavelength increases strongly and can become almost twice the one predicted by the linear stability analysis in the vicinity of C_1. This explains why all spatio-temporal dynamics slightly above C_2 exhibit periods and wavelengths much greater than those predicted by the linear stability analysis. By choosing appropriate initial conditions, multistability between some of these patterns can be realized as indicated. In general, all mixed modes disappear in favor of either a Hopf or a Turing mode if the parameters α and D are sufficiently far from the codimension-two Turing–Hopf points. Furthermore, it should be noted that the asymptotic patterns are in general preceded by transient spatio-temporal chaos when random initial conditions are used, as discussed in Section 7.3.1, and that the transient times are often so long that it is difficult to observe the asymptotic state.

The behavior found in the reaction–diffusion model (7.31) and (7.32) is quite similar to that obtained for the Brusselator model of chemical active media (Nicolis and Prigogine 1977):

$$\partial_t X = A - (B + 1)X + X^2 Y + D_x \nabla^2 X,$$

$$\partial_t Y = BX - X^2 Y + D_y \nabla^2 Y, \tag{7.63}$$

where X and Y are the concentrations of two chemical species, A and B are reaction coefficients, and D_x and D_y are diffusion constants. B is the control parameter. Here the spatio-temporal dynamics near the codimension-two Turing–Hopf point can be classified in the A versus σ/σ_c control parameter space, where $\sigma = D_x/D_y$, i.e. in the "time scale" versus "space scale" plane. A straightforward comparison between the two models can be achieved if we consider the following equivalences:

$$\sqrt{\alpha} = A, \tag{7.64}$$

$$D = 1/\sigma. \tag{7.65}$$

The corresponding thresholds of instabilities and other characteristics of the two systems are given in Table 7.2. An important difference between the two models is that, in the Brusselator, once $(A, \sigma/\sigma_c)$ is fixed, one has only one codimension-two Turing–Hopf point whereas in the semiconductor model two Turing–Hopf points exist for given values of α and D/D_c, since two values of j_0 correspond to a given value of γ. However, in the Brusselator model Hopf and Turing modes, localized structures, mixed modes and subharmonic Turing–Hopf modes, including spatio-temporal chaos, have also been found (De Wit *et al.* 1993, 1996, Petrov *et al.* 1995, Lima *et al.* 1996).

7.3.4   The Karhunen–Loève decomposition of spatio-temporal modes

In the preceding sections we have demonstrated that the reaction–diffusion system considered gives rise to rather complex spatio-temporal patterns. From visual inspection of the space–time plots it is clear that temporal and spatial degrees of freedom are simultaneously excited and intrinsically coupled in a complex way. It is therefore desirable to develop a systematic procedure to extract the relevant spatial and temporal eigenmodes from the numerical data and quantify the *spatio-temporal complexity*. Such a method of data reduction is provided by the *Karhunen–Loève decomposition*, which is also called the method of empirical orthogonal functions, proper orthogonal decomposition, principal component analysis, singular spectrum analysis, or singular value decomposition (Sirovich 1987, 1989a, Vautard and Ghil 1989, Deane *et al.* 1991, Caponeri and Ciliberto 1992, Holmes *et al.* 1996). It is a pattern-recognition algorithm that uses linear correlations in space to find coherent structures, the *empirical orthogonal functions* (or *empirical eigenmodes*) in spatio-temporal data. With the help of this analysis a few relevant spatial "patterns" can be extracted and we will use these eigenmodes as basis functions to separate the spatial and temporal dynamics (Meixner *et al.* 1997a).

Let the current density $j(x, t)$ be the quantity under consideration. Then the Karhunen–Loève expansion (or proper orthogonal decomposition)

$$j_p(x, t) = \sum_{k=1}^{p} a_k(t)\, w_k \phi_k(x) \tag{7.66}$$

Table 7.2. *Comparison between the semiconductor activator–inhibitor system and the Brusselator chemical reaction–diffusion system*

	Semiconductor	Brusselator
Bifurcation parameter	γ	B
Ratio of diffusion coefficients	$1/D$	$D_x/D_y = \sigma$
Hopf threshold	$\gamma^{\mathrm{H}} = \mathcal{T} + \alpha$	$B^{\mathrm{H}} = 1 + A^2$
Hopf frequency ω_{c}	$\sqrt{\alpha \mathcal{T}}$	A
Turing threshold	$\gamma^{\mathrm{T}} = \left(\sqrt{\mathcal{T}} + \sqrt{\dfrac{\alpha}{D}}\right)^2$	$B^{\mathrm{T}} = (1 + A\sqrt{\sigma})^2$
Turing wave number k_{c}^2	$\sqrt{\dfrac{\alpha \mathcal{T}}{D}}$	$\dfrac{A}{\sqrt{D_x D_y}}$
Codimension-two Turing–Hopf point	$D_{\mathrm{c}} = \left(\sqrt{\dfrac{\mathcal{T}}{\alpha}} + \sqrt{\dfrac{\mathcal{T}}{\alpha} + 1}\right)^2$	$\sigma_{\mathrm{c}} = [(\sqrt{1 + A^2} - 1)/A^2]^2$

yields a good approximation to $j(x, t)$ in the sense of a minimum mean squared error for a given number of expansion terms p. The spatial functions $\phi_k(x)$ are called *eigenmodes*, whereas the temporal functions $a_k(t)$ are the *amplitudes* of the corresponding modes. The first eigenmode which fits best is determined by solving the variational problem

$$w_1 = \max_{\phi_1} \left\{ \lim_{T \to \infty} \frac{1}{T} \int_0^T (\phi_1, j)^2 \, dt \right\} \tag{7.67}$$

with $(\phi_1, j) = \int_0^L \phi_1(x) j(x, t) \, dx$ being a scalar product with respect to space. The other eigenmodes are determined analogously, taking into account orthogonality and normalization.

Replacing the continuous variables x and t by discrete values with, say, M steps in time and N steps in space, eq. (7.66) becomes

$$j_p(x_j, t_i) = \sum_{k=1}^{p} a_k(t_i) w_k \phi_k(x_j). \tag{7.68}$$

A closer look reveals that this has the structure of a matrix equation of the form

$$J = A \cdot W \cdot \Phi, \tag{7.69}$$

where the $M \times N$ matrix A holds the M discretized values of the mode amplitudes $a_k(t_i)$ as columns, and the $N \times N$ matrix Φ contains the N eigenmodes as rows each consisting of the N values of $\phi_k(x_j)$. The $N \times N$ diagonal matrix W contains the eigenvalues w_k. The approximation J_p of J is then calculated by using only the first p columns of A and the first p rows of Φ, i.e. by setting the remaining $N - p$ eigenvalues approximately equal to zero.

Indeed, such a matrix decomposition is guaranteed to exist by a mathematical theorem for any $M \times N$ matrix J, where M is greater than N. In practice, numerical algorithms may be used to find this unique decomposition (Press *et al.* 1992).

The discrete Karhunen–Loève decomposition (7.68) is different from the continuous one (7.66) insofar as complete data set of the $M \times N$ matrix J can be reconstructed exactly by using all N eigenmodes.

Note that this variational problem is equivalent to the eigenvalue problem for the spatial co-variance matrix

$$C_{jj'} = \frac{1}{T} \int_0^T j(x_j, t) j(x_{j'}, t) \, dt. \tag{7.70}$$

The eigenvectors $\phi_k(x_j)$ of $C_{jj'}$ ($k = 1, \ldots, N$) then represent an optimal basis for expansion of $j(x, t)$ in the sense that they maximize the projected mean $\langle (\phi_k, j)^2 \rangle = (1/T) \int_0^T (\phi_k, j)^2 \, dt$, i.e. for a given accuracy the minimum number of eigenvectors is needed in the expansion. The eigenvalues w_k corresponding to the eigenmodes

ϕ_k of the co-variance matrix are equal to the variance of the kth mode amplitude $\sigma_k^2 \equiv \langle a_k(t)^2 \rangle$.

Instead of the eigenvalues w_k the normalized eigenvalues

$$w_k^* = \frac{w_k}{\sum_{i=1}^{N} w_i} \tag{7.71}$$

are used in the following. They are a measure of how much the corresponding eigenmodes ϕ_k contribute to the pattern as a whole.

The Karhunen–Loève expansion has been widely used in the analysis of nonlinear hydrodynamic flows (Deane *et al.* 1991) and self-organized pattern formation in surface chemical reactions (Graham *et al.* 1995, Arndt *et al.* 1997). It has also been applied to the complex spatio-temporal dynamics of semiconductor laser arrays (Hess and Schöll 1994, Merbach *et al.* 1995) and to the reaction–diffusion system studied in this Chapter (Meixner *et al.* 1997*a*). For spatio-temporal spiking in the reaction–diffusion system (7.4) and (7.5) with global coupling studied in Section 7.2, it was found that this can be described in terms of just a few eigenmodes. The empirical eigenmodes for periodic (Fig. 7.5(c)) and chaotic (Fig. 7.5(d)) spiking look very similar. The individual eigenvalues for these two data sets are also quite similar. In both cases the first two eigenmodes cover about 87%–89% of the patterns in terms of the relative accumulated eigenvalues. The first mode describes the spikes, whereas the small-amplitude oscillations between the spikes are determined mainly by the second mode. It can be deduced that Fig. 7.5(d) corresponds to the case of primarily temporal chaos.

In Figs. 7.40–7.46 the results of the Karhunen–Loève decomposition for the reaction–diffusion system (7.31) and (7.32) are shown. The Karhunen–Loève expansion has been performed by use of a singular value decomposition algorithm. This algorithm is mainly based on matrix manipulation (Press *et al.* 1992). In the following the spatial average $\int j(x, t)\, dx/L$ was subtracted from the data sets at every sampled time. For the expansion this acts like choosing the zeroth eigenmode to be constant with a time-dependent mode amplitude. All higher modes then have a vanishing spatial mean value. The spatio-temporal patterns are analyzed for a selection of values of the parameters α, D, and j_0, and typical random initial conditions. In all cases the system first undergoes a chaotic transient, which provides patterns that are irregular in time and space. After an abrupt transition an asymptotic, periodic pattern is approached.

Depending upon the parameters, the asymptotic patterns near a codimension-two Turing–Hopf point are pure Hopf or Turing modes, coexisting localized structures, or mixed modes (spatio-temporal spiking), as discussed in Section 7.3.3. For pure Hopf or Turing patterns the Karhunen–Loève decomposition yields, as expected, only one dominant mode that determines the system's behavior completely. In the Hopf case this is a spatially homogeneous mode that oscillates in time. Pure Turing patterns are fully described by one inhomogeneous spatially periodic mode with constant amplitude.

Figure 7.40 shows the analysis of a pattern that represents a coexisting Turing–Hopf structure with a localized front between the two phases. The right-hand part of the pattern is described by the first eigenmode that has a Turing shape to the right and vanishes to the left. This mode exhibits no time dependency. There is almost no contribution to the Turing pattern from the other modes. The second most important eigenmode is constant to the left and vanishes to the right. (Note that the spatial average has been subtracted from the quantity plotted in Fig. 7.40.) It is oscillating periodically in time, and, in conjunction with the third mode, essentially generates the Hopf pattern. All higher modes describe more or less details of the system's behavior at the interface of the two patterns. As can be seen from the relative contribution of the eigenvalues w_i^* (upper right-hand side), the first three modes alone provide a reconstruction of 97% of the pattern.

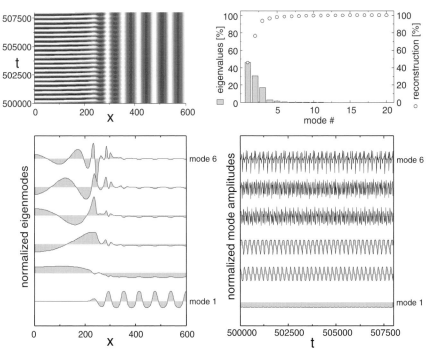

Figure 7.40. Karhunen–Loève decomposition of a spatially coexisting Hopf and Turing pattern. The upper left-hand plot shows the density plot of j in the time-versus-space diagram. The lower left-hand part shows the first six Karhunen–Loève eigenmodes versus space x. The lower right-hand part shows the corresponding normalized mode amplitudes versus time t. The upper right-hand part shows the relative eigenvalues w_i^* (full columns) and accumulated eigenvalues $\sum_{i=1}^{n} w_i^*$ (open circles) of the first 20 eigenmodes. ($L = 600$, $\alpha = 0.02$, $D = 8.0$, $j_0 = 3.1$, and $T = 0.05$). (After Meixner *et al.* (1997a).)

A more complex localized pattern is shown in Fig. 7.41. It consists in two coexisting spiking domains separated by a one-period Turing pattern. The two spiking domains have a frequency ratio of $3 : 2$. Here the Karhunen–Loève decomposition yields three main modes that contain about 80% of the system's dynamics. The first eigenmode describes the Turing stripe and is almost time-independent. The next two modes supply the information for the left- and the right-hand spiking regions, respectively. The frequency ratio of $3 : 2$ is reflected in the respective mode amplitudes. All higher modes describe the interaction of the two sides with a relative frequency of six.

Although the pattern shown in Fig. 7.42 does not look simple, it consists essentially in only one single eigenmode that reconstructs more than 96% of the spatio-temporal dynamics. The eigenmode is periodic in space, i.e. it has a fixed wave number k_0 and its mode amplitude is oscillating in time with a fixed frequency ω_0. It thus describes a mixed Turing–Hopf mode of which k_0 and ω_0 are the Turing wave number and Hopf frequency, respectively.

A more complex mixed Turing–Hopf pattern is given by the periodic spatio-temporal spiking discussed in the earlier sections of this chapter (Fig. 7.43). The Karhunen–Loève decomposition yields three main eigenmodes reconstructing together about 80% of the pattern. Two of these modes (numbers 1 and 3) have a spatial

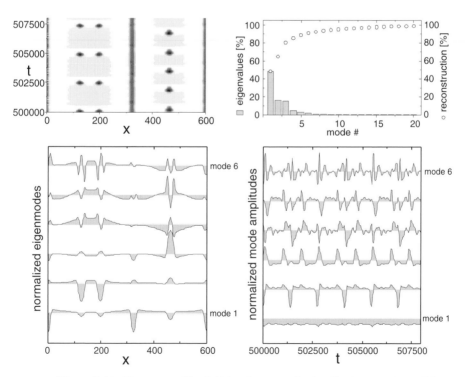

Figure 7.41. The same as Fig. 7.40 but for a complex localized pattern ($L = 600$, $\alpha = 0.02$, $D = 12.5$, $j_0 = 1.18$, and $\mathcal{T} = 0.05$) (after Meixner *et al.* (1997a)).

periodicity with wave number k_0 and oscillate in time with a fixed frequency ω_0. The other mode (number 2) has a wave number $k_0/2$ and is oscillating with only half the frequency, $\omega_0/2$. Thus the spatio-temporal dynamics is indeed associated with a subharmonic instability of the pure modes near a codimension-two Turing–Hopf point, and can be explained by invoking the interaction of the Turing mode (wave number k_0) and the Hopf mode (frequency ω_0) and their subharmonics $k_0/2$ and $\omega_0/2$, cf. Section 7.3.3. We see that the Karhunen–Loève decomposition provides a more profound understanding of the spiking mode.

If the Karhunen–Loève decomposition is applied to an arbitrary time interval in the transient regime, one finds eigenmodes without spatial and temporal periodicity (Fig. 7.44). Choosing different time intervals produces completely different sets of eigenmodes. It is obvious from the relative eigenvalues that even ten modes are not able to provide an acceptable reconstruction of the spatio-temporal patterns. Here the Karhunen–Loève decomposition obviously fails, and one may take this as another indication that here spatio-temporal chaos involving many degrees of freedom is indeed excited. Nevertheless, useful information can be gained from the Karhunen–Loève decomposition by using a scaling property of the number of relevant modes, as will be explained in Section 7.3.5.

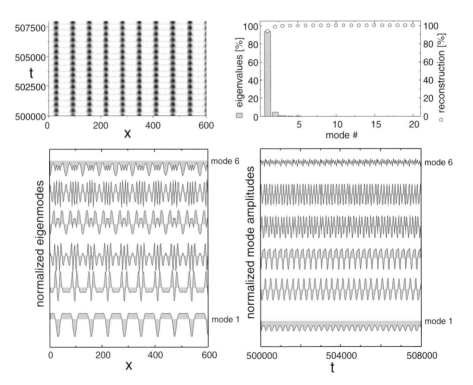

Figure 7.42. The same as Fig. 7.40 but for a mixed Turing–Hopf mode ($L = 600$, $\alpha = 0.038$, $D = 8.0$, $j_0 = 1.25$, and $\mathcal{T} = 0.05$) (after Meixner *et al.* (1997a)).

Also, one can always observe in the transient chaotic regime that all eigenmodes have comparable eigenvalues, which decrease only very slowly with increasing mode index. This indicates that there exist no preferred modes during the chaotic transient that might be responsible for the gradual emergence of an asymptotic pattern. These findings are also consistent with the argument in Section 7.3.1 which postulated that the transient patterns are essentially uncorrelated in space and time, and therefrom inferred an exponential increase of the mean transient times with the system's size L.

Let us now analyze the transition from transient spatio-temporal chaos to the asymptotic periodic spiking behavior (Fig. 7.45). The Karhunen–Loève decomposition shows that the transition from chaotic to asymptotic behavior is quite abrupt and takes place within only a few time periods defined by the oscillation of the most dominant mode. The mode amplitudes of those three modes governing the asymptotic subharmonic spatio-temporal spiking (cf. Fig. 7.43) begin to grow suddenly and become dominant over all other modes even if they have been almost nonexistent in terms of their relative eigenvalues toward the end of the chaotic phase (Fig. 7.45). All other modes decay or perform oscillations with very small amplitudes reflecting the absolute values of their eigenvalues. The abrupt transition is characteristic of a chaotic repeller. There is no continuous gradual growth of the asymptotic eigenmodes until

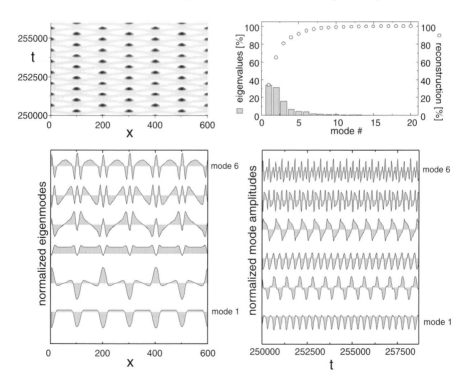

Figure 7.43. The same as Fig. 7.40 but for a subharmonic Turing–Hopf pattern (spatio-temporal spiking) ($L = 600$, $\alpha = 0.02$, $D = 8.0$, $j_0 = 1.21$, and $\mathcal{T} = 0.05$) (after Meixner *et al.* (1997a)).

dominance over the transient modes is gained, but rather the system over its whole length at a certain time coincidentally locks into a periodic solution in space and time.

A more efficient way to monitor the emergence of the asymptotic modes during the transition is to directly calculate the projections of the system's state onto a set of eigenmodes found in an analysis performed in the asymptotic regime after the transition. This allows one to examine the temporal evolution of the pure modes, unspoilt by transition effects as well as by nonasymptotic eigenmodes (Fig. 7.46).

Although the space–time patterns look like the asymptotic state immediately after the transition, it may take a fairly long time, e.g. 100 000 time steps, before there is no noticeable change, neither in the higher eigenmodes nor in the eigenvalues of the most dominant modes. For this reason all asymptotic patterns examined above were taken from data sets recorded for at least 500 000 time steps after the transition from spatio-temporal chaos to asymptotic behavior.

In conclusion, the Karhunen–Loève decomposition proves to be a powerful tool for extracting detailed quantitative information on complex space–time data. For asymptotic patterns the Karhunen–Loève decomposition shows that these are usually determined by only a few characteristic eigenmodes. Patterns consisting in modes with

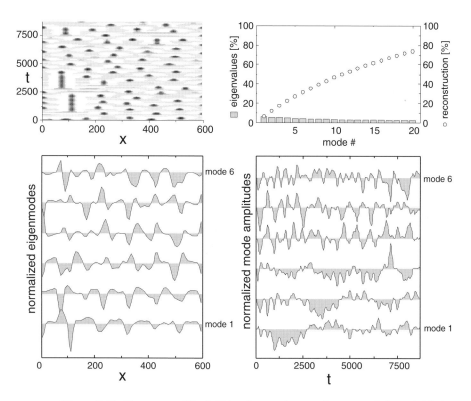

Figure 7.44. The same as Fig. 7.40 but for transient spatio-temporal chaos (with the same parameters as those in Fig. 7.43) (after Meixner *et al.* (1997*a*)).

one wave number k_0 and one frequency ω_0 as well as patterns with three frequencies with a ratio of $1:2:3$, and a subharmonic spatio-temporal spiking pattern with two wave numbers k_0 and $k_0/2$ and two frequencies ω_0 and $\omega_0/2$ have been found. The decomposition of patterns recorded during transient spatio-temporal chaos in terms of their eigenmodes reveals no preferred modes. The transition from transient chaotic to asymptotic periodic behavior occurs abruptly, and all asymptotic modes begin to emerge simultaneously as soon as the system locks into the asymptotic state.

7.3.5 Local characterization of extensive chaos

In this section we use the Karhunen–Loève decomposition (KLD) to compute a characteristic quantity of spatio-temporal chaos, viz. the KLD correlation length ξ_{KLD}. We show that it is a sensitive measure of spatial dynamic inhomogeneities in the extensive chaos regime. It reveals substantial spatial nonuniformity of the dynamics at the boundaries and can also detect slow spatial variations in system parameters.

Results of the previous sections have indicated that states of spatio-temporal chaos exhibit a lack of correlation both in space and in time. Such states are found in

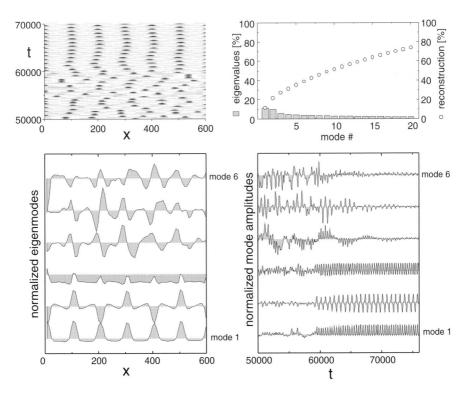

Figure 7.45. The same as Fig. 7.40 but for the transition from transient spatio-temporal chaos to asymptotic periodic spiking (with the same parameters as those in Fig. 7.43) (after Meixner *et al.* (1997*a*)).

many physical systems, including reacting and diffusing chemical flows (Gray and Scott 1984), convective transport of heat (Morris *et al.* 1993, 1996), and charge transport in semiconductors (Niedernostheide 1995, Wacker *et al.* 1995*a*). These states may be characterized using two-point correlation functions (Tufillaro *et al.* 1989) and dimension correlation lengths (O'Hern *et al.* 1996), though both are global measures of the dynamics that assume uniform homogeneous chaos. However, averages of spatio-temporal chaos in experiments (Rudroff and Rehberg 1997, Gluckman *et al.* 1995, Ning *et al.* 1993) and simulation (Eguíluz *et al.* 1999) have demonstrated that the dynamics can be strongly effected by the boundaries. Furthermore, system parameters can often vary spatially (an example being the concentration of a chemical species in a spatially extended reaction–diffusion system), so a means of quantifying parametric variations in the system can be important in making comparisons with experiment. The Karhunen–Loève decomposition (KLD) correlation length (Zoldi and Greenside 1997) was recently used to characterize extensive chaos *locally* in small subsystems of the larger dynamics on the basis of the extensive growth of the KLD dimension (Holmes *et al.* 1996). Since the KLD correlation length is defined in a subsystem, it allows a dynamic measure of nonuniformities in space. Furthermore, this length scale was shown to be an independent quantity that behaves similarly to the dimension correlation length (Zoldi and Greenside 1997) but has a parametric dependency other than the two-point correlation length ξ_2 in coupled map

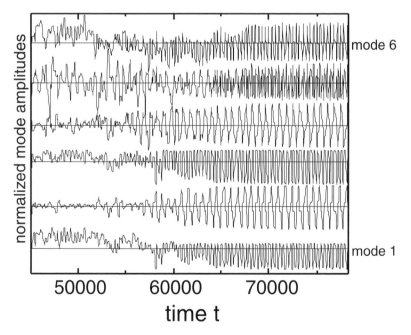

Figure 7.46. Normalized mode amplitudes calculated by projecting the transient spatio-temporal dynamics onto a set of asymptotic Karhunen–Loève eigenmodes (with the same parameters as those in Fig. 7.43) (after Meixner *et al.* (1997*a*)).

lattices (Zoldi and Greenside 1997) and convection data (Zoldi *et al.* 1998). These results suggest that the KLD correlation length is a useful independent length scale in spatio-temporal chaotic systems.

In this section, we present an application of the KLD correlation length to the reaction–diffusion model (7.31) and (7.32) describing charge transport in a semi-conductor with bistable current–voltage characteristics (Meixner *et al.* 2000*b*). This model has an advantage over the Kuramoto–Sivashinsky equation from which the KLD correlation length was originally computed in that there are tunable parameters in the partial differential equations, making it a more general model of spatio-temporal chaos. For a sufficiently large system size, the model exhibits long transients that are extensively chaotic, i.e. the number of positive Lyapunov exponents grows in proportion to the system's volume (Meixner *et al.* 1997*a*). If one defines the KLD dimension D_{KLD} as the number of KLD eigenmodes needed to approximate the space–time data with a certain accuracy, this number also scales extensively with the subsystem's volume. A KLD local correlation length ξ_{KLD}

$$\xi_{KLD} = \left(\frac{D_{KLD}}{V} \right)^{-1/d}, \tag{7.72}$$

where d is the spatial dimensionality of the system, is then derived from the rate of growth of the KLD dimension with the subsystem's volume V. This KLD correlation length is used to quantify dynamic inhomogeneity near boundaries. We demonstrate the existence of a long-range spatial nonuniformity in the KLD correlation length similar to the average patterns of spatio-temporal chaos (Eguíluz *et al.* 1999). Furthermore, we vary a system parameter in the model and confirm that the KLD correlation length can detect parametric changes. Finally, we demonstrate that the KLD correlation length computed for small subsystems is proportional to the two-point correlation length computed over the entire volume of the system.

The Karhunen–Loève decomposition, introduced in Section 7.3.4, is a classical statistical method of representing complex space–time data $j(t_i, x_j)$ by a minimum number of spatial and temporal eigenmodes (Holmes *et al.* 1996). This decomposition proceeds by organizing the discretized data into a space–time matrix,

$$J_{ij} = j(t_i, x_j) - \langle j(t_i, x_j) \rangle, \tag{7.73}$$

where $\langle j(t_i, x_j) \rangle$ is the space–time average of the current density $j(t_i, x_j)$. The space–time matrix is of dimensions $T \times X$, where T is the number of observation times t_i, and X is the number of observation sites x_j within the subsystem. A singular value decomposition of this matrix provides an optimal 2-norm variance decomposition of the space–time matrix $\mathbf{J}$ in the sense that the expansion of

$$J_{ij} \approx \sum_{k=1}^{p} a_k(t_i) \sigma_k^2 \phi_k(x_j) \tag{7.74}$$

in terms of spatial eigenmodes $\phi_k(x_j)$ and normalized mode amplitudes $a_k(t_i)$ has a minimum squared error for a fixed number of expansion terms p. The weightings of the various terms in the expansion are given by their variances σ_k^2, which correspond to the eigenvalues of the positive semidefinite covariance matrix $\mathbf{J}^{\mathrm{T}}\mathbf{J}$, ordered in decreasing size, $\sigma_1^2 \geq \sigma_2^2 \geq \cdots \geq \sigma_X^2$, and $\phi_k(x_j)$ are the eigenvectors of $\mathbf{J}^{\mathrm{T}}\mathbf{J}$. The KLD dimension (Sirovich 1989b) of the matrix J_{ij}

$$
D_{\mathrm{KLD}} = \max \left\{ p: \frac{\sum_{k=1}^{p} \sigma_k^2}{\sum_{k=1}^{X} \sigma_k^2} \leq f \right\} \tag{7.75}
$$

represents the number of linear eigenmodes needed to approximate some fraction $0 < f < 1$ of the total variance of the data (Zoldi and Greenside 1997, Sirovich 1989b, Sirovich and Deane 1991, Ciliberto and Nicolaenko 1991, Vautard and Ghil 1989).

The relation of the KLD decomposition to Fourier analysis depends on the symmetries of the matrix $\mathbf{J}$ (Holmes $et\ al.$ 1996). If the data are homogeneous, i.e. periodic in time or translationally invariant in space, then the autocorrelation matrix $\mathbf{J}^T\mathbf{J}$ becomes translationally invariant and then the KLD modes are Fourier modes. In most physical situations, the dynamics will not be homogeneous or space-translationally invariant due to the boundaries and varying system parameters. Therefore, in spatio-temporally chaotic systems the KLD modes most likely will not trivially correspond to Fourier modes of the data.

The KLD $local$ correlation length is based on the computation of $D_{\mathrm{KLD}}(x_j)$ for concentric subsystems of volume V centered at the point x_j in space. The dimension $D_{\mathrm{KLD}}(x_j)$ typically depends on the point x_j and so provides a measure of spatial dynamic inhomogeneity. For extensive chaotic systems the KLD dimension $D_{\mathrm{KLD}}(x_j)$ will increase linearly with the subsystem's volume V with a slope δ_{KLD}. This indicates that the KLD dimension density $\delta_{\mathrm{KLD}} = D_{\mathrm{KLD}}/V$ is a more useful measure, since it is an intensive property of the subsystem. To derive a characteristic length scale, the KLD correlation length ξ_{KLD} is defined to be $\delta_{\mathrm{KLD}}^{-1/d}$ with spatial dimensionality d. The advantage of the KLD correlation length over the dimension correlation length (Egolf and Greenside 1994) and the two-point correlation length (Tufillaro $et\ al.$ 1989) is that it is computed directly from data in small localized spatial subsystems of larger space–time data sets. This locality has allowed the detection of smooth spatial dynamic nonuniformities of a system parameter of a coupled map lattice (Zoldi and Greenside 1997) and of experimental spatial inhomogeneities in convection data (Zoldi $et\ al.$ 1998). Furthermore, in the study of an Ising-like phase transition, the KLD correlation length ξ_{KLD} was shown (Zoldi and Greenside 1997) to have a critical parametric dependency other than the commonly computed two-point correlation length, indicating, that, at least for some cases ξ_{KLD} is an independent length scale of spatio-temporal chaos. It is not self-evident whether these properties hold for systems described by partial differential equations with boundaries and tunable parameters, in which continuity smoothes out inhomogeneities.

Here we consider the spatially extended reaction–diffusion system of activator–inhibitor type (7.31) and (7.32) studied in the previous sections. We consider a system in a range of sizes $L = 1000$ to $L = 2750$ and parameter values corresponding to transient spatio-temporal chaos. We record the spatial field $j(x, t) - \langle j(x, t) \rangle$ at each mesh point in a concentric subsystem of length S at intervals of 5 time units. To eliminate the influence of initial conditions, the first 250 time units are discarded. The resulting KLD dimension D_{KLD} is shown as a function of the subsystem's size S in Fig. 7.47. For most cases in the following, the KLD dimension D_{KLD} is computed with a relative total variance corresponding to $f = 99.999\%$. This results in keeping roughly a fifth of the number of eigenmodes of the covariance matrix $J^{\text{T}}J$. The KLD correlation length $\xi_{\text{KLD}} = (D_{\text{KLD}}/S)^{-1}$ is computed from the linear growth of the KLD dimension D_{KLD} with the subsystem's size S, and at least five subsystem sizes S are chosen to extract the slope δ_{KLD}. It has been checked that it agrees to good accuracy with the slope determined by varying the *total* system size L. When the total system is made larger, the change in D_{KLD} is equivalent to adding more homogeneous center subsystems. The qualitative behavior of ξ_{KLD} remains similar for fractions f from 85% to 99.9999%. For fractions $f < 85\%$ the scaling of D_{KLD} becomes coarse due to the small number of modes required to satisfy smaller fractions of data variance.

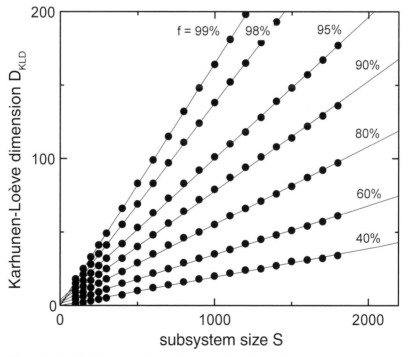

Figure 7.47. The Karhunen–Loève decomposition dimension D_{KLD} versus the subsystem size S for various fractions f of the reconstruction ($L = 2200$, $\Delta x = 0.5$, and $T = 4000$ time snapshots). The numerical parameters are $\mathcal{T} = 0.05$, $j_0 = 1.218$, $D = 8$, and $\alpha = 0.02$. (After Meixner *et al.* (2000*b*).)

First we study a system of size $L = 1000$ to perform a high-resolution ($\Delta x = 0.25$) investigation of the dynamic inhomogeneity induced by the boundaries. This is motivated by calculations of mean patterns of spatio-temporally chaotic systems exhibiting complicated patterns that persist over a large distance into the interior of the system (Gluckman *et al.* 1995, Ning *et al.* 1993). In our reaction–diffusion model, for the parameter regime considered, the two-point correlation length ξ_2, which is computed from the inverse full width at half maximum of the main peak of the power spectrum $P(k) \sim \exp[-(4 \ln 2)\, \xi_2^2 (k - k_0)^2]$ with respect to the wave vector k, is approximately 4.0 and so the system's length is approximately equal to $250\xi_2$. We compute various KLD correlation lengths ξ_{KLD} centered at a distance x from the boundary, with $0 < x < 50\xi_2$. Figure 7.48 demonstrates that the KLD correlation length ξ_{KLD} oscillates and decays to its bulk value of approximately $\xi_{KLD} \approx 2$ over the range of x from 0 to $50\xi_2 \approx 200$. The oscillations indicate that the system of size $L = 1000$ is still relatively far from the limit of homogeneous extensive chaos. These oscillations are fingerprints of interference effects induced by the Neumann system boundaries.

A second problem often encountered in experiments is that system parameters can vary nonuniformly in space or vary over time. To evaluate whether the KLD correlation length can detect small spatial parameter changes in partial differential equations we vary the parameter $\mathcal{T}$ spatially in the $L = 1000$ system. Physically this could be achieved, for example, by modulating the layer thickness of the semiconductor structure and thus the tunneling rate. We wish to test whether, in a more strongly coupled spatio-temporally chaotic system, the KLD correlation length is able to detect changes in parameters, similar to what was shown for weakly coupled

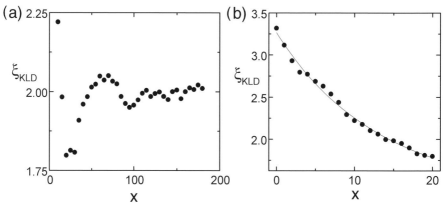

Figure 7.48. (a) The local Karhunen–Loève decomposition correlation length ξ_{KLD} versus position x measured from the boundary of the system. (b) The same data close to the boundary with a higher spatial resolution. Subsystems of sizes $S = 3, 4, 5, 6, 7,$ and 8, and a reconstruction of $f = 99.999\%$ are used (system size $L = 1000$, $\Delta x = 0.25$, and $T = 2000$ time snapshots, with the same parameters as those in Fig. 7.47). (After Meixner *et al.* (2000b).)

systems of iterated chaotic maps (Zoldi and Greenside 1997). The parameter $\mathcal{T}$ is modulated periodically according to $\mathcal{T} = 0.05 + 0.0025 \sin(2\pi x/1000)$ (this corresponds to a 5% change in a system parameter). Figure 7.49 shows that the KLD correlation length ξ_{KLD} has a sinusoidal form and is clearly able to predict the parametric dependency of the variation in $\mathcal{T}$. The noise in the data could be reduced by performing additional averages over time. The local nature of the KLD allows a determination of the dynamic inhomogeneity.

Next we consider whether the KLD correlation length has a parametric dependency other than the more commonly used two-point correlation length. For the case of a lattice of weakly coupled one-dimensional iterated maps it was shown by Zoldi and Greenside (1997) that, near a nonequilibrium Ising-like phase transition, the KLD correlation length ξ_{KLD} varies in a similar way to the fractal dimension correlation length $\xi_\delta = (D/V)^{-1/d}$, where D is the fractal dimension, but differently from the two-point correlation length. Here we consider a system of length $L = 2750$ and vary $\mathcal{T}$ from 0.048 to 0.052. We have been unable to extend this range of $\mathcal{T}$ because the dynamics qualitatively changes outside this window of $\mathcal{T}$ and becomes nonchaotic. Note that the two-point correlation length ξ_2 is computed over the *entire* system size $L = 2750$. Figure 7.50 demonstrates that, within the errors in

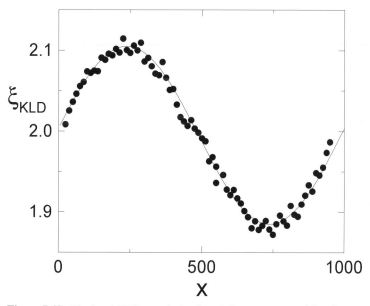

Figure 7.49. The local KLD correlation length ξ_{KLD} versus position for a nonuniform modulation of the parameter $\mathcal{T} = 0.05 + 0.0025 \sin(2\pi x/1000)$ over a system size $L = 1000$. Subsystems of size $S = 15, 20, 25$, and 30, and a reconstruction of $f = 99.999\%$ are used to compute the KLD correlation length ($\Delta x = 0.5$ and $T = 4000$ time snapshots sampled every five time units, with the same parameters as those in Fig. 7.47). The continuous line is a guide to the eye. (After Meixner *et al.* (2000*b*).)

calculating the KLD correlation lengths ξ_{KLD} and the two-point correlation lengths ξ_2, they are proportional. Therefore we suggest that, in the reaction–diffusion model in this range of $\mathcal{T}$, the two-point and KLD correlation lengths are related. Given the relationship between the KLD correlation length and the fractal dimension correlation length (Zoldi and Greenside 1997), we expect that the fractal dimension correlation length is also proportional to the two-point correlation length. This is similar to the proportionality relationship between the dimension correlation length and the two-point correlation length found for the magnitude of the order parameter of the complex Ginzburg–Landau equation (Egolf and Greenside 1994). However, we can not exclude the possibility that, near parameter values at which the dynamics changes critically and abruptly, the KLD correlation length may exhibit different behavior.

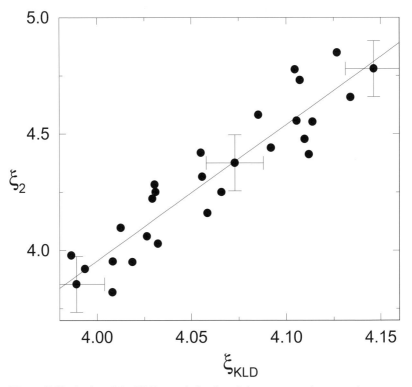

Figure 7.50. A plot of the KLD correlation length ξ_{KLD} versus the two-point correlation length ξ_2 corresponding to the range of $\mathcal{T}$ from 0.048 to 0.052. The system's size is $L = 2750$. To compute the KLD correlation length ξ_{KLD}, $T = 4000$ snap-shots separated by five time units are used in subsystems of sizes $S = 45, 60, 75,$ and 90 with a reconstruction of $f = 99.999\%$ ($\Delta x = 0.5$). In computing the two-point correlation length, the entire system size of $L = 2750$ is used and the power spectra are averaged over 2000 realizations. Error estimates for ξ_{KLD} and ξ_2 are $\Delta\xi_2 = 0.12$ and $\Delta\xi_{KLD} = 0.015$. The line is a guide to the eye and is of the form $\xi_2 = -19.5 + 5.9\xi_{KLD}$. The parameters are the same as those in Fig. 7.47. (After Meixner *et al.* (2000*b*).)

Moreover, since the KLD correlation length is much more readily accessible from small data sets than are the other two types of correlation lengths, it appears to be more suitable for quantitative investigations.

In conclusion, the KLD correlation length can be utilized to characterize dynamic inhomogeneities of spatio-temporal chaos in a system of partial differential equations of reaction–diffusion type. It has been confirmed that the KLD correlation length is able to detect dynamic inhomogeneities both in space and in time due to boundaries or variations in a system parameter. The KLD correlation length ξ_{KLD} can be computed for small localized subsystems, allowing quantification of spatial dynamic nonuniformities. Furthermore, the KLD correlation length is based solely on spatio-temporal data, and therefore can be easily applied both to experiments and to computer simulation. An analysis of such experimental data would provide potentially very interesting insights into inhomogeneities in the physical system.

References

Abakumov, V. N., Perel, V. I. and Yassievich, I. N. (1978), 'Capture of carriers by attractive centers in semiconductors', *Sov. Phys. Semicond.* **12**, 1.

Abe, Y. (1988), 'Impact ionization and chaotic states in narrow-gap semiconductors under a strong magnetic field', *Solid State Electron.* **31**, 795.

Abe, Y., ed. (1989), *Nonlinear and chaotic transport phenomena in semiconductors*, Special issue of *Appl. Phys. A* **48**, 93.

Adler, D., Shur, M. S., Silver, M. and Ovshinsky, S. R. (1980), 'Threshold switching in chalcogenide-glass thin films', *J. Appl. Phys.* **51**, 3289.

Aguado, R., Platero, G., Moscoso, M. and Bonilla, L. L. (1997), 'Microscopic model for sequential tunneling in semiconductor multiple quantum wells', *Phys. Rev. B* **55**, 16 053.

Alekseev, A., Bose, S., Rodin, P. and Schöll, E. (1998), 'Stability of current filaments in a bistable semiconductor system with global coupling', *Phys. Rev. E* **57**, 2640.

Aliev, K. M., Bashirov, R. I. and Gadzhialiev, M. M. (1994), 'Experimental study of the intrinsic effects in doped bismuth telluride by electrical methods', *Semiconductors* **28**, 899. [*Fiz. Tekhn. Poluprovodn.* **28**, 899 (1994).]

Amann, A., Wacker, A., Bonilla, L. L. and Schöll, E. (2001), Field domains in semiconductor superlattices: Dynamic scenarios of multistable switching, in N. Miura, ed., *Proceedings of the 25th International Conference on the Physics of Semiconductors (ICPS-25), Osaka 2000*, Springer, Berlin.

Andronov, A., Leontovich, E. A., Gordon, I. and Maier, A. G. (1971), *Theory of Bifurcations of Systems on a Plane*, Israel program for scientific translation, Jerusalem.

Andronov, A., Leontovich, E. A., Gordon, I. and Maier, A. G. (1973), *Qualitative Theory of Second-Order Dynamic Systems*, Wiley, New York.

Aoki, K. (1997), *Int. J. Bifurc. Chaos* **7**, 1059.

Aoki, K. (1999*a*), Pattern dynamics of a current density filament during impact ionization avalanche in n-GaAs, in D. Gershoni, ed., *Proceedings of the 24th International Conference on The Physics of Semiconductors (ICPS-24), Jerusalem, Israel*, World Scientific, Singapore.

Aoki, K. (1999*b*), 'Photoluminescence from two-dimensional electron gas in Al-GaAs/GaAs heterojunctions under high electric fields', *Physica B* **272**, 146.

Aoki, K. (2000), *Nonlinear Dynamics and Chaos in Semiconductors*, Institute of Physics Publishing, Bristol.

Aoki, K. and Fukui, S. (1998), 'Nucleation process of a filamentary current during impact ionization avalanche in n-GaAs', *J. Phys. Soc. Japan* **67**, 1106.

Aoki, K. and Fukui, S. (1999), 'Dynamics of filamentary currents in Corbino discs under magnetic fields', *Physica B* **272**, 274.

Aoki, K. and Kondo, T. (1991), 'Complexity of the magnetic-field-induced cross-over instability of a filamentary current in n-GaAs', *Phys. Lett. A* **154**, 281.

Aoki, K. and Yamamoto, K. (1983), 'Firing wave instability of the current filaments in a semiconductor. An analogy with neurodynamics', *Phys. Lett. A* **98**, 72.

Aoki, K. and Yamamoto, K. (1989), 'Bifurcation phenomena in a periodically driven current filament and a conjecture on the turbulent patterns by computer simulations', *Appl. Phys. A* **48**, 161.

Aoki, K., Kobayashi, T. and Yamamoto, K. (1981), 'Periodic oscillations and turbulence of hot-carrier plasma at 4.2K in n-GaAs', *J. Physique Colloque C* **C7**, 51.

Aoki, K., Mugibayashi, N. and Yamamoto, K. (1986), 'New bifurcation routes to chaos of a driven current filament in semiconductors simulated by numerical computation', *Physica Scripta* **T14**, 76.

Aoki, K., Yamamoto, K. and Mugibayashi, N. (1988), 'Chaotic behavior of a driven current filament observed in high-purity n-GaAs at 4.2 K', *J. Phys. Soc. Japan* **57**, 26.

Aoki, K., Yamamoto, K., Mugibayashi, N. and Schöll, E. (1989), 'Complex dynamical behavior and chaos in the Hess oscillator', *Solid State Electron.* **32**, 1149.

Aoki, K., Kawase, Y., Yamamoto, K. and Mugibayashi, N. (1990*a*), 'Current instability and chaos induced during low-temperature avalanche breakdown in n-GaAs under longitudinal magnetic field', *J. Phys. Soc. Japan* **59**, 20.

Aoki, K., Rau, U., Peinke, J., Parisi, J. and Huebener, R. P. (1990*b*), 'Observation of a large-scale sheetlike current filament in a thin n-GaAs layer', *J. Phys. Soc. Japan* **59**, 420.

Arndt, J., Herzel, H. P., Bose, S., Falcke, M. and Schöll, E. (1997), 'Quantification of transients using empirical orthogonal functions', *Chaos, Solitons and Fractals* **8**, 1911.

Arnold, D., Hess, K. and Iafrate, G. J. (1988), 'Electron transport in heterostructure hot-electron diodes', *Appl. Phys. Lett.* **53**, 373.

Arnold, D., Hess, K., Higman, T., Coleman, J. J. and Iafrate, G. J. (1989), 'Dynamics of heterostructure hot-electron diodes', *J. Appl. Phys.* **66**, 1423.

Asche, M., Gribnikov, Z. S., Ivastchenko, V. M., Kostial, H. and Mitin, V. V. (1982), 'Conductivity and transverse fields in n-Si for currents in the 110 and 100 planes',

Physica Status Solidi (b) **114**, 429.

Asche, M., Kostial, H., Ivastchenko, V. M. and Mitin, V. V. (1984), '$j-B$ characteristics of S-type in the presence of multivalued electron distribution between the equivalent valleys in Si', *J. Phys. C* **17**, 6781.

Baker, G. L. and Gollub, J. P. (1990), *Chaotic Dynamics: An Introduction*, Cambridge University Press, Cambridge.

Balkan, N., ed. (1998), *Hot Electrons in Semiconductors: Physics and Devices*, Oxford University Press, Oxford.

Balkan, N. and Ridley, B. (1989), *Superlattices and Microstructures* **5**, 539.

Balkan, N., Ridley, B. K. and Vickers, A. J., eds (1993), *Negative Differential Resistance and Instabilities in Two-Dimensional Semiconductors*, Plenum Press, New York.

Balkan, N., da Cunha, A., O'Brien, A., Teke, A., Gupta, R., Straw, A. and Ç. Arikan, M. (1996), Hot electron light emitting semiconductor heterojunction devices (HELLISH) – type 1 and type 2, in K. Hess, J. P. Leburton and U. Ravaioli, eds, *Proceedings of the 9th International Conference on Hot Carriers in Semiconductors, Chicago*, Plenum Press, New York.

Balkarei, Y. and Elinson, M. (1991), *Sov. J. Techn. Phys.* **17**, 487. [*Pis'ma Zh. Tekhn. Fiz.* **17**, 73 (1991)].

Balkarei, Y., Evtikhov, M. and Elinson, M. (1987), *Sov. Techn. Phys.* **32**, 127. [*Zh. Tekhn. Fiz.* **57**, 209 (1987)].

Bar-Joseph, I. (1999), 'Spatial and temporal studies of a gated electron gas'. Invited talk, HCIS-11, Kyoto 1999.

Bär, M., Falcke, M., Hildebrand, M., Neufeld, M., Engel, H. and Eiswirth, M. (1994), 'Chemical turbulence and standing waves in a surface reaction model: The influence of global coupling and wave instabilities', *Int. J. Bifurc. Chaos* **4**, 499.

Barnett, A. M. (1970), in *Semiconductors and Semimetals*, Vol. 6, Academic Press, New York.

Bass, F. G., Bochkov, V. S. and Gurevich, Y. G. (1970), 'The effect of sample dimension on current voltage characteristic of media with ambiguous field dependence on electron temperature', *Sov. Phys. JETP* **31**, 972. [*Zh. Éksp. Teor. Fiz.* **58**, 1814 (1970)].

Bastard, G. (1988), *Wave Mechanics Applied to Semiconductor Heterostructures*, Les Editions de Physique, Les Ulis.

Battogtokh, D. and Mikhailov, A. (1996), 'Controlling turbulence in the complex Ginzburg–Landau equation', *Physica D* **90**, 84.

Baumann, H., Symanczyk, R., Radehaus, C., Purwins, H. and Jäger, D. (1987), 'Properties of solitary current filaments in silicon pin diodes', *Phys. Lett. A* **123**, 421.

Beck, C. and Schlögl, F. (1993), *Thermodynamics of Chaotic Systems*, Cambridge University Press, Cambridge.

Beenakker, C. W. J. and van Houten, H. (1991), Quantum transport in semiconductor nanostructures, in H. Ehrenreich and D. Turnbull, eds, *Solid-State Physics*, Vol. 44, Academic Press, New York.

Belkov, V. V., Hirschinger, J., Novák, V., Niedernostheide, F.-J., Ganichev, S. D. and Prettl, W. (1991), 'Pattern formation in semiconductors', *Nature* **397**, 398.

Belyantsev, A. M., Ignatov, A. A., Piskarev, V. I., Sinitsyn, M. A., Shashkin, V. I., Yavich, B. S. and Yakovlev, M. L. (1986), 'New nonlinear high-frequency effects and S-shaped negative differential conductivity in multilayer heterostructures', *JETP Lett.* **43**, 437.

Belyantsev, A. M., Gavrilenko, V. I., Ignatov, A. A., Piskarev, V. I., Shashkin, V. I. and Andronov, A. A. (1988), 'Ultrafast S-type NDC and self-oscillations under vertical transport in multilayer heterostructures', *Solid State Electron.* **31**, 379.

Belyantsev, A. M., Demidov, E. V. and Romanov, Y. A. (1993), Electron transport and oscillatory instability in the heterostructure hot-electron diode, in *Proceedings of the 8th Vilnius Symposium on Ultrafast Phenomena in Semiconductors*, Academia, Vilnius, p. 31.

Benettin, G., Galgani, L., Giorgilli, A. and Strelcyn, J. (1980), 'Lyapunov characteristic exponents for smooth dynamical systems; a method for computing them all', *Meccanica* **15**, 9 and 21.

Bergé, P., Pomeau, Y. and Vidal, C. (1987), *Order Within Chaos*, Wiley, New York.

Bergmann, M. J., Teitsworth, S. W., Bonilla, L. L. and Cantalapiedra, I. R. (1996), 'Solitary-wave conduction in p-type Ge under time-dependent voltage bias', *Phys. Rev. B* **53**, 1327.

Bielawski, S., Derozier, D. and Glorieux, P. (1994), 'Controlling unstable periodic orbits by a delayed continuous feedback', *Phys. Rev. E* **49**, R971.

Bimberg, D., Grundmann, M. and Ledentsov, N. (1999), *Quantum Dot Heterostructures*, Wiley, New York.

Bleich, M. E. and Socolar, J. E. S. (1996), 'Stability of periodic orbits controlled by time-delay feedback', *Phys. Lett. A* **210**, 87.

Bleich, M. E., Hochheiser, D., Moloney, J. V. and Socolar, J. E. S. (1997), 'Controlling extended systems with spatially filtered, time-delay feedback', *Phys. Rev. E* **55**, 2119.

Bludau, W. and Wagner, E. (1976), 'Impact ionization of excitons in GaAs', *Phys. Rev. B* **13**, 5410.

Boccaletti, S., Maza, D., Mancini, H., Genesio, R. and Arecchi, F. T. (1997), 'Control of defects and spacelike structures in delayed dynamical systems', *Phys. Rev. Lett.* **79**, 5246.

Böer, K. W. (1959), 'Inhomogene Feldverteilung in CdS-Einkristallen im Bereich hoher Feldstärken', *Z. Phys.* **155**, 184.

Böer, K. W. (1961), 'Feld- und Strominhomogenitäten bei hohen elektrischen Belastungen in Isolatoren und Photoleitern', *Festkörperprobleme – Advances in Solid State Physics* **1**, 38.

Böer, K. W. and Döhler, G. (1969), 'Influence of boundary conditions on high-field domains in Gunn diodes', *Phys. Rev.* **186**, 793.

Böer, K. W. and Quinn, P. L. (1966), 'Inhomogeneous field distribution in homogeneous semiconductors having an N-shaped negative differential conductivity', *Phys. Status Solidi* **17**, 307.

Böer, K. W. and Voss, P. (1968a), 'Stationary anode-adjacent high-field domains in cadmium sulfide', *Phys. Status Solidi* **28**, 355.

Böer, K. W. and Voss, P. (1968b), 'Stationary high-field domains in the range of negative differential conductivity in CdS single crystals', *Phys. Rev.* **171**, 899.

Böer, K. W. and Williges, A. (1961), 'Feld- und Strominhomogenitäten in Si-Einkristallen', *Phys. Status Solidi* K72.

Böer, K. W., Hänsch, H. J. and Kümmel, U. (1959), 'Anwendung elektro-optischer Effekte zur Analyse des elektrischen Leitungsvorganges in CdS-Einkristallen', *Z. Phys.* **155**, 170.

Böer, K. W., Hänsch, H. J., Kümmel, U., Lange, H. and Nebauer, E. (1961a), 'Inhomogene Stromdichteverteilung bei hohen elektrischen Leistungen in CdS-Einkristallen; Vorprozesse des Wärmedurchschlags', *Phys. Status Solidi* **1**, 169.

Böer, K. W., Jahne, E. and Nebauer, E. (1961b), *Phys. Status Solidi* **1**, 231.

Bonch-Bruevich, V. L., Zvyagin, I. P. and Mironov, A. G. (1975), *Domain Electrical Instabilities in Semiconductors*, Consultants Bureau, New York.

Bonilla, L. L. (1995), Dynamics of electric field domains in superlattices, in F.-J. Niedernostheide, ed., *Nonlinear Dynamics and Pattern Formation in Semiconductors and Devices*, Springer, Berlin, chapter 1, pp. 1–20.

Bonilla, L. L., Cantalapiedra, I. R., Bergmann, M. J. and Teitsworth, S. W. (1994a), 'Onset of current oscillations in extrinsic semiconductors under dc voltage bias', *Semicond. Sci. Technol.* **9**, 599.

Bonilla, L. L., Galán, J., Cuesta, J. A., Martínez, F. C. and Molera, J. M. (1994b), 'Dynamics of electric field domains and oscillations of the photocurrent in a simple superlattice model', *Phys. Rev. B* **50**, 8644.

Bonilla, L. L., Higuera, F. J. and Venakides, S. (1994c), 'The Gunn effect: Instability of the steady state and stability of the solitary wave in long extrinsic semiconductors', *SIAM J. Appl. Math.* **54**, 1521.

Bonilla, L. L., Bulashenko, O. M., Galán, J., Kindelan, M. and Moscoso, M. (1996), 'Dynamics of electric-field domains and chaos in semiconductor superlattices', *Solid State Electron.* **40**, 161.

Bonilla, L. L., Cantalapiedra, I. R., Gomila, G. and Rubí, J. M. (1997a), 'Asymptotic analysis of the Gunn effect with realistic boundary conditions', *Phys. Rev. E* **56**, 1500.

Bonilla, L. L., Hernando, P. J., Herrero, M. A., Kindelan, M. and Velázquez, J. J. L. (1997b), 'Asymptotics of the trap-dominated Gunn effect in p-type Ge', *Physica D* **108**, 168.

Borisov, V. I., Sablikov, V. A., Chmil, A. I. and Borisova, I. V. (1999), 'Origin of current instability in GaAs/AlGaAs heterostructures', *Physica E* **8** no. 4.

Bose, S., Wacker, A. and Schöll, E. (1994), 'Bifurcation scenarios of spatio-temporal spiking in semiconductor devices', *Phys. Lett. A* **195**, 144.

Bose, S., Rodin, P. and Schöll, E. (2000), 'Competing spatial and temporal instabilities in a globally coupled bistable semiconductor system near a codimension-two bifurcation', *Phys. Rev. E* **62**, 1778.

Bosnell, J. R. and Thomas, C. B. (1972), *Solid State Electron.* **15**, 1261.

Braiman, Y. and Goldhirsch, I. (1991), 'Taming chaotic dynamics with weak periodic perturbations', *Phys. Rev. Lett.* **66**, 2545.

Brandl, A. and Prettl, W. (1991), 'Chaotic fluctuations and formation of a current filament in n-type GaAs', *Phys. Rev. Lett.* **66**, 3044.

Brandl, A., Geisel, T. and Prettl, W. (1987), 'Oscillations and chaotic current flow in n-GaAs', *Europhys. Lett.* **3**, 401.

Brandl, A., Völcker, M. and Prettl, W. (1989), 'Reconstruction of the spatial structure of current filaments in n-GaAs', *Appl. Phys. Lett.* **55**, 238.

Brandl, A., Kröninger, W., Prettl, W. and Obermair, G. (1990), 'Hall voltage collapse at filamentary current flow causing chaotic fluctuations in n-GaAs', *Phys. Rev. Lett.* **64**, 212.

Brown, E. R., Sollner, T. C. L. G., Parker, C. D., Goodhue, W. D. and Chen, C. L. (1989), 'Oscillations up to 420 Ghz in GaAs/AlAs resonant tunneling diodes', *Appl. Phys. Lett.* **55**, 1777.

Bude, J., Hess, K. and Iafrate, G. J. (1992), 'Impact ionization in semiconductors: effects of high electric fields and high scattering rates', *Phys. Rev. B* **45**, 10958.

Bulashenko, O. M. and Bonilla, L. L. (1995), 'Chaos in resonant-tunneling superlattices', *Phys. Rev. B* **52**, 7849.

Bulashenko, O. M., Garcia, M. C. and Bonilla, L. L. (1996), 'Chaotic dynamics of electric-field domains in periodically driven superlattices', *Phys. Rev. B* **53**, 10008.

Bulashenko, O. M., Luo, K. J., Grahn, H. T., Ploog, K. H. and Bonilla, L. L. (1999), 'Multifractal dimension of chaotic attractors in a driven semiconductor superlattice', *Phys. Rev. B* **60**, 5694.

Bumeliene, S. B., Požela, J., Pyragas, K. A. and Tamaševičius, A. V. (1985), 'Chaotic behavior of hot electron plasma in Ni-compensated Ge', *Physica B* **134**, 293.

Busse, F. H. and Müller, S. C., eds. (1998), *Evolution of Spontaneous Structures in Dissipative Continuous Systems*, Springer, Berlin.

Butcher, P. N. (1965), 'Theory of stable domain propagation in the Gunn effect', *Phys. Lett.* **19**, 546.

Büttiker, M. and Thomas, H. (1977), 'Current instability and domain propagation due to Bragg scattering', *Phys. Rev. Lett.* **38**, 78.

Büttiker, M. and Thomas, H. (1978), 'Travelling dipole domains and fluctuations for a current instability due to Bragg scattering', *Solid State Electron.* **21**, 95.

Cantalapiedra, I. R., Bonilla, L. L., Bergmann, M. J. and Teitsworth, S. W. (1993), 'Solitary-wave dynamics in extrinsic semiconductors under dc voltage bias', *Phys. Rev. B* **48**, 12278.

Cao, J. C. and Lei, X. L. (1999), 'Hydrodynamic balance-equation analysis of spatiotemporal domains and negative differential conductance in a voltage-biased GaAs superlattice', *Phys. Rev. B* **59**, 2199.

Capasso, F., Mohammed, K. and Cho, A. Y. (1986), *IEEE Trans. Electron Dev.* **34**, 297 and 1853.

Caponeri, M. and Ciliberto, S. (1992), 'Thermodynamic aspects of the transition to spatiotemporal chaos', *Physica D* **58**, 365.

Carpio, A., Bonilla, L. L., Wacker, A. and Schöll, E. (2000), 'Wavefronts may move upstream in semiconductor superlattices', *Phys. Rev. E* **61**, 4866.

Cenys, A., Lasiene, G. and Pyragas, K. (1992), 'Spatiotemporal chaos in Gunn diodes in presence of impact ionization', *Solid State Electron.* **35**, 975.

Cheianov, V., Rodin, P. and Schöll, E. (2000), 'On transverse coupling in bistable resonant-tunneling structures', *Phys. Rev. B* **62**, 9966.

Cheng, M. and Chang, H.-C. (1992), *Phys. Fluids A* **4**, 505.

Chien, C. and Westgate, C. (1980), *The Hall Effect and its Applications*, Plenum Press, New York.

Choi, K. K., Levine, B. F., Malik, R. J., Walker, J. and Bethea, C. G. (1987), 'Periodic negative conductance by sequential resonant tunneling through an expanding high-field superlattice domain', *Phys. Rev. B* **35**, 4172.

Christen, T. (1994*a*), 'The velocity of current filaments in weak magnetic fields', *Z. Naturforsch.* **49a**, 847.

Christen, T. (1994*b*), 'Wave instability in semiconductors without negative differential conductivity', *Phys. Rev. B* **49**, 16 423.

Christen, T. (1995), 'Complex Ginzburg–Landau equation for nonlinear travelling waves in extrinsic semiconductors', *Z. Phys. B* **97**, 473.

Ciliberto, S. and Nicolaenko, B. (1991), 'Estimating the number of degrees of freedom in spatially extended systems', *Europhys. Lett.* **14**, 303.

Clauss, W., Rau, U., Peinke, J., Parisi, J., Kittel, A., Bayerbach, M. and Huebener, R. (1991), 'Dynamics of current filaments in p-type germanium under the influence of a transverse magnetic field', *J. Appl. Phys.* **70**, 232.

Claussen, J. C., Mausbach, T., Piel, A. and Schuster, H. G. (1998), 'Memory difference control of unknown unstable fixed points: Drifting parameter condition and delayed measurement', *Phys. Rev. E* **58**, 7256.

Coleman, P. D., Freeman, J., Morkoç, H., Hess, K., Streetman, B. G. and Keever, M. (1982), 'Demonstrations of a new oscillator based on real-space transfer in heterojunctions', *Appl. Phys. Lett.* **40**, 493.

Coon, D. D., Ma, S. N. and Perera, A. G. U. (1987), 'Farey-fraction frequency modulation in the neuronlike output of silicon p–i–n diodes at 4.2 K', *Phys. Rev. Lett.* **58**, 1139.

Cooper, D. P. and Schöll, E. (1995), 'Tunable real space transfer oscillator by delayed feedback control of chaos', *Z. Naturforsch.* **50a**, 117.

Crandall, R. S. (1970), 'Low-temperature non-Ohmic electron transport in GaAs', *Phys. Rev. B* **1**, 730.

Cross, M. C. and Hohenberg, P. C. (1993), 'Pattern formation outside of equilibrium', *Rev. Mod. Phys.* **65**, 851.

Crutchfield, J. P. and Kaneko, K. (1988), 'Are attractors relevant to turbulence?', *Phys. Rev. Lett.* **60**, 2715.

Datta, S. (1995), *Electronic Transport in Mesoscopic Systems*, Cambridge University Press, Cambridge.

De Wit, A., Dewel, G. and Borckmans, P. (1993), 'Chaotic Turing–Hopf mixed mode', *Phys. Rev. E* **48**, 4191.

De Wit, A., Lima, D., Dewel, G. and Borckmans, P. (1996), 'Spatiotemporal dynamics near a codimension-two point', *Phys. Rev. E* **54**, 261.

Deane, A. E., Kevrekides, I. G., Karniadakis, G. E. and Orszag, S. A. (1991), *Phys. Fluids A* **3**, 2337.

Derrick, W. R. (1972), *Introductory Complex Analysis and Applications*, Academic Press, New York, London.

Ding, M., Yang, W., In, V., Ditto, W. L., Spano, M. L. and Gluckman, B. (1996), 'Controlling chaos in high dimensions: Theory and experiment', *Phys. Rev. E* **53**, 4334.

Ditto, W. L. and Showalter, K. (1997), 'Introduction: Control and synchronisation of chaos', *Chaos* **7**, 509.

Ditto, W. L., Rauseo, S. N. and Spano, M. L. (1990), 'Experimental control of chaos', *Phys. Rev. Lett.* **65**, 3211.

Dohmen, R. (1991), PhD thesis, University of Münster.

Döttling, R. and Schöll, E. (1992), 'Oscillatory bistability of real-space transfer in semiconductor heterostructures', *Phys. Rev. B* **45**, 1935.

Döttling, R. and Schöll, E. (1993), 'Front and domain propagation in semiconductor heterostructures', *Physica D* **67**, 418.

Döttling, R. and Schöll, E. (1994), 'Domain formation in modulation-doped GaAs/Al$_x$Ga$_{1-x}$As heterostructures', *Solid State Electron.* **37**, 685.

Dubitskij, A. L., Kerner, B. S. and Osipov, V. V. (1986), 'Spontaneous formation of local impact ionization regions in homogeneous semiconductors subjected to weak electric fields', *Sov. Phys. Semicond.* **20**, 755.

D'yakonov, M. and Levinstein, M. (1978), 'Theory of propagation of the turned-on state in a thyristor', *Sov. Phys. Semicond.* **12**, 426. [*Fiz. Tekhn. Poluprovodn.* **12**, 729 (1978)].

Eaves, L. (1998), 'Quantum Hall effect breakdown: analogies with fluid dynamics', *Physica B* **256**, 47.

Eberle, W., Hirschinger, J., Margull, U., Prettl, W., Novák, V. and Kostial, H. (1996), 'Visualization of current filaments in n-GaAs by photoluminescence quenching', *Appl. Phys. Lett.* **68**, 3329.

Ebert, U. and van Saarloos, W. (1998), 'Universal algebraic relaxation of fronts propagating into an unstable state and implications for moving boundary approximations', *Phys. Rev. Lett.* **80**, 1650.

Ebert, U., van Saarloos, W. and Caroli, C. (1996), 'Streamer propagation as a pattern formation problem: Planar fronts', *Phys. Rev. Lett.* **77**, 4178.

Ebert, U., van Saarloos, W. and Caroli, C. (1997), 'Propagation and structure of planar streamer fronts', *Phys. Rev. E* **55**, 1530.

Egolf, D. A. and Greenside, H. S. (1994), 'Relation between fractal dimension and spatial correlation length for extensive chaos', *Nature* **369**, 129.

Eguíluz, V. M., Alstrøm, P., Hernández-García, E. and Piro, O. (1999), 'Average patterns of spatiotemporal chaos: a boundary effect', *Phys. Rev. E* **59**, 2822.

Eisenstein, J. P. and Störmer, H. L. (1990), 'The fractional quantum Hall effect', *Science* **248**, 1510.

Elmer, F. J. (1990), 'Nonlocal dynamics of domains and domain walls in dissipative systems', *Phys. Rev. A* **41**, 4174.

Elmer, F. J. (1992), 'Limit cycles of the ballast resistor caused by intrinsic instabilities', *Z. Phys. B* **87**, 377.

Engel, H., Niedernostheide, F.-J., Purwins, H. G. and Schöll, E., eds (1996), *Self-Organization in Activator–Inhibitor–Systems: Semiconductors, Gas Discharge, and Chemical Active Media*, Wissenschaft und Technik Verlag, Berlin.

Esaki, L. and Chang, L. L. (1974), 'New transport phenomenon in a semiconductor superlattice', *Phys. Rev. Lett.* **33**, 495.

Etemadi, G. and Palmier, J. F. (1993), 'Effect of interface roughness on non-linear vertical transport in GaAs/AlAs superlattices', *Solid Stat. Commun.* **86**, 739.

Falcke, M. and Engel, H. (1994), 'Influence of global coupling through the gas phase on the dynamics of CO oxidation on Pt(110)', *Phys. Rev. E* **50**, 1353.

Falcke, M. and Neufeld, M. (1997), 'Traveling pulses in anisotropic media with global coupling', *Phys. Rev. E* **56**, 635.

Falcke, M., Engel, H. and Neufeld, M. (1995), 'Cluster formation, standing waves, and stripe patterns in oscillatory active media with local and global coupling', *Phys. Rev. E* **52**, 763.

Feigenbaum, M. (1978), 'Quantitative universality for a class of nonlinear transformations', *J. Statist. Phys.* **19**, 25.

Feiginov, M. N. and Volkov, V. A. (1998), 'Self-excitation of 2D plasmons in resonant tunneling diodes', *JETP Lett.* **68**, 633.

Feistel, R. and Ebeling, W. (1989), *Evolution of Complex Systems*, Kluwer, Dordrecht.

Ferry, D. K. and Goodnick, S. M. (1997), *Transport in Nanostructures*, Cambridge University Press, Cambridge.

Ferry, D. K., Grubin, L., Jacoboni, C. and Jauho, A. P., eds (1995), *Quantum Transport in Ultrasmall Devices*, Plenum Press, New York.

Ferry, D. K., Gardner, C. and Ringhofer, C., eds (1998), *Proceedings of the 4th International Workshop on Computational Electronics, Tempe, Az.*, VLSI Design, Vol. 6.

Fife, P. (1983), Deterministic (continuous and discrete) mathematics of nonlinear problems, in *Current Topics in Reaction–Diffusion Systems, Nonequilibrium Coop. Phenom. in Phys. and Related Fields*, Plenum Press, New York.

Fleischmann, R., Geisel, T. and Ketzmerick, R. (1994), 'Quenched and negative Hall effect in periodic media: application to antidot superlattices', *Europhys. Lett.* **25**, 219.

Franceschini, G., Bose, S. and Schöll, E. (1999), 'Control of chaotic spatiotemporal spiking by time-delay autosynchronisation', *Phys. Rev. E* **60**, 5426.

Fujii, K., Ohyama, T. and Otsuka, E. (1989), 'Magnetic field dependence of spontanious oscillation in n-InSb', *Appl. Phys. A* **48**, 189.

Fujimura, K. and Renardy, Y. (1995), 'The 2 : 1 steady/Hopf mode interaction in the two-layer Bénard problem', *Physica D* **85**, 25.

Gaa, M. and Schöll, E. (1996), 'Traveling carrier density waves in n-GaAs at low-temperature impurity breakdown', *Phys. Rev. B* **54**, 16 733.

Gaa, M., Kunz, R. E. and Schöll, E. (1996a), 'Dynamics of nascent current filaments in low-temperature impurity breakdown', *Phys. Rev. B* **53**, 15 971.

Gaa, M., Kunz, R. E. and Schöll, E. (1996b), Spatio-temporal dynamics of filament formation induced by impurity impact ionization in GaAs, in K. Hess, J. P. Leburton and U. Ravaioli, eds, *Proceedings of the 9th International Conference on Hot Carriers in Semiconductors*, Plenum, New York, pp. 347–351.

Gaa, M., Kunz, R. E., Schöll, E., Eberle, W., Hirschinger, J. and Prettl, W. (1996*c*), 'Spatial structure of impact-ionization induced current filaments in n-GaAs films', *Semicond. Sci. Technol.* **11**, 1646.

Gafiichuk, V. V., Datsko, B. I., Kerner, B. S. and Osipov, V. V. (1990), 'Microplasmas in perfectly homogeneous p–i–n structures', *Sov. Phys. Semicond.* **24**, 455.

Gajewski, H. (1985), 'On existence, uniqueness and asymptotic behavior of solutions of the basic equations for carrier transport in semiconductors', *Z. Angew. Math. Mech.* **2**, 101.

Gajewski, H. and Gärtner, K. (1992), 'On the iterative solution of van Roosbroeck's equations', *Z. Angew. Math. Mech.* **72**, 19.

Gajewski, H., Heinemann, B., Nürnberg, R., Langmach, H., Telschow, G. and Zarachias, K. (1991), 'Manual of the two-dimensional semi-conductor analysis package (Tosca)' (unpublished).

Gierer, A. and Meinhardt, H. (1972), *Kybernetik* **12**, 30.

Glansdorff, P. and Prigogine, I. (1971), *Thermodynamic Theory of Structure, Stability, and Fluctuations*, Wiley, New York.

Glavin, B., Kochelap, V. and Mitin, V. (1997), 'Patterns in bistable resonant-tunneling structures', *Phys. Rev. B* **56**, 13 346.

Glendinning, P. (1994), *Stability, Instability and Chaos*, Cambridge University Press, Cambridge.

Glicksman, M. (1971), *Plasmas in Solids*, Vol. XXVI of *Solid State Physics*, Academic Press, New York, p. 275.

Gluckman, B. J., Arnold, C. B. and Gollub, J. P. (1995), 'Statistical studies of chaotic wave patterns', *Phys. Rev. E* **51**, 1128.

Godlewski, M. K., Fronc, K., Gajewska, M., Chen, W. M. and Monemar, B. (1992), *Semicond. Sci. Technol.* **B7**, 483.

Goldhirsch, I. and Orszag, S. (1987), 'Stability and Lyapunov stability of dynamical systems: a differential approach and a numerical method', *Physica D* **27**, 311.

Goldman, V. J., Tsui, D. C. and Cunningham, J. E. (1987), 'Observation of intrinsic bistability in resonant-tunneling structures', *Phys. Rev. Lett.* **58**, 1256.

Gorbatyuk, A. V. and Niedernostheide, F.-J. (1996), 'Analytical model for a nonlinear current feedback mechanism in a self-organizing semiconductor system', *Physica D* **99**, 339.

Gorbatyuk, A. V. and Niedernostheide, F.-J. (1999), 'Mechanism of spatial current-density instabilities in ppnpn structures', *Phys. Rev. B* **59**, 13 157.

Gorbatyuk, A. V. and Rodin, P. B. (1990), 'Mechanism for spatially periodic stratification of current in a thyristor', *Sov. J. Techn. Phys.* **16**, 519. [*Pis'ma Zh. Tekhn. Fiz.* **16**, 519 (1990)].

Gorbatyuk, A. V. and Rodin, P. B. (1992*a*), 'Effect of distributed-gate control on current filamentation in thyristors', *Solid State Electron.* **35**, 1359.

Gorbatyuk, A. V. and Rodin, P. B. (1992*b*), 'Spontaneous crowding of current in GTO', *Sov. J. Communication Technol. Electron.* **37**, 97. [*Radiotekhn. Elektron.* **37**, 910 (1992)].

Gorbatyuk, A. V. and Rodin, P. B. (1992*c*), 'Spontaneous current filamentation in a semiconductor system with nonlocal transverse coupling', *Russian Microelectron.* **21**, 172.

Gorbatyuk, A. V. and Rodin, P. B. (1994), Turing's instability as a failure mechanism of GTO, in *Proceedings of the 6th International Symposium on Power Semiconductor Devices, Davos, Switzerland*, p. 227.

Gorbatyuk, A. V. and Rodin, P. B. (1995), 'Turing's instability and space periodic current filamentation in a thyristor', *J. Communication Technol. Electron.* **40**, 49. [*Radiotekhn. Elektron.* **40**, 1876 (1994)].

Gorbatyuk, A. V. and Rodin, P. B. (1997), 'Current filamentation in a bistable semiconductor system with two global constraints', *Z. Phys. B* **104**, 45.

Gorbatyuk, A. V., Linijchuk, I. A. and Svirin, A. V. (1989), 'Space-periodical destruction of dynamically overloaded thyristor', *Sov. J. Techn. Phys.* **15**, 224. [*Pis'ma Zh. Tekhn. Fiz.* **15**, 42 (1989)].

Graham, M. D., Kevrekides, I. G., Hudson, J. L., Veser, G., Krischer, K. and Imbihl, R. (1995), *Chaos, Solitons and Fractals* **5**, 1817.

Graham, R. and Haken, H. (1968), 'Quantum theory of light propagation in a fluctuating laser-active medium', *Z. Phys.* **213**, 420.

Grahn, H. T. (1995*a*), Electric field domains, in H. T. Grahn, ed., *Semiconductor Superlattices, Growth and Electronic Properties*, World Scientific, Singapore, chapter 5.

Grahn, H. T., ed. (1995*b*), *Semiconductor Superlattices, Growth and Electronic Properties*, World Scientific, Singapore.

Grahn, H. T., Haug, R. J., Müller, W. and Ploog, K. (1991), 'Electric-field domains in semiconductor superlattices: A novel system for tunneling between 2d systems', *Phys. Rev. Lett.* **67**, 1618.

Grahn, H. T., Kastrup, J., Ploog, K., Bonilla, L. L., Galán, J., Kindelan, M. and Moscoso, M. (1995), 'Self-oscillations of the current in doped semiconductor superlattices', *Japan. J. Appl. Phys.* **34**, 4526.

Grassberger, P. and Procaccia, I. (1983*a*), *Physica D* **9**, 189.

Grassberger, P. and Procaccia, I. (1983*b*), 'Characterization of strange attractors', *Phys. Rev. Lett.* **50**, 346.

Gray, P. and Scott, S. K. (1984), *Chem. Eng. Sci.* **39**, 1087.

Grebogi, C., Ott, E. and Yorke, J. A. (1983), 'Fractal basin boundaries, long-lived chaotic transients, and unstable–unstable pair bifurcation', *Phys. Rev. Lett.* **50**, 935.

Grenzer, J., Schomburg, E., Lingott, I., Ignatov, A. A., Renk, K. F., Pietsch, U., Zeimer, U., Melzer, B. J., Ivanov, S., Schaposchnikov, S., Kop'ev, P. S., Pavel'ev, D. G. and Koschurinov, Y. (1998), 'X-ray characterization of an Esaki-Tsu superlattice and transport properties', *Semicond. Sci. Technol.* **13**, 733.

Gribnikov, Z. S. (1973), *Sov. Phys. Semicond.* **6**, 1204.

Gribnikov, Z. S. and Mel'nikov, V. I. (1966), 'Injection and extraction of hot electrons in n–n heterojunctions with rapid Maxwellization of the electron gas', *Sov. Phys. Solid State* **7**, 2364.

Gribnikov, Z. S., Hess, K. and Kosinovsky, G. (1995), 'Nonlocal and nonlinear transport in semiconductors: Real-space transfer effects', *J. Appl. Phys.* **77**, 1337.

Grigoriev, R. O., Cross, M. C. and Schuster, H. G. (1997), 'Pinning control of spatiotemporal chaos', *Phys. Rev. Lett.* **79**, 2795.

Grossmann, S. and Thomae, S. (1977), 'Invariant distributions and stationary correlation functions of one-dimensional discrete processes', *Z. Naturforsch.* **32a**, 1353.

Guckenheimer, J. and Holmes, P. (1983), *Nonlinear Oscillations, Dynamical Systems, and Bifurcations of Vector Fields*, Applied mathematical sciences 42, Springer, Berlin.

Gunn, J. B. (1963), *Solid Stat. Commun.* **1**, 88.

Gunn, J. B. (1964), 'Instabilities of current in III–V semiconductors', *IBM J. Res. Develop.* **8**, 141.

Haken, H. (1983), *Synergetics*, 2nd edn, Springer, Berlin.

Haken, H. (1987), *Advanced Synergetics*, 2nd edn, Springer, Berlin.

Haken, H., Schulz, C.-D. and Schindel, M. (1994), in F. G. Boebel and T. Wagner, eds, *Proceedings of the First Conference of Applied Synergetics and Synergetical Engineering*, Fraunhofer Society, Erlangen.

Hall, K., Christini, D. J., Tremblay, M., Collins, J. J., Glass, L. and Billete, J. (1997), 'Dynamic control of cardiac alterans', *Phys. Rev. Lett.* **78**, 4518.

Hapke-Wurst, I., Zeitler, U., Schumacher, H. W., Haug, R. J., Pierz, K. and Ahlers, F. J. (1999), 'Size determination of InAs quantum dots using magneto-tunneling experiments'. *Semicond. Sci. Technol.* **14**, L41.

Haug, H., ed. (1988), *Optical Nonlinearities and Instabilities in Semiconductors*, Academic Press, New York.

Haug, H. and Jauho, A.-P. (1996), *Quantum Kinetics in Transport and Optics of Semiconductors*, Springer, Berlin.

Haug, H. and Koch, S. W. (1993), *Quantum Theory of the Optical and Electronic Properties of Semiconductors*, 2nd edn, World Scientific, Singapore.

Heidemann, G., Bode, M. and Purwins, H. (1993), 'Fronts between Hopf and Turing-type domains in a two-component reaction–diffusion system', *Phys. Lett. A* **177**, 225.

Held, G., Jeffries, C. and Haller, E. (1984), 'Observation of chaotic behavior in an electron–hole plasma in Ge', *Phys. Rev. Lett.* **52**, 1037.

Helgesen, P. and Finstad, T. G. (1990), Sequential resonant and non-resonant tunneling in GaAs/AlGaAs multiple quantum well structures: High field domain formation, in O. Hansen, ed., *Proceedings of the 14th Nordic Semiconductor Meeting*, University of Århus, Århus, p. 323.

Helm, M., England, P., Colas, E., DeRosa, F. and Allen, S. J. Jr. (1989), 'Intersubband emission from semiconductor superlattices excited by sequential resonant tunneling', *Phys. Rev. Lett.* **63**, 74.

Hempel, H., Schebesch, I. and Schimansky-Geier, L. (1998), 'Traveling pulses in reaction–diffusion systems under global constraints', *Eur. Phys. J. B* **2**, 399.

Hendriks, P., Zwaal, E. A. E., Dubois, J. G. A., Blom, F. P. and Wolter, J. H. (1991),

'Electric field induced parallel conduction in GaAs/AlGaAs heterostructures', *J. Appl. Phys.* **69**, 302.

Hess, K. (1988), *Advanced Theory of Semiconductor Devices*, Prentice Hall, Englewood Cliffs, New Jersey.

Hess, K., Morkoç, H., Shichijo, H. and Streetman, B. G. (1979), 'Negative differential resistance through real-space electron transfer', *Appl. Phys. Lett.* **35**, 469.

Hess, K., Higman, T. K., Emanuel, M. A. and Coleman, J. J. (1986), 'New ultrafast switching mechanism in semiconductor heterostructures', *J. Appl. Phys.* **60**, 3775.

Hess, O. (1993), *Spatio-Temporal Dynamics of Semiconductor Lasers*, Dissertation TU-Berlin, Wissenschaft und Technik Verlag, Berlin.

Hess, O. and Kuhn, T. (1996), 'Spatio-temporal dynamics of semiconductor lasers: Theory, modelling and analysis', *Prog. Quantum Electron.* **20**, 84.

Hess, O. and Schöll, E. (1994), 'Spatio-temporal dynamics in twin-stripe semiconductor lasers', *Physica D* **70**, 165.

Hess, O., Merbach, D., Herzel, H.-P. and Schöll, E. (1994), 'Bifurcations of a 3-torus in a twin-stripe semiconductor laser model', *Phys. Lett. A* **194**, 289.

Higman, T. K., Higman, J. M., Emanuel, M. A., Hess, K. and Coleman, J. J. (1987), 'Theoretical and experimental analysis of the switching mechanism in heterostructure hot-electron diodes', *J. Appl. Phys.* **62**, 1495.

Higuera, F. J. and Bonilla, L. L. (1992), 'Gunn instability in finite samples of GaAs – II. Oscillatory states in long samples', *Physica D* **57**, 161.

Hill, A. and Stewart, I. (1991), *Dynam. Stab. Syst.* **6**, 149.

Hirsch, M., Kittel, A., Flätgen, G., Huebener, R. P. and Parisi, J. (1994), *Phys. Lett. A* **186**, 157.

Hirschinger, J., Eberle, W., Prettl, W., Niedernostheide, F.-J. and Kostial, H. (1997*a*), 'Self-organized current-density patterns and bifurcations in n-GaAs with a circular contact symmetry', *Phys. Lett. A* **236**, 249.

Hirschinger, J., Niedernostheide, F.-J., Prettl, W., Novák, V. and Kostial, H. (1997*b*), 'Self-organized current filament patterns and bifurcations in n-GaAs Corbino disks', *Phys. Status Solidi* **204**, 477.

Hirschinger, J., Niedernostheide, F.-J., Prettl, W., Novák, V., Cukr, M., Oswald, J. and Kostial, H. (1997*c*), 'Current filamentation in n-GaAs samples with different contact geometries', *Acta Techn. ČSAV* **42**, 661.

Hirschinger, J., Kostial, H. and Prettl, W. (1998), 'Visualization of lateral movement of current filaments in n-GaAs', *Solid Stat. Commun.* **106**, 187.

Hockney, R. and Eastwood, J. (1981), *Computer Simulation Using Particles*, McGraw-Hill, New York.

Hofbeck, K., Grenzer, J., Schomburg, E., Ignatov, A. A., Renk, K. F., Pavel'ev, D. G., Koschurinov, Y., Melzer, B., Ivanov, S., Schaposchnikov, S. and Kop'ev, P. S. (1996), 'High-frequency self-sustained current oscillation in an Esaki-Tsu superlattice monitored via microwave emission', *Phys. Lett. A* **218**, 349.

Holden, A. V., ed. (1986), *Chaos*, Manchester University Press, Manchester.

Holmes, P., Lumley, J. and Berkooz, G. (1996), *Turbulence, Coherent Structures, Dynamical Systems and Symmetry*, Cambridge University Press, Cambridge.

Hübler, A. and Lüscher, E. (1989), 'Resonant stimulation and control of nonlinear oscillators', *Naturwissenschaften* **76**, 67.

Hüpper, G. and Schöll, E. (1991), 'Dynamic Hall effect as a mechanism for self-sustained oscillations and chaos in semiconductors', *Phys. Rev. Lett.* **66**, 2372.

Hüpper, G., Schöll, E. and Reggiani, L. (1989), 'Global bifurcation and hysteresis of self-generated oscillations in a microscopic model of nonlinear transport in p-Ge', *Solid State Electron.* **32**, 1787.

Hüpper, G., Schöll, E. and Rein, A. (1992), 'Nonlinear and chaotic oscillations in semiconductors under the influence of a transverse magnetic field: The dynamic Hall effect', *Mod. Phys. Lett. B* **6**, 1001.

Hüpper, G., Pyragas, K. and Schöll, E. (1993*a*), 'Complex dynamics of current filaments in the low temperature impurity breakdown regime of semiconductors', *Phys. Rev. B* **47**, 15 515.

Hüpper, G., Pyragas, K. and Schöll, E. (1993*b*), 'Complex spatio-temporal dynamics of current filaments in crossed electric and magnetic fields', *Phys. Rev. B* **48**, 17 633.

Hyman, J. M., Nicolaenko, B. and Zaleski, S. (1986), 'Order and compexity in the Kuramoto–Sivashinsky model of weakly turbulent interfaces', *Physica D* **23**, 265.

Ignatov, A. A., Piskarev, V. I. and Shashkin, V. I. (1985), 'Instability (formation of domains) of an electric field in multilayer quantum structures', *Sov. Phys. Semicond.* **19**, 1345. [*Fiz. Tekhn. Poluprovodn.* **19**, 1283 (1985)].

Ignatov, A. A., Dodin, E. P. and Shashkin, V. I. (1991), 'Transient response theory of semiconductor superlattices: Connection with Bloch oscillations', *Mod. Phys. Lett. B* **5**, 1087.

Itskevich, I. E., Ihn, T., Thornton, A., Henini, M., Foster, T. J., Moriarty, P., Nogaret, A., Beton, P. H., Eaves, L. and Main, P. C. (1996), 'Resonant magnetotunneling through individual self-assembled InAs quantum dots', *Phys. Rev. B* **54**, 16 401.

Jacoboni, C. and Lugli, P. (1989), *The Monte Carlo Method for Semiconductor Device Simulation*, Springer, Vienna.

Jäger, D., Baumann, H. and Symanczyk, R. (1986), 'Experimental observation of spatial structures due to current filament formation in silicon pin diodes', *Phys. Lett. A* **117**, 141.

Joosten, H. P., Noteborn, H. J. M. F., Kaski, K. and Lenstra, D. (1991), 'The stability of the self-consistently determined current of a double-barrier resonant-tunneling diode', *J. Appl. Phys.* **70**, 3141.

Just, W. (1999), Principles of time delayed feedback control, in H. G. Schuster, ed., *Handbook of Chaos Control*, Wiley-VCH, Weinheim.

Just, W., Bernard, T., Ostheimer, M., Reibold, E. and Benner, H. (1997), 'Mechanism of time-delayed feedback control', *Phys. Rev. Lett.* **78**, 203.

Just, W., Reckwerth, D., Möckel, J., Reibold, E. and Benner, H. (1998), 'Delayed feedback control of periodic orbits in autonomous systems', *Phys. Rev. Lett.* **81**, 562.

Just, W., Reibold, E., Benner, H., Kacperski, K., Fronczak, P. and Holyst, J. (1999), 'Limits of time-delayed feedback control', *Phys. Lett. A* **254**, 158.

Kahn, A. M., Mar, D. J. and Westervelt, R. M. (1991), 'Spatial measurements of moving space-charge domains in ultrapure Ge', *Phys. Rev. B* **43**, 9740.

Kahn, A. M., Mar, D. J. and Westervelt, R. M. (1992*a*), 'Dynamics of space-charge domains in ultrapure Ge', *Phys. Rev. Lett.* **46**, 369.

Kahn, A. M., Mar, D. J. and Westervelt, R. M. (1992*b*), 'Spatial measurements near the instability threshold in ultrapure Ge', *Phys. Rev. B* **45**, 8342.

Kantz, H. and Schreiber, T. (1997), *Nonlinear Time Series Analysis*, Cambridge University Press, Cambridge.

Kaplan, J. L. and Yorke, J. A. (1979), Chaotic behavior of multidimensional difference equations, in H. O. Peitgen and H. O. Walter, eds, *Functional Differential Equations and Approximations of Fixed Points*, Lecture notes in mathematics Vol. 730', Springer, Berlin.

Karel, F., Oswald, J., Pastrňák, J. and Petřček, O. (1992), 'Impurity breakdown and electric-field-dependent luminescence in MBE and VPE GaAs layers', *Semicond. Sci. Technol.* **7**, 203.

Kastalsky, A. A. (1973), *Phys. Status Solidi* **15**, 599.

Kastalsky, A., Milshtein, M., Shantharama, L. G., Harbison, J. and Florez, L. (1989), 'New features of real-space hot-electron transfer in the NERFET', *Solid State Electron.* **32**, 1841.

Kastrup, J., Grahn, H. T., Ploog, K., Prengel, F., Wacker, A. and Schöll, E. (1994), 'Multistability of the current–voltage characteristics in doped GaAs–AlAs superlattices', *Appl. Phys. Lett.* **65**, 1808.

Kastrup, J., Klann, R., Grahn, H. T., Ploog, K., Bonilla, L. L., Galán, J., Kindelan, M., Moscoso, M. and Merlin, R. (1995), 'Self-oscillations of domains in doped GaAs–AlAs superlattices', *Phys. Rev. B* **52**, 13 761.

Kastrup, J., Prengel, F., Grahn, H. T., Ploog, K. and Schöll, E. (1996), 'Formation times of electric field domains in doped GaAs-AlAs superlattices', *Phys. Rev. B* **53**, 1502.

Kastrup, J., Hey, R., Ploog, K. H., Grahn, H. T., Bonilla, L. L., Kindelan, M., Moscoso, M., Wacker, A. and Galán, J. (1997), 'Electrically tunable GHz oscillations in doped GaAs–AlAs superlattices', *Phys. Rev. B* **55**, 2476.

Kawamura, Y., Wakita, K., Asahi, H. and Kurumada, K. (1986), 'Observation of room temperature current oscillation in InGaAs/InAlAs MQW pin diodes', *Japan. J. Appl. Phys.* **25**, L928.

Kaya, I. I., Nachtwei, G., von Klitzing, K. and Eberl, K. (1998), 'Spatial evolution of hot-electron relaxation in quantum Hall conductors', *Phys. Rev. B* **58**, R7536.

Kazarinov, R. F. and Suris, R. A. (1971), 'Possibility of the amplification of electromagnetic waves in a semiconductor with a superlattice', *Sov. Phys. Semicond.* **5**, 707.

Kazarinov, R. F. and Suris, R. A. (1972), 'Electric and electromagnetic properties of semiconductors with a superlattice', *Sov. Phys. Semicond.* **6**, 120. [*Fiz. Tekhn. Poluprov.* **6**, 148 (1972)].

Keener, J. P. (1976), *Studies Appl. Math.* **55**, 187.

Keever, M., Shichijo, H., Hess, K., Banarjee, S., Witkowski, L. and Morkoç, H.

(1981), 'Measurements of hot-electron conduction and real-space transfer in GaAs–$Al_xGa_{1-x}As$ heterojunction layers', *Appl. Phys. Lett.* **38**, 36.

Kehrer, B., Quade, W. and Schöll, E. (1995*a*), 'Monte Carlo simulation of impact-ionization-induced breakdown and current filamentation in δ-doped GaAs', *Phys. Rev. B* **51**, 7725.

Kehrer, B., Quade, W. and Schöll, E. (1995*b*), Monte Carlo simulation of low temperature impurity breakdown and current filamentation in δ-doped GaAs, in D. J. Lockwood, ed., *Proceedings of the 22nd International Conference on the Physics of Semiconductors, Vancouver 1994*, World Scientific, Singapore.

Kerner, B. S. and Osipov, V. V. (1976), *Sov. Phys. JETP* **44**, 807.

Kerner, B. S. and Osipov, V. V. (1980), *Sov. Phys. JETP* **52**, 112.

Kerner, B. S. and Osipov, V. (1982), 'Pulsating 'heterophase' regions in nonequilibrium systems', *Sov. Phys. JETP* **56**, 1275. [*Sov. Fiz. JETP* **56** 1275 (1982)].

Kerner, B. S. and Osipov, V. (1989), 'Autosolitons', *Sov. Phys. Usp.* **157**, 101.

Kerner, B. S. and Osipov, V. V. (1994), *Autosolitons*, Kluwer, Dordrecht.

Kerner, B. S., Litvin, D. P. and Sankin, V. S. (1987), *Sov. Phys. Techn. Phys. Lett.* **13**, 342.

Kidachi, H. (1980), 'On mode interactions in reaction diffusion equations with nearly degenerate bifurcations', *Prog. Theor. Phys.* **63**, 1152.

Kittel, A., Parisi, J. and Pyragas, K. (1995), 'Delayed feedback control of chaos by self-adapted delay time', *Phys. Lett. A* **198**, 433.

Knap, W., Jezewski, M., Lusakowski, J. and Kuszko, W. (1988), *Solid State Electron.* **31**, 813.

Knight, B. W. and Peterson, G. A. (1967), 'Theory of the Gunn effect', *Phys. Rev.* **155**, 393.

Kogan, S. M. (1968), *Sov. Phys. JETP* **27**, 656.

Kolodzey, J., Laskar, J., Higman, T., Emanuel, M., Coleman, J. and Hess, K. (1988), 'Microwave frequency operation of the heterostructure hot-electron diode', *IEEE Electron Dev. Lett.* **9**, 272.

Kometer, K., Zandler, G. and Vogl, P. (1992), 'Lattice gas cellular-automaton method for semiclassical transport in semiconductors', *Phys. Rev. B* **46**, 1382.

Korotkov, A. N., Averin, D. V. and Likharev, K. K. (1993), 'Single-electron quantization of electric field domains in slim semiconductor superlattices', *Appl. Phys. Lett.* **62**, 3282.

Kostial, H., Asche, M., Hey, R., Ploog, K. and Koch, F. (1993*a*), *Japan. J. Appl. Phys.* **32**, 491.

Kostial, H., Ihn, T., Kleinert, P., Hey, R., Asche, M. and Koch, F. (1993*b*), 'Field-induced real-space transfer in Delta-doped GaAs', *Phys. Rev. B* **47**, 4485.

Kostial, H., Asche, M., Hey, R., Ploog, K., Kehrer, B., Quade, W. and Schöll, E. (1995*a*), 'Low temperature breakdown and current filamentation in n-type GaAs with homogeneous and partially ordered Si doping', *Semicond. Sci. Technol.* **10**, 775.

Kostial, H., Ploog, K., Hey, R. and Böbel, F. (1995*b*), 'Current filament dynamics in n-GaAs', *J. Appl. Phys.* **78**, 4560.

Kozhevnikov, M., Ashkinadze, B. M., Cohen, E. and Arza, R. (1995), *Phys. Rev. B* **52**,

4855.

Kroemer, H. (1964), 'Theory of the Gunn effect', *Proceedings of the IEEE* **52**, 1736.

Kroemer, H. (1968), *IEEE Trans. Electron Dev.* **15**, 819.

Krotkus, A., Reklaitis, A., Geizutis, A. and Asche, M. (1998), 'Voltage switching and oscillations in a single barrier heterostructure hot-electron diode', *J. Appl. Phys.* **84**, 3980.

Krotkus, A., Reklaitis, A., Geizutis, A. and Asche, M. (1999), 'S-type negative differential conductivity and voltage switching due to the avalanche in semiconductor heterostructures', *Semicond. Sci. Technol.* **14**, 341.

Kuhn, T., Hüpper, G., Quade, W., Rein, A., Schöll, E., Varani, L. and Reggiani, L. (1993), 'Microscopic analysis of noise and nonlinear dynamics in p-type germanium', *Phys. Rev. B* **48**, 1478.

Kukuk, B., Reil, F., Niedernostheide, F.-J. and Purwins, H.-G. (1996), in H. Engel, F.-J. Niedernostheide, H. G. Purwins and E. Schöll, eds, *Selforganization in Activator–Inhibitor-Systems*, Wissenschaft and Technik Verlag, Berlin, pp. 62–66.

Kunihiro, K., Gaa, M. and Schöll, E. (1997), 'Formation of current filaments in n-type GaAs under crossed electric and magnetic fields', *Phys. Rev. B* **55**, 2207.

Kunz, R. E. and Schöll, E. (1992), 'Globally coupled dynamics of breathing current filaments in semiconductors', *Z. Phys. B* **89**, 289.

Kunz, R. E. and Schöll, E. (1996), 'Dynamics of stochastically induced and spatially inhomogeneous impurity breakdown in semiconductors', *Z. Phys. B* **99**, 185.

Kunz, R. E., Schöll, E., Gajewski, H. and Nürnberg, R. (1996), 'Low-temperature impurity breakdown in semiconductors: an approach towards efficient device simulation', *Solid State Electron.* **39**, 1155.

Kuramoto, Y. (1988), *Chemical Oscillations, Waves and Turbulence*, Springer, Berlin.

Kwok, S. H., Grahn, H. T., Ramsteiner, M., Ploog, K., Prengel, F., Wacker, A., Schöll, E., Murugkar, S. and Merlin, R. (1995), 'Non-resonant carrier transport through high-field domains in semiconductor superlattices', *Phys. Rev. B* **51**, 9943.

Landsberg, P. T. (1991), *Recombination in Semiconductors*, Cambridge University Press, Cambridge.

Landsberg, P. T., ed. (1992), *Handbook on Semiconductors, Vol. I*, 2nd edn, Elsevier, Amsterdam.

Landsberg, P. T. and Pimpale, A. (1976), 'Recombination-induced nonequilibrium phase transitions in semiconductors', *J. Phys. C* **9**, 1243.

Landsberg, P. T., Robbins, D. J. and Schöll, E. (1978), 'Threshold switching as a generation–recombination induced non-equilibrium phase transition', *Physica Status Solidi (a)* **50**, 423.

Landsberg, P. T., Schöll, E. and Shukla, P. (1988), 'A simple model for the origin af chaos in semiconductors', *Physica D* **30**, 235.

Lax, M. (1960), 'Cascade capture of electrons in solids', *Phys. Rev.* **119**, 1502.

Leadbeater, M. L., Eaves, L., Henini, M., Hughes, O. H., Hill, G. and Pate, M. A. (1989), 'Inverted bistability in the current–voltage characteristics of a resonant tunneling device', *Solid State Electron.* **32**, 1467.

Lentine, A. L. and Miller, D. A. B. (1993), 'Evolution of the SEED technology: bistable logic gates to optoelectronic smart pixels', *IEEE J. Quant. Electron.* **29**, 655.

Lima, D., DeWit, A., Dewel, G. and Borckmans, P. (1996), 'Chaotic spatially subharmonic oscillations', *Phys. Rev. E* **53**, 1305.

Loeb, L. B. (1965), *Science* **148**, 1417.

Lorenz, E. N. (1963), 'Deterministic nonperiodic flow', *J. Atmos. Sci.* **20**, 130.

Lourenço, C. and Babloyantz, A. (1994), 'Control of chaos in networks with delay: A model for synchronisation of cortical tissue', *Neural Comput.* **6**, 1141.

Lueder, H., Schottky, W. and Spenke, E. (1936), 'Zur technischen Beherrschung des Wärmedurchschlags', *Naturwissenschaften* **24**, 61.

Luo, K. J., Grahn, H. T. and Ploog, K. H. (1998*a*), 'Relocation time of the domain boundary in weakly coupled GaAs/AlAs superlattices', *Phys. Rev. B* **57**, 6838.

Luo, K. J., Grahn, H. T., Ploog, K. H. and Bonilla, L. L. (1998*b*), 'Explosive bifurcation to chaos in weakly coupled semiconductor superlattices', *Phys. Rev. Lett.* **81**, 1290.

Luryi, S. and Pinto, M. R. (1991), 'Broken symmetry and the formation of hot-electron domains in real-space-transfer transistors', *Phys. Rev. Lett.* **67**, 2351.

Madelung, O. (1957), *Encyclopedia of Physics*, Vol. XX, Springer, Berlin, chapter electrical conductivity II.

Magyari, E. (1982), 'Travelling kinks in Schlögl's second model for nonequilibrium phase transitions', *J. Phys. A* **15**, L139.

Mahan, G. D. (1990), *Many-Particle Physics*, Plenum, New York.

Manneville, P. (1990), *Dissipative Structures and Weak Turbulence*, Academic Press, Boston.

Mantegna, R. N. and Stanley, H. E. (1999), *Econophysics: An Introduction*, Cambridge University Press, Cambridge.

Maracas, G. N., Porod, W., Johnson, D. A., Ferry, D. K. and Goronkin, H. (1985), *Physica B* **134**, 276.

Maracas, G. N., Johnson, D. A., Puechner, A., Edwards, J. L., Myhajlenko, S., Goronkin, H. and Tsui, R. (1989), *Solid State Electron.* **32**, 1887.

Margull, U. (1996), Nichtlineare Dynamik räumlicher Strukturen in n-GaAs unter dem Einfluß des Magnetfeldes, PhD thesis, Universität Regensburg.

Markovich, P. A. (1986), *The Stationary Semiconductor Device Equations*, Springer, Vienna.

Martin, A. D., Lerch, M. L. F., Simmonds, P. E. and Eaves, L. (1994), 'Observation of intrinsic tristability in a resonant tunneling structure', *Appl. Phys. Lett.* **64**, 1248.

Mayer, K., Gross, R., Parisi, J., Peinke, J. and Huebener, R. (1987), 'Spatially resolved observation of current filament dynamics in semiconductors', *Solid State Commun.* **63**, 55.

Mayer, K., Parisi, J. and Huebener, R. (1988), 'Imaging of self-generated multifilamentary current patterns in GaAs', *Z. Phys. B* **71**, 171.

Mazouz, N., Flätgen, G. and Krischer, K. (1997), 'Tuning the range of spatial coupling in electrochemical systems: From local via nonlocal to global coupling', *Phys. Rev. E* **55**, 2260.

McCumber, D. E. and Chynoweth, A. G. (1966), 'Theory of negative-conductance amplification and of Gunn instabilities in two-valley semiconductors', *IEEE Trans. Electron Dev.* **13**, 4.

Meixner, M., Bose, S. and Schöll, E. (1997a), 'Analysis of chaotic patterns near a codimension-2 Turing–Hopf point in a reaction–diffusion model', *Physica D* **109**, 128.

Meixner, M., De Wit, A., Bose, S. and Schöll, E. (1997b), 'Generic spatio-temporal dynamics near Turing–Hopf codimension-two bifurcations', *Phys. Rev. E* **55**, 6690.

Meixner, M., Rodin, P. and Schöll, E. (1998a), 'Accelerated, decelerated and oscillating fronts in a globally coupled bistable semiconductor system', *Phys. Rev. E* **58**, 2796.

Meixner, M., Rodin, P. and Schöll, E. (1998b), 'Fronts in a bistable medium with two global constraints: Oscillatory instability and large-amplitude limit-cycle motion', *Phys. Rev. E* **58**, 5586.

Meixner, M., Rodin, P., Schöll, E. and Wacker, A. (1999), Dynamics and stability of lateral current density patterns in resonant-tunneling structures, in *Proceedings of the 7th International Symposium on Nanostructure: Physics and Technology*, Ioffe Institute, Sankt Petersburg, pp. 280–3.

Meixner, M., Rodin, P., Schöll, E. and Wacker, A. (2000a), 'Lateral current density fronts in globally coupled bistable semiconductors with S- or Z-shaped current voltage characteristic', *Eur. Phys. J. B* **13**, 157.

Meixner, M., Zoldi, S., Bose, S. and Schöll, E. (2000b), 'Karhunen–Loève local characterization of spatio-temporal chaos in a reaction–diffusion system', *Phys. Rev. E* **61**, 1382.

Mel'nikov, D. and Podlivaev, A. (1998), 'Lateral traveling wave as a type of transient process in a resonant tunneling structure', *Semiconductors* **32**, 206.

Merbach, D., Hess, O., Herzel, H. and Schöll, E. (1995), 'Injection-induced bifurcations of transverse spatiotemporal patterns in semiconductor laser arrays', *Phys. Rev. B* **52**, 1571.

Merbach, D., Schöll, E. and Gutowski, J. (1999), 'Simulation of ZnSe-based self-electrooptic effect devices', *J. Appl. Phys.* **85**, 7051.

Merlin, R., Kwok, S. H., Norris, T. B., Grahn, H. T., Ploog, K., Bonilla, L. L., Galán, J., Cuesta, J. A., Martínez, F. C. and Molera, J. M. (1995), Dynamics of resonant tunneling domains in superlattices: Theory and experiment, in D. J. Lockwood, ed., *Proceedings of the 22nd International Conference on the Physics of Semiconductors, Vancouver 1994*, World Scientific, Singapore, p. 1039.

Mikhailov, A. S. (1989), in A. V. Gaponov-Grekhov and M. I. Rabinovich, eds, *Nonlinear Waves, Dynamics and Evolution*, Springer, Berlin.

Mikhailov, A. S. (1992), 'Stable autonomous pacemakers in the enlarged Ginzburg–Landau model', *Physica D* **55**, 99.

Mikhailov, A. S. (1994), *Foundations of Synergetics Vol. I*, 2nd edn, Springer, Berlin.

Mikhailov, A. S. and Loskutov, A. Y. (1996), *Foundations of Synergetics Vol. II*, 2nd edn, Springer, Berlin.

Miller, D. and Laikhtman, B. (1994), 'Theory of high-field-domain structures in superlattices', *Phys. Rev. B* **50**, 18 426.

Miller, D. A. B. (1990), 'Quantum-well self-electro-optic effect devices', *Opt. Quant. Electron.* **22**, S61.

Minarsky, A. M. and Rodin, P. B. (1997), 'Transverse instability and inhomogeneous dynamics of superfast impact ionization waves in diode structures', *Solid State Electron.* **41**, 813.

Mitkov, I., Kladko, K. and Pearson, J. E. (1998), 'Tunable pinning of burst waves in extended systems with discrete sources', *Phys. Rev. Lett.* **81**, 5453.

Mityagin, Y. A. and Murzin, V. N. (1996), 'Current hysteresis and the formation condition for electric-field domains in lightly doped superlattices', *JEPT Lett.* **64**, 155. [*Pis. Zh. Éksp. Teor. Fiz.* **64**, 146 (1996)].

Mityagin, Y. A., Murzin, V. N., Efimov, Y. A. and Rasulova, G. K. (1997), 'Sequential excited-to-excited states resonant tunneling and electric field domains in long period superlattices', *Appl. Phys. Lett.* **70**, 3008.

Montroll, E. W. and Shuler, K. E. (1958), *Adv. Chem. Phys.* **1**, 361.

Mori, H. and Kuramoto, Y. (1998), *Dissipative Structures and Chaos*, Springer, Berlin.

Morris, S. W., Bodenschatz, E., Cannell, D. S. and Ahlers, G. (1993), *Phys. Rev. Lett.* **71**, 2026.

Morris, S. W., Bodenschatz, E., Cannell, D. S. and Ahlers, G. (1996), 'The spatio-temporal structure of spiral-defect chaos', *Physica D* **97**, 164.

Mosekilde, E., Feldberg, R., Knudsen, C. and Hindsholm, M. (1990), 'Mode locking and spatiotemporal chaos in periodically driven Gunn diodes', *Phys. Rev. B* **41**, 2298.

Mosekilde, E., Thomson, J. S., Knudsen, C. and Feldberg, R. (1993), 'Phase diagrams for periodically driven Gunn-diodes', *Physica D* **66**, 143.

Moss de Oliveira, S., de Oliveira, P. M. C. and Stauffer, D. (1999), *Evolution, Money, War and Computers*, Teubner, Stuttgart.

Münkel, M., Kaiser, F. and Hess, O. (1997), 'Stabilization of spatiotemporally chaotic semiconductor laser arrays by means of delayed optical feedback', *Phys. Rev. E* **56**, 3868.

Murray, J. D. (1993), *Mathematical Biology*, Vol. 19 of *Biomathematics Texts*, 2nd edn, Springer, Berlin.

Naber, H. and Schöll, E. (1990), 'Mode-locking of self-generated oscillations in a semiconductor model for low-temperature impurity breakdown', *Z. Phys. B* **78**, 305.

Nakajima, H. (1997), 'On analytical properties of delayed feedback control of chaos', *Phys. Lett. A* **232**, 207.

Nakamura, K. (1989), 'Microscopic simulation of chaotic current fluctuations in a two-valley semiconductor', *Phys. Lett. A* **138**, 396.

Narihiro, M., Yusa, G., Nakamura, Y., Noda, T. and Sakaki, H. (1997), 'Resonant tunneling of electrons via 20 nm scale InAs quantum dot and magnetotunneling spectroscopy of its electronic states', *Appl. Phys. Lett.* **70**, 105.

Naundorf, H., Gupta, R. and Schöll, E. (1998), 'A model for hot electron light emission from semiconductor heterostructures', *Semicond. Sci. Technol.* **13**, 548.

Nicolis, G. and Prigogine, I. (1977), *Self-Organization in Non-Equilibrium Systems*, Wiley, New York.

Niedernostheide, F.-J., ed. (1995), *Nonlinear Dynamics and Pattern Formation in Semiconductors and Devices*, Springer, Berlin.

Niedernostheide, F. J. and Kleinkes, M. (1999), 'Spatiotemporal dynamics of current-density filaments in a periodically driven multilayered semiconductor device', *Phys. Rev. B* **59**, 7663.

Niedernostheide, F.-J., Arps, M., Dohmen, R., Willebrand, H. and Purwins, H.-G. (1992*a*), 'Spatial and spatio-temporal patterns in pnpn semiconductor devices', *Physica status solidi (b)* **172**, 249.

Niedernostheide, F.-J., Kerner, B. S. and Purwins, H.-G. (1992*b*), 'Spontaneous appearance of rocking localized current filaments in a nonequilibrium distributive system', *Phys. Rev. B* **46**, 7559.

Niedernostheide, F.-J., Kreimer, M., Schulze, H.-J. and Purwins, H.-G. (1993), 'Periodic and irregular spatio-temporal bahaviour of current density filaments in Si p^+-n^+-p-n^- diodes', *Phys. Lett. A* **180**, 113.

Niedernostheide, F.-J., Ardes, M., Or-Guil, M. and Purwins, H.-G. (1994*a*), 'Spatio-temporal behavior of localized current filaments in p-n-p-n diodes: Numerical calculations and comparison with experimental results', *Phys. Rev. B* **49**, 7370.

Niedernostheide, F.-J., Kreimer, M., Kukuk, B., Schulze, H.-J. and Purwins, H.-G. (1994*b*), 'Travelling current density filaments in multilayered silicon devices', *Phys. Lett. A* **191**, 285.

Niedernostheide, F.-J., Brillert, C., Kukuk, B., Purwins, H.-G. and Schulze, H.-J. (1996*a*), 'Frequency-locked, quasiperiodic, and chaotic motions of current-density filaments in a semiconductor device', *Phys. Rev. B* **54**, 14 012.

Niedernostheide, F.-J., Schulze, H., Bose, S., Wacker, A. and Schöll, E. (1996*b*), 'Spiking in a semiconductor device: Experiments and comparison with a model', *Phys. Rev. E* **54**, 1253.

Niedernostheide, F.-J., Or-Guil, M., Kleinkes, M. and Purwins, H.-G. (1997), 'Dynamical behavior of spots in a nonequilibrium distributive active medium', *Phys. Rev. E* **55**, 4107.

Niedernostheide, F.-J., Hirschinger, J., Prettl, W., Novák, V. and Kostial, H. (1998), 'Oscillations of current filaments in n-GaAs caused by a magnetic field', *Phys. Rev. B* **58**, 4454.

Ning, L., Hu, Y., Ecke, R. E. and Ahlers, G. (1993), 'Spatial and temporal averages in chaotic patterns', *Phys. Rev. Lett.* **71**, 2216.

Northrup, D. C., Thornton, P. R. and Tresize, K. E. (1964), *Solid State Electron.* **7**, 17.

Noteborn, H. J. M. F., Joosten, H. P., Kaski, K. and Lenstra, D. (1993), 'Alternative for the quantum-inductance model in double-barrier resonant-tunneling', *Superlattices Microstructures* **13**, 153.

Nougier, J. P., Vaissière, J. and Gasquet, D. (1981), 'Determination of transient regime of hot carriers in semiconductors, using the relaxation time approximations', *J. Appl. Phys.* **52**, 825.

Novák, V. and Prettl, W. (1995), Current filamentation in dipolar electric fields, in F.-J. Niedernostheide, ed., *Nonlinear Dynamics and Pattern Formation in Semiconductors*, Springer, Berlin.

Novák, V., Wimmer, C. and Prettl, W. (1995), 'Impurity-breakdown-induced current filamentation in a dipolar electric field', *Phys. Rev. B* **52**, 9023.

Novák, V., Hirschinger, J., Niedernostheide, F.-J., Prettl, W., Cukr, M. and Oswald, J. (1998*a*), 'Direct experimental observation of the Hall angle in the low-temperature breakdown regime of n-GaAs', *Phys. Rev. B* **58**, 13 099.

Novák, V., Hirschinger, J., Prettl, W. and Niedernostheide, F.-J. (1998*b*), 'Current filamentation in point contact geometry and its 2D stationary model', *Semicond. Sci. Technol.* **13**, 756.

O'Hern, C., Egolf, D. and Greenside, H. (1996), 'Lyapunov spectral analysis of an nonequilibrium Ising-like transition', *Phys. Rev. E* **53**, 3374.

Ohta, T. (1989), 'Decay of metastable rest state in excitable reaction–diffusion system', *Prog. Theor. Phys. Suppl.* **99**, 425.

Ohta, T., Ito, A. and Tetsuka, A. (1990), 'Self-organization in an excitable reaction–diffusion system: Synchronization of oscillatory domains in one dimension', *Phys. Rev. A* **42**, 3225.

Ohtani, N., Egami, N., Grahn, H.-T., Ploog, K. H. and Bonilla, L. (1998), 'Transition between static and dynamic electric-field domain formation in weakly coupled GaAs/AlAs superlattices', *Phys. Rev. B* **58**, 7528.

Oshio, K. (1998), 'Bifurcation and chaos of current oscillations in semiconductors with NDC', *J. Phys. Soc. Japan* **67**, 2538.

Osipov, V. V. and Kholodnov, V. A. (1971), 'Current filamentation in a long diode', *Sov. Phys. Semicond.* **4**, 1033. [*Fiz. Tekhn. Poluprovodn* **4**, 1216 (1970)].

Osipov, V. V. and Kholodnov, V. A. (1973), *Microelectronica* **2**, 529. [in Russian.]

Ott, E. (1993), *Chaos in Dynamical Systems*, Cambridge University Press, Cambridge.

Ott, E., Grebogi, C. and Yorke, J. A. (1990), 'Controlling chaos', *Phys. Rev. Lett.* **64**, 1196.

Patra, M., Schwarz, G. and Schöll, E. (1998), 'Bifurcation analysis of stationary and oscillating field domains in semiconductor superlattices with doping fluctuations', *Phys. Rev. B* **57**, 1824.

Peinke, J., Mühlbach, A., Huebener, R. P. and Parisi, J. (1985), 'Spontaneous oscillations and chaos in p-germanium', *Phys. Lett.* **108 A**, 407.

Peinke, J., Rau, U., Clauss, W., Richter, R. and Parisi, J. (1989), 'Critical dynamics near the onset of spontaneous oscillations in p-germanium', *Europhys. Lett.* **9**, 743.

Peinke, J., Parisi, J., Rössler, O. and Stoop, R. (1992), *Encounter with Chaos*, Springer, Berlin.

Pelcé, P. (1988), *Dynamics of Curved Fronts*, Academic Press, San Diego.

Perraud, J.-J., De Wit, A., Dulos, E., De Kepper, P., Dewel, G. and Borckmans, P. (1993), 'One-dimensional spiral: novel asynchronous chemical wave sources', *Phys. Rev. Lett.* **71**, 1271.

Petersen, K. E. and Adler, D. (1976), 'On-state of amorphous threshold switches', *J. Appl. Phys.* **47**, 256.

Petrov, V., Crowley, M. J. and Showalter, K. (1994), 'Tracking unstable periodic orbits in the Belousov–Zhabotinsky reaction', *Phys. Rev. Lett.* **72**, 2955.

Petrov, V., Metens, S., Borckmans, P., Dewel, G. and Showalter, K. (1995), 'Tracking unstable Turing patterns through mixed-mode spatiotemporal chaos', *Phys. Rev. Lett.* **75**, 2895.

Piazza, F., Christianen, P. C. M. and Maan, J. C. (1997), 'Propagating high-electric-field domains in semi-insulating GaAs: Experiment and theory', *Phys. Rev. B* **55**, 15 591.

Pickin, W. (1978), *Solid State Electron.* **21**, 309 and 1299.

Pierre, T., Bonhomme, G. and Atipo, A. (1996), 'Controlling the chaotic regime of nonlinear ionization waves using time-delay autsynchronisation method', *Phys. Rev. Lett.* **76**, 2290.

Pikovsky, A., Rosenblum, M. and Kurths, J. (1996), 'Synchronisation in a population of globally coupled oscillators', *Europhys. Lett.* **34**, 165.

Politi, A., Livi, R., Oppo, G.-L. and Kapral, R. (1993), 'Unpredictable behaviour in stable systems', *Europhys. Lett.* **22**, 571.

Požela, J. (1981), *Plasma and Current Instabilities in Semiconductors*, Pergamon, Oxford.

Prengel, F. and Schöll, E. (1999), 'Quantum kinetics of intersubband impact ionization in quantum wires', *Semicond. Sci. Technol.* **14**, 379.

Prengel, F., Wacker, A. and Schöll, E. (1994), 'Simple model for multistability and domain formation in semiconductor superlattices', *Phys. Rev. B* **50**, 1705. (See also *ibid.* **52**, 11 518 (1995).)

Prengel, F., Wacker, A., Schwarz, G., Schöll, E., Kastrup, J. and Grahn, H. T. (1995), Dynamics of domain formation in semiconductor superlattices, *Lithuanian J. Phys.* **35**, 404.

Prengel, F., Patra, M., Schwarz, G. and Schöll, E. (1996), Nonlinear dynamics of field domains in weakly disordered superlattices, in M. Scheffler and R. Zimmermann, eds, *Proceedings of the 23rd International Conference on the Physics of Semiconductors, Berlin 1996*, Vol. 3, World Scientific, Singapore, pp. 1667–70.

Press, W. H., Flannery, B. P., Teukolsky, S. A. and Vettering, W. T. (1992), *Numerical Recipes in C*, 2nd edn, Cambridge University Press, Cambridge.

Purwins, H. G., Klempt, G. and Berkemeier, J. (1987), in 'Festkörperprobleme', Vol. 27, Vieweg, Braunschweig, p. 27.

Purwins, H.-G., Radehaus, C. and Berkemeier, J. (1988), 'Experimental investigation of spatial pattern formation in physical systems of activator inhibitor type', *Z. Naturforsch.* **43a**, 17.

Pyragas, K. (1992), 'Continuous control of chaos by self-controlling feedback', *Phys. Lett. A* **170**, 421.

Pyragas, K. and Tamaševičius, A. (1993), 'Experimental control of chaos by delayed self-controlling feedback', *Phys. Lett. A* **180**, 99.

Pytte, E. and Thomas, H. (1969), 'Soft modes, critical fluctuations, and optical properties for a two-valley model of Gunn-instability semiconductors', *Phys. Rev.* **179**, 431.

Quade, W., Hüpper, G., Schöll, E. and Kuhn, T. (1994*a*), 'Monte Carlo simulation of the nonequilibrium phase transition in p-type Ge at impurity breakdown', *Phys. Rev. B* **49**, 13 408.

Quade, W., Schöll, E., Rossi, F. and Jacoboni, C. (1994*b*), 'Quantum theory of impact ionization in coherent high-field semiconductor transport', *Phys. Rev. B* **50**, 7398.

Radehaus, C., Kardell, K., Baumann, H., Jäger, D. and Purwins, H.-G. (1987), 'Pattern formation in S-shaped negative differential conductivity material', *Z. Phys. B* **65**, 515.

Radehaus, C., Dohmen, R., Willebrand, H. and Niedernostheide, F.-J. (1990), 'Model for current patterns in physical systems with two charge carriers', *Phys. Rev. A* **42**, 7426.

Radehaus, C. V. and Willebrand, H. (1995), in F.-J. Niedernostheide, ed., *Nonlinear Dynamics and Pattern Formation in Semiconductors and Devices*, Springer, Berlin, p. 250.

Rau, U., Clauss, W., Kittel, A., Lehr, M., Bayerbach, M., Parisi, J., Peinke, J. and Huebener, R. (1991), 'Classification of sponaneous oscillations at the onset of avalanche breakdown in p-type germanium', *Phys. Rev. B* **43**, 2255.

Rauch, C., Strasser, G., Unterrainer, K., Boxleitner, W., Gornik, E. and Wacker, A. (1998), 'Transition between coherent and incoherent electron transport in GaAs/GaAlAs superlattices', *Phys. Rev. Lett.* **81**, 3495.

Redmer, R., Madureira, J. R., Fitzer, N., Goodnick, S. M., Schattke, W. and Schöll, E. (2000), 'Field effect on the impact ionization rate in semiconductors', *J. Appl. Phys.* **87**, 781.

Reed, M. and Simon, B. (1972), *Analysis of Operators*, Vol. 4 of *Methods of Modern Mathematical Physics*, Academic Press, New York.

Reggiani, L. and Mitin, V. (1989), 'Recombination and ionization processes at impurity centres in hot-electron semiconductor transport', *Rev. Nuov. Cim.* **12**, 1.

Rein, A., Hüpper, G. and Schöll, E. (1993), 'Fluctuations and critical scaling in type-I intermittency', *Europhys. Lett.* **21**, 7.

Reklaitis, A., Stasch, R., Asche, M., Hey, R., Krotkus, A. and Schöll, E. (1997), 'Dynamics of the single barrier heterostructure hot electron diode', *J. Appl. Phys.* **82**, 1706.

Reznik, D. and Schöll, E. (1993), 'Oscillation modes, transient chaos and its control in a modulation-doped semiconductor double-heterostructure', *Z. Phys. B* **91**, 309.

Richter, R., Peinke, J., Clauss, W., Rau, U. and Parisi, J. (1991), 'Evidence of type-III intermittency in the electric breakdown of p-type germanium', *Europhys. Lett.* **14**, 1.

Richter, R., Rau, U., Kittel, A., Heinz, G., Peinke, J., Parisi, J. and Huebener, R. (1992), *Z. Naturforsch.* **46a**, 1012.

Ridley, B. K. (1963), 'Specific negative resistance in solids', *Proceedings of the Phys. Soc.* **82**, 954.

Ridley, B. K. (1983), 'Lucky-drift-mechanism for impact ionisation in semiconductors', *J. Phys. C* **16**, 3373.

Ridley, B. K. (1987), *Semicond. Sci. Technol.* **2**, 116.

Ridley, B. K. (1991), 'Hot electrons in low-dimensional structures', *Rep. Prog. Phys.* **54**, 169.

Ridley, B. K. and Watkins, T. B. (1961), 'The possibility of negative resistance effects in semiconductors', *Proceedings of the Phys. Soc.* **78**, 293.

Robbins, D. J. and Landsberg, P. T. (1980), 'Impact ionisation and Auger recombination involving traps in semiconductors', *J. Phys. C* **13**, 2425.

Robbins, D. J., Landsberg, P. T. and Schöll, E. (1981), 'Threshold switching as a generation–recombination induced non-equilibrium phase transition (II)', *Phys-*

ica Status Solidi (a) **65**, 353.

Rossi, F., Brunetti, R. and Jacoboni, C. (1992), Quantum transport, in J. Shah, ed., *Hot Carriers in Semiconductor Nanostructures: Physics and Applications*, Academic Press, Boston, p. 153.

Rössler, O. E. (1976), *Phys. Lett. A* **56**, 397.

Rudroff, S. and Rehberg, I. (1997), 'Pattern formation and spatiotemporal chaos in the presence of boundaries', *Phys. Rev. E* **55**, 2742.

Rudzick, O. and Schöll, E. (1995), 'Influence of weak disorder on oscillatory current in-stabilities in modulation-doped multilayer heterostructures', *Semicond. Sci. Technol.* **10**, 143.

Ruelle, D. (1989), *Chaotic Evolution and Strange Attractors*, Cambridge University Press, Cambridge.

Rühle, W. W., Heberle, A. P., Alexander, M. G. W., Nido, M. and Köhler, K. (1991), 'Time-resolved optical investigation of tunneling of carriers through single $Al(x)Ga(1-x)As$ barriers', *Physica Scripta* **T39**, 278.

Ruwisch, D., Bode, M., Schulze, H.-J. and Niedernostheide, F.-J. (1996), in J. Parisi, S. Müller and W. Zimmermann, eds, *Nonlinear Physics of Complex Systems*, Lecture Notes in Physics, Vol, 476, Springer, Berlin, p. 194.

Ryabushkin, O. and Bader, V. (1991), in *Proceedings of the International Conference on Optical Science and Engineering (Optics for Computer)*, The Hague, p. 1505.

Sablikov, V. A., Ryabushkin, O. A. and Polyakov, S. V. (1996), 'On the mechanism of low-temperature impurity breakdown', *Semiconductors* **30**, 660.

Sacks, H. K. and Milnes, A. G. (1970), *Int. J. Electron.* **28**, 565.

Sakamoto, R., Akai, K. and Inoue, M. (1989), 'Real space transfer and hot-electron transport properties in III–V semiconductor heterostructures', *IEEE Trans. Electron Dev.* **36**, 2344.

Samuilov, V. A. (1995), in F.-J. Niedernostheide, ed., *Nonlinear Dynamics and Pattern Formation in Semiconductors and Devices*, Springer, Berlin.

Sattinger, D. H. (1973), *Topics in Stability and Bifurcation Theory*, Springer, Berlin.

Schimansky-Geier, L., Zülicke, C. and Schöll, E. (1991), 'Domain formation due to Ostwald ripening in bistable systems far from equilibrium', *Z. Phys. B* **84**, 433.

Schimansky-Geier, L., Zülicke, C. and Schöll, E. (1992), 'Growth of domains under global constraints', *Physica A* **188**, 436.

Schimansky-Geier, L., Hempel, H., Bartussek, R. and Zülicke, C. (1995), 'Analysis of domain-solutions in reaction–diffusion systems', *Z. Phys. B* **96**, 417.

Schlögl, F. (1972), 'Chemical reaction models for non-equilibrium phase transitions', *Z. Phys.* **253**, 147.

Schlögl, F., Escher, C. and Berry, R. S. (1983), 'Fluctuations in the interface between two phases', *Phys. Rev. A* **27**, 2698.

Schmolke, R., Schöll, E., Nägele, M. and Gutowski, J. (1995), 'Nonlinear dynamics of optical switching fronts in CdS', *Physica Status Solidi (b)* **187**, 631.

Schöll, E. (1982*a*), 'Bistability and nonequilibrium phase transitions in a semiconductor recombination model with impact ionization of donors', *Z. Phys. B* **46**, 23.

Schöll, E. (1982*b*), 'Current layers and filaments in a semiconductor model with an impact ionization induced instability', *Z. Phys. B* **48**, 153.

Schöll, E. (1983), 'Stability of generation–recombination induced dissipative structures in semiconductors', *Z. Phys. B* **52**, 321.

Schöll, E. (1985), 'Mechanisms for chaos in semiconductors induced by impact ionization', *Physica B* **134**, 271.

Schöll, E. (1986*a*), 'Equal areas rules for filamentation in SNDC elements', *Solid State Electron.* **29**, 687.

Schöll, E. (1986*b*), 'Impact ionization mechanism for self-generated chaos in semiconductors', *Phys. Rev. B* **34**, 1395.

Schöll, E. (1987), *Nonequilibrium Phase Transitions in Semiconductors*, Springer, Berlin.

Schöll, E. (1988), Nonlinear dynamics and breathing of current filaments during impact ionization breakdown, in W. Zawadzki, ed., *Proceedings of the 19th International Conference Physics of Semiconductors, Warsaw 1988*, Institute of Physics, Polish Academy of Science, pp. 1407–10.

Schöll, E. (1989), 'Instabilities in semiconductors including chaotic phenomena', *Physica Scripta* **T29**, 152.

Schöll, E. (1992), Nonlinear dynamics, phase transitions and chaos in semiconductors, in P. T. Landsberg, ed., *Handbook on Semiconductors*, 2nd edn, Vol. 1, North Holland, Amsterdam.

Schöll, E. (1998*a*), Impact phenomena and nonlinear spatiotemporal dynamics of hot electrons in semiconductors, in N. Balkan, ed., *Hot Electrons in Semiconductors: Physics and Devices*, Oxford University Press, Oxford, chapter 9, pp. 209–231.

Schöll, E., ed. (1998*b*), *Theory of Transport Properties of Semiconductor Nanostructures*, Vol. 4 of *Electronic Materials Series*, Chapman and Hall, London.

Schöll, E. (1999), 'Nonlinear spatio-temporal dynamics in semiconductors', *Braz. J. Phys.* **29**, 627.

Schöll, E. and Aoki, K. (1991), 'Novel mechanism of a real-space transfer oscillator', *Appl. Phys. Lett.* **58**, 1277.

Schöll, E. and Drasdo, D. (1990), 'Nonlinear dynamics of breathing current filaments in n-GaAs and p-Ge', *Z. Phys. B* **81**, 183.

Schöll, E. and Landsberg, P. T. (1979), *Proceedings of the Roy. Soc. A* **365**, 495.

Schöll, E. and Landsberg, P. T. (1988), 'Generalised equal areas rules for spatially extended systems', *Z. Phys. B* **72**, 515.

Schöll, E. and Pyragas, K. (1993), 'Tunable semiconductor oscillator based on self-control of chaos in the dynamic Hall effect', *Europhys. Lett.* **24**, 159.

Schöll, E., Parisi, J., Röhricht, B., Peinke, J. and Huebener, R. P. (1987), 'Spatial correlations of chaotic oscillations in the post-breakdown regime of p-Ge', *Phys. Lett. A* **119**, 419.

Schöll, E., Hüpper, G. and Rein, A. (1992), 'Dynamic Hall effect of hot electrons as a novel mechanism for current oscillations, chaos and intermittency', *Semicond. Sci. Technol.* **7**, B480.

Schöll, E., Schwarz, G., Patra, M., Prengel, F. and Wacker, A. (1996), Oscillatory instabilities and field domain formation in imperfect superlattices, in K. Hess, J. P. Leburton and U. Ravaioli, eds, *Proceedings of the 9th International Conference on Hot Carriers in Semiconductors, Chicago 1995*, Plenum, New York, pp. 177–181.

Schöll, E., Niedernostheide, F.-J., Parisi, J., Prettl, W. and Purwins, H. (1998*a*), Formation of spatio-temporal structures in semiconductors, in F. H. Busse and S. C. Müller, eds, *Evolution of Spontaneous Structures in Dissipative Continuous Systems*, Springer, Berlin, pp. 446–494.

Schöll, E., Schwarz, G. and Wacker, A. (1998*b*), 'Nonlinear and oscillatory electronic transport in superlattices as a probe of structural imperfections', *Physica B* **249**, 961.

Schomburg, E., Brandl, S., Hofbeck, K., Blomeier, T., Grenzer, J., Ignatov, A. A., Renk, K. F., Pavel'ev, D. G., Koschurinov, Y., Ustinov, V., Zhukov, A., Kovsch, A., Ivanov, S. and Kopev, P. S. (1998), 'Generation of millimeter waves with a GaAs/AlAs superlattice oscillator', *Appl. Phys. Lett.* **72**, 1498.

Schomburg, E., Henini, M., Chamberlain, J. M., Steenson, D. P., Brandl, S., Hofbeck, K., Renk, K. F. and Wegscheider, W. (1999), 'Self-sustained current oscillation above 100 GHz in a GaAs/AlAs superlattice', *Appl. Phys. Lett.* **74**, 2179.

Schuster, H. G. (1988), *Deterministic Chaos*, 2nd edn, VCH Verlagsgesellschaft, Weinheim.

Schuster, H. G. (1999), *Handbook of Chaos Control*, Wiley-VCH, Weinheim.

Schwarz, G. (1995), Felddomänen in imperfekten Übergitterstrukturen, Master's thesis, TU Berlin.

Schwarz, G. and Schöll, E. (1996), 'Field domains in semiconductor superlattices', *Physica Status Solidi (b)* **194**, 351.

Schwarz, G. and Schöll, E. (1997), 'Simulation of current filaments in semiconductors with point contacts and Corbino disks', *Acta Technica ČSAV* **42**, 669.

Schwarz, G., Prengel, F., Schöll, E., Kastrup, J., Grahn, H. T. and Hey, R. (1996*a*), 'Electric field domains in intentionally perturbed semiconductor superlattices', *Appl. Phys. Lett.* **69**, 626.

Schwarz, G., Wacker, A., Prengel, F., Schöll, E., Kastrup, J., Grahn, H. T. and Ploog, K. (1996*b*), 'Influence of imperfections and weak disorder on domain formation in superlattices', *Semicond. Sci. Technol.* **11**, 475.

Schwarz, G., Patra, M., Prengel, F. and Schöll, E. (1998), 'Multistable current–voltage characteristics as fingerprints of growth-related imperfections in semiconductor superlattices', *Superlattices Microstructures* **23**, 1353.

Schwarz, G., Lehmann, C. and Schöll, E. (2000*a*), 'Self-organized symmetry-breaking current filamentation and multistability in Corbino disks', *Phys. Rev. B* **61**, 10 194.

Schwarz, G., Lehmann, C., Reimann, A., Schöll, E., Hirschinger, J., Prettl, W. and Novák, V. (2000*b*), 'Current filamentation in n-GaAs thin films with different contact geometries', *Semicond. Sci. Technol.* **15**, 593.

Seiler, D., Littler, C., Justice, R. and Milonni, P. (1985), 'Nonlinear oscillations and chaos in n-InSb', *Phys. Lett. A* **108**, 462.

Selberherr, S. (1984), *Analysis and Simulation of Semiconductor Devices*, Springer, Vienna, New York.

Shaw, M. P. and Gastman, I. J. (1971), 'Circuit controlled current instabilities in S-shaped negative differential conductivity elements', *Appl. Phys. Lett.* **19**, 243.

Shaw, M. P., Grubin, H. L. and Solomon, P. (1979), *The Gunn–Hilsum Effect*, Academic Press, New York.

Shaw, M. P., Mitin, V. V., Schöll, E. and Grubin, H. L. (1992), *The Physics of Instabilities in Solid State Electron Devices*, Plenum Press, New York.

Shichijo, H., Hess, K. and Streetman, B. (1980), *Solid State Electron.* **23**, 817.

Shimada, I. and Nagashima, T. (1979), 'A numerical approach to ergodic problem of dissipative dynamical systems', *Prog. Theor. Phys* **61**, 1605.

Shimada, Y. and Hirakawa, K. (1997), 'Sequential resonant magnetotunneling through Landau levels in GaAs/AlGaAs multiple quantum well structures', *Phys. Status Solidi* (b) **204**, 427.

Shockley, W. (1961), *Solid State Electron.* **2**, 35.

Shraiman, B. I. (1986), 'Order, disorder, and phase turbulence', *Phys. Rev. Lett.* **57**, 325.

Simmendinger, C. and Hess, O. (1996), 'Controlling delay-induced chaotic behavior of a semiconductor laser with optical feedback', *Phys. Lett. A* **216**, 97.

Sirovich, L. (1987), *Quart. Appl. Math.* **45**, 561.

Sirovich, L. (1989*a*), *Physica D* **58**, 365.

Sirovich, L. (1989*b*), 'Chaotic dynamics of coherent structures', *Physica D* **37**, 126.

Sirovich, L. and Deane, A. E. (1991), *J. Fluid Mech.* **222**, 251.

Smoller, J. (1983), *Shock Waves and Reaction–Diffusion Equations*, Springer, New York.

Socolar, J. E. S., Sukow, D. W. and Gauthier, D. J. (1994), 'Stabilizing unstable periodic orbits in fast dynamical systems', *Phys. Rev. E* **50**, 3245.

Song, X., Seiler, D. and Loloee, M. (1989), 'Nonlinear oscillations and chaotic behavior due to impact ionization of shallow donors in InSb', *Appl. Phys. A* **48**, 137.

Spangler, J. and Prettl, W. (1994), 'Self-generated and externally driven current oscillations in n-GaAs', *Physica Scripta* **T55**, 25.

Spangler, J., Margull, U. and Prettl, W. (1992), 'Regular and chaotic oscillations in n-type GaAs in transverse and longitudinal magnetic fields', *Phys. Rev. B* **45**, 12 137.

Spangler, J., Finger, B., Wimmer, C., Eberle, W. and Prettl, W. (1994), 'Magnetic-field-induced lateral displacements of current filaments in n-GaAs', *Semicond. Sci. Technol.* **9**, 373.

Spinnewyn, J., Strauven, H. and Verbeke, O. (1989), 'Complementary influence of electric and magnetic fields on the sample voltage oscillations of p-type ultrapure germanium', *Z. Phys. B* **75**, 159.

Stasch, R., Hey, R., Asche, M., Wacker, A. and Schöll, E. (1996), 'Temperature persistent bistability and threshold switching in a single barrier heterostructure hot-electron diode', *J. Appl. Phys.* **80**, 3376.

Steuer, H., Wacker, A. and Schöll, E. (1999), 'Complex behavior due to electron heating in superlattices exhibiting high-frequency current oscillations', *Physica B* **272**, 202.

Steuer, H., Wacker, A., Schöll, E., Ellmauer, M., Schomburg, E. and Renk, K. F. (2000), 'Thermal breakdown, bistability, and complex high-frequency current oscillations due to carrier heating in superlattices', *Appl. Phys. Lett.* **76**, 2059.

Stillman, G. E., Wolfe, C. M. and Dimmock, J. O. (1977), *Semiconductors and Semimetals*, Vol. 12, Academic Press, New York.

Straw, A., Balkan, N., O'Brien, A., da Cunha, A., Gupta, R. and Arikan, M. C. (1995), *Superlattices and Microstructures* **18**, 33.

Sukow, D. W., Bleich, M. E., Gauthier, D. J. and Socolar, J. E. S. (1997), 'Controlling chaos in a fast diode resonator using time-delay autosynchronisation: Experimental observations and theoretical analysis', *Chaos* **7**, 560.

Suzuki, T., Nomoto, K., Taira, K. and Hase, I. (1997), 'Tunneling spectroscopy of InAs wetting layer and self-assembled quantum dots: Resonant tunneling through two- and zero-dimensional electronic states', *Japan. J. Appl. Phys.* **36**, 1917.

Svirezhev, Y. M. (1987), *Nonlinear Waves, Dissipative Structures and Catastrophies in Ecology*, Nauka, Moscow. [in Russian.]

Swinney, H. and Gollub, J. P. (1984), *Hydrodynamic instabilities*, Springer, Berlin.

Swinney, H. L. and Krinsky, V. I., eds (1991), Waves and patterns in chemical and biological media, *Physica* D **49**, 1.

Symanczyk, R., Gaelings, S. and Jäger, D. (1991*a*), 'Observation of spatio-temporal structures due to current filaments in Si pin diodes', *Phys. Lett. A* **160**, 397.

Symanczyk, R., Jäger, D. and Schöll, E. (1991*b*), 'Equivalent cirquit model for current filamentation in p–i–n diodes', *Appl. Phys. Lett.* **59**, 105.

Sze, S. M. (1998), *Modern Semiconductor Device Physics*, Wiley, New York.

Tang, J. Y. and Hess, K. (1982), 'Investigation of near ballistic transport following high energy injection', *IEEE Trans. Electron Dev.* **29**, 1906.

Teitsworth, S. and Westervelt, R. (1984), 'Chaos and broadband noise in extrinsic photoconductors', *Phys. Rev. Lett.* **53**, 2587.

Teitsworth, S. W., Westervelt, R. M. and Haller, E. E. (1983), 'Nonlinear oscillations and chaos in electrical breakdown in Ge', *Phys. Rev. Lett.* **51**, 825.

Tél, T. (1990), Transient chaos, in H. Bai-lin, ed., *Directions in Chaos*, Vol. 3, World Scientific, Singapore, p. 149 and 211.

Thomas, H., ed. (1992), *Nonlinear Dynamics in Solids*, Springer, Berlin.

Thompson, J. M. T. and Stewart, H. B. (1986), *Nonlinear Dynamics and Chaos: Geometrical Methods for Engineers and Scientists*, Wiley, Chichester.

Tolstikhin, V. I. (1986), 'Transverse transport in multilayer heterostructures under carrier-heating conditions', *Sov. Phys. Semicond.* **20**, 1375.

Tominaga, H. and Mori, H. (1994), 'Hopf bifurcations and chaos in the S-shaped I–V characteristic systems', *Prog. Theor. Phys.* **91**, 1081.

Troger, H. and Steindl, A. (1991), *Nonlinear Stability and Bifurcation Theory*, Springer, Vienna.

Tsemekhman, V., Tsemekhman, K., Wexler, C., Han, J. H. and Thouless, D. J. (1997), 'Theory of the breakdown of the quantum Hall effect', *Phys. Rev. B* **55**, R10 201.

Tsu, R. and Esaki, L. (1973), 'Tunneling in a finite superlattice', *Appl. Phys. Lett.* **22**, 562.

Tufillaro, N. B., Ramshankar, R. and Gollub, J. P. (1989), 'Order–disorder transition in capillary ripples', *Phys. Rev. Lett.* **62**, 422.

Turing, A. M. (1952), 'The chemical basis of morphogenesis', *Phil. Trans. Roy. Soc.* **237**,

37.

van Kampen, N. G. (1985), 'Elimination of fast variables', *Phys. Rep.* **124**, 70.

Varlamov, I. V. and Osipov, V. V. (1970), 'Filamentation of the current in multilayered structures', *Sov. Phys. Semicond.* **3**, 803. [*Fiz. Tekn. Poluprovodn.* **3**, 950 (1969)].

Varlamov, I. V., Osipov, V. V. and Poltoratsky, E. A. (1970), 'Current filamentation in a four-layer structure', *Sov. Phys. Semicond.* **3**, 978. [*Fiz. Tekhn. Poluprovodn* **3**, 1162 (1969)].

Vashchenko, V., Kerner, B., Osipov, V. and Sinkevich, V. (1990), *Sov. Phys. Semicond.* **24**, 1045.

Vashchenko, V. A., Vodakov, Y. A., Gafiichuk, V. V., Datsko, B. I., Kerner, B. S., Litvin, D. P., Osipov, V. V., Roenkov, A. D. and Sankin, V. S. (1991), 'Formation of pulsating thermal–diffusion autosolitons and turbulence in a nonequilibrium electron–hole plasma', *Sov. Phys. Semicond.* **25**, 1020.

Vashchenko, V. A., Martynov, J. B. and Sinkevich, V. F. (1997), 'Simulation of multiple filaments in AgAs structures', *Solid State Electron.* **41**, 75.

Vautard, R. and Ghil, M. (1989), 'Singular spectrum analysis in nonlinear dynamics with applications to paleoclimatic time series', *Physica D* **35**, 395.

Vickers, A. J., Straw, A. and Roberts, J. S. (1989), 'High frequency current oscillations in GaAs/AIGaAs single quantum wells', *Semicond. Sci. Technol.* **4**, 743.

Volkov, A. F. and Kogan, S. M. (1967), 'Nonuniform current distribution in semiconductors with negative differential conductivity', *Sov. Phys. JETP* **25**, 1095. [*Zh. Éksp. Teor. Fiz.* **52**, 1647 (1967)].

Volkov, A. F. and Kogan, S. M. (1969), 'Physical phenomena in semiconductors with negative differential conductivity', *Sov. Phys. Usp.* **11**, 881. [*Usp. Phys. Nauk* **96**, 633 (1968)].

von Klitzing, K. (1990), 'Ten years quantum Hall effect', *Festkörperprobleme – Advances in Solid State Physics* **30**, 25.

Wacker, A. (1993), Nichtlineare Dynamik bei senkrechtem Ladungstransport in einer Halbleiter-Heterostruktur, PhD thesis, Berlin.

Wacker, A. (1994), 'Dynamical behavior in a quantum-dot structure', *Phys. Rev. B* **49**, 16 785.

Wacker, A. (1998), Vertical transport and domain formation in multiple quantum wells, in E. Schöll, ed., *Theory of Transport Properties of Semiconductor Nanostructures*, Chapman and Hall, London, chapter 10.

Wacker, A. and Hu, B. Y.-K. (1999), 'Theory of transmission through disorderd superlattices', *Phys. Rev. B* **60**, 16 039.

Wacker, A. and Jauho, A.-P. (1998a), 'Impact of interface roughness on perpendicular transport and domain formation in superlattices', *Superlattices and Microstructures* **23**, 297.

Wacker, A. and Jauho, A.-P. (1998b), 'Quantum transport: The link between standard approaches in superlattices', *Phys. Rev. Lett.* **80**, 369.

Wacker, A. and Schöll, E. (1991), 'Oscillatory instability in the heterostructure hot-electron diode', *Appl. Phys. Lett.* **59**, 1702.

Wacker, A. and Schöll, E. (1992), 'Spatio-temporal dynamics of vertical charge transport in a semiconductor heterostructure', *Semicond. Sci. Technol.* **7**, 1456.

Wacker, A. and Schöll, E. (1994*a*), 'Spiking at vertical electrical transport in a heterostructure device', *Semicond. Sci. Technol.* **9**, 592.

Wacker, A. and Schöll, E. (1994*b*), 'Spiking in an activator–inhibitor model for elements with S-shaped negative differential conductivity', *Z. Phys. B* **93**, 431.

Wacker, A. and Schöll, E. (1995), 'Criteria for stability in bistable electrical devices with S- or Z-shaped current voltage characteristic', *J. Appl. Phys.* **78**, 7352.

Wacker, A., Bose, S. and Schöll, E. (1995*a*), 'Transient spatio-temporal chaos in a reaction–diffusion model', *Europhys. Lett.* **31**, 257.

Wacker, A., Prengel, F. and Schöll, E. (1995*b*), Theory of multistability and domain formation in semiconductor superlattices, in D. J. Lockwood, ed., *Proceedings of the 22nd International Conference on the Physics of Semiconductors, Vancouver 1994*, Vol. 2, World Scientific, Singapore, p. 1075.

Wacker, A., Schwarz, G., Prengel, F., Schöll, E., Kastrup, J. and Grahn, H. T. (1995*c*), 'Probing growth-related disorder by high-field transport in semiconductor superlattices', *Phys. Rev. B* **52**, 13 788.

Wacker, A., Moscoso, M., Kindelan, M. and Bonilla, L. L. (1997), 'Current–voltage characteristic and stability in resonant-tunneling n-doped semiconductor superlattices', *Phys. Rev. B* **55**, 2466.

Walgraef, D. (1997), *Spatio-temporal Pattern Formation*, Springer, New York.

Wang, W., Kiss, I. Z. and Hudson, J. L. (2000), 'Experiments on arrays of globally coupled chaotic electrochemical oscillators: Synchronization and clustering', *Chaos* **10**, 248.

Weidlich, W. and Haag, G. (1984), *Concepts and Models of a Quantitative Sociology*, Springer, Berlin. Springer Series in Synergetics Vol. 14.

Weiss, C. O. and Vilaseca, R. (1991), *Dynamics of Lasers*, VCH Verlagsgesellschaft, Weinheim.

Westervelt, R. and Teitsworth, S. (1985), 'Nonlinear transient response of extrinsic Ge far-infrared photoconductors', *J. Appl. Phys.* **57**, 5457.

Wierschem, A., Niedernostheide, F.-J., Gorbatyuk, A. and Purwins, H.-G. (1995), *Scanning* **17**, 106.

Willebrand, H., Hünteler, T., Niedernostheide, F.-J., Dohmen, R. and Purwins, H. (1992), 'Periodic and turbulent behavior of solitary in distributed active media', *Phys. Rev. A* **45**, 8766.

Wu, J. C., Wybourne, M. N., Berven, C., Goodnick, S. M. and Smith, D. D. (1992), 'Negative differential conductivity observed in a lateral double constriction device', *Appl. Phys. Lett.* **61**, 2425.

Yamada, K., Takara, N., Imada, H., Miura, N. and Hamagushi, C. (1988), *Solid State Electron.* **31**, 809.

Yano, H., Goto, N. and Ohno, Y. (1992), 'Relationship between the chaotic oscillation and side-gating effect S-type negative differential resistance in GaAs MESFETs', *Semicond. Sci. Technol.* **B7**, 491.

Zabrodskij, A. G. and Shlimak, I. S. (1975), *Sov. Phys. Solid State* **16**, 1528.

Zel'dovich, Y. and Frank-Kamenetskij, D. (1938), *Zh. Éksp. Khim.* **12**, 100. [in Russian.]

Zhang, Y., Yang, X., Liu, W., Zhang, P. and Jiang, D. (1994), 'New formation mechanism of electric field domain due to $\Gamma - X$ sequential tunneling in GaAs/AlAs superlattices', *Appl. Phys. Lett.* **65**, 1148.

Zhang, Y., Kastrup, J., Klann, R., Ploog, K. and Grahn, H. T. (1996), 'Synchronization and chaos induced by resonant tunneling in GaAs/AlAs superlattices', *Phys. Rev. Lett.* **77**, 3001.

Zhang, Y., Klann, R., Ploog, K. and Grahn, H. T. (1997), 'Observation of bistability in GaAs/AlAs superlattices', *Appl. Phys. Lett.* **70**, 2825.

Zoldi, S. and Greenside, H. (1997), 'Karhunen–Loève decomposition of extensive chaos', *Phys. Rev. Lett.* **78**, 1687.

Zoldi, S. and Greenside, H. (1998), 'Spatially localized unstable periodic orbits of a high-dimensional chaotic system', *Phys. Rev. E* **57**, R2511.

Zoldi, S., Liu, J., Bajaj, K. M. S., Greenside, H. S. and Ahlers, G. (1998), 'Extensive scaling and nonuniformity of the Karhunen–Loeve decomposition for the spiral-defect chaos state', *Phys. Rev. E* **58**, R6903.

Zoldi, S., Franceschini, G., Bose, S. and Schöll, E. (2000), Stabilizing unstable periodic orbits in reaction–diffusion systems by global time-delayed feedback control, in *Proceedings of Equadiff 99*, World Scientific, Singapore.

Zongfu, J. and Benkum, M. (1991), 'Period doubling and chaos in the Gunn effect', *Phys. Rev. B* **44**, 11 072.

Zykov, V. S., Mikhailov, A. S. and Müller, S. C. (1997), 'Controlling spiral waves in confined geometries by global feedback', *Phys. Rev. Lett.* **78**, 3398.

Index